Control Engineering

Series Editor

William S. Levine
University of Maryland

Lyapunov-Based Control of Mechanical Systems

Marcio S. de Queiroz
Darren M. Dawson
Siddharth P. Nagarkatti
Fumin Zhang

With 89 Figures

Springer Science+Business Media, LLC

Marcio S. de Queiroz
Department of Mechanical Engineering
Louisiana State University
Baton Rouge, LA 70803-6413
USA

Darren M. Dawson
Siddharth P. Nagarkatti
Fumin Zhang
Department of Electrical and
Computer Engineering
Clemson University
Clemson, SC 29634-0915
USA

Library of Congress Cataloging-in-Publication Data
Lyapunov-based control of mechanical systems / Marcio S. de Queiroz.
p. cm.
Includes bibliographical references and index.
ISBN 978-1-4612-7108-6
1. Automatic control. 2. Control theory. 3. Lyapunov functions. I. Queiroz, Marcio S. de
TJ213 .L92 2000
629.8—dc21 00-029233
CIP

Printed on acid-free paper.

Originally published by Birkhäuser Boston in 2000
Softcover reprint of the hardcover 1st edition 2000

ISBN 978-1-4612-7108-6 ISBN 978-1-4612-1352-9 (eBook)
DOI 10.1007/978-1-4612-1352-9

SPIN 10682927

Typeset by the authors in TeX.

9 8 7 6 5 4 3 2 1

To my parents,

José and Nancy de Queiroz

M. S. Q.

To my daughter,

Jacklyn Dawson

D. M. D.

To my parents,

Usha and Pravin Nagarkatti

S. P. N.

To my grandmother,

Huanxu Xie

F. Z.

Contents

Preface

The design of nonlinear controllers for mechanical systems has been an extremely active area of research in the last two decades. From a theoretical point of view, this attention can be attributed to their interesting dynamic behavior, which makes them suitable benchmarks for nonlinear control theoreticians. On the other hand, recent technological advances have produced many real-world engineering applications that require the automatic control of mechanical systems.

Often, Lyapunov-based techniques are utilized as the mechanism for developing different nonlinear control structures for mechanical systems. The allure of the Lyapunov-based framework for mechanical system control design can most likely be assigned to the fact that Lyapunov function candidates can often be crafted from physical insight into the mechanics of the system. That is, despite the nonlinearities, couplings, and/or the flexible effects associated with the system, Lyapunov-based techniques can often be used to analyze the stability of the closed-loop system by using an energy-like function as the Lyapunov function candidate. In practice, the design procedure often tends to be an iterative process that results in the death of many trees. That is, the controller and energy-like function are often constructed in concert to foster an advantageous stability property and/or robustness property. Fortunately, over the last 15 years, many system theory and control researchers have labored in this area to produce various design tools that can be applied in a variety of situations. Concurrent with this theoretical progress, many control research groups around the world have developed real-time computer platforms and the associated hardware to facilitate the implementation of nonlinear control strategies for many different types of applications (e.g., electric motors, magnetic bearings, robotics, flexible structures, etc.).

The simultaneous progress of Lyapunov-based control techniques and real-time computer system technology promulgated the research delineated in this book. It is our intent to illustrate, in a modest way, how Lyapunov-based techniques can be used to design nonlinear controllers for mechanical systems. In the first half of the book, we focus our attention on the design of controllers for rigid mechanical systems, while in the latter half of the book, we design boundary controllers for systems that exhibit vibrations owing to structural flexibilities. Experimental results are provided for almost every controller to illustrate the feasibility of the theoretical concepts. The book is organized as follows. In Chapter 1, we provide the motivation for the control design problems tackled in subsequent chapters. In Chapter 2, we visit a problem common to many mechanical systems — the compensation for friction effects. Specifically, we present the design of several friction compensation controllers for position setpoint and tracking applications of single-input single-output, rigid mechanical systems. Some of the control design issues in this chapter (e.g., parametric uncertainty, modular control/adaptation law design, and unmeasurable state variables) will serve as a preview for the topics presented in Chapters 3 and 4 for multi-input multi-output (MIMO), rigid mechanical systems. In Chapter 3, we design adaptive tracking controllers for MIMO systems using full-state feedback. Since velocity measurements in mechanical systems are often noisy or inaccurate, several solutions to the MIMO, output feedback, tracking problem are given in Chapter 4. In Chapters 5 through 7, we turn our attention to flexible mechanical systems with the intention of illustrating how Lyapunov techniques, which normally are applied to systems modeled by ordinary differential equations, can be extended to systems described by partial differential equations. In these chapters, we design both model-based and adaptive boundary controllers, with the former serving as a stepping stone for the latter. Chapter 5 presents boundary controllers for string-like systems, while Chapter 6 deals with the design of boundary controllers for flexible beams. In Chapter 7, we focus on flexible systems that model some typical, real-world engineering applications. Specifically, we design boundary control strategies for an axially moving string system, a flexible link robot manipulator, and a flexible rotor system. A significant portion of the mathematical background necessary to follow the control designs and analyses are combined in Appendix A. With the exception of Chapter 1, each chapter includes detailed proofs. While a few of the proofs in some of the chapters appear to be slightly repetitive, we believe this will allow the reader to follow each chapter with a certain level of independence.

All of the material contained in this book (unless noted otherwise) has resulted from the authors' research over the last seven years. The material is intended for audiences with an undergraduate background in control theory and linear systems. Some knowledge of nonlinear systems theory would also be useful; however, we do not believe that it is necessary. This book is mainly aimed at researchers and graduate students in the area of control applications.

We would like to acknowledge and express our sincere gratitude to Professor Chris Rahn for introducing us to the boundary control philosophy and for his technical support in the flexible structures area. Special appreciation is also due to the following past and present graduate students of the Department of Electrical and Computer Engineering at Clemson University whose hard work throughout the past seven years made this book a reality: Manish Agarwal, Keith Anderson, Tim Burg, Hüseyin Canbolat, Nick Costescu, Bret Costic, Warren Dixon, Matthew Feemster, Deepak Haste, Ser-Yong Lim, Mike Mercier, Praveen Vedagarhba, and Erkan Zergeroğlu.

Baton Rouge, Louisiana
Clemson, South Carolina

Marcio de Queiroz
Darren Dawson
Siddharth Nagarkatti
Fumin Zhang

1 Introduction

1.1 Lyapunov-Based Control

The synthesis of controllers basically involves two steps: (i) the *design* step, where the goal is to construct a control algorithm for a given system to satisfy certain performance specifications; and (ii) the *analysis* step; where the goal is to verify the closed-loop system behavior (e.g., stability properties) once the controller has been designed. For a nonlinear system, these two steps become interdependent in the sense that the design and analysis are the result of an iterative procedure. The nuances of this iterative procedure become explicitly apparent if one exploits *Lyapunov's stability theory*[1] (i.e., the so-called Direct Method) for the control design and analysis. Roughly speaking, Lyapunov's Direct Method is a mathematical interpretation of the physical property that if a system's total energy is dissipating, then the states of the system will ultimately travel to an equilibrium point [6]. Simply stated, this property can be explored by constructing a scalar, energy-related function for the system (e.g., $V(t)$, where this function usually contains the closed-loop system states[2]) and then investigating its time variation denoted by $\dot{V}(t)$. If $\dot{V}(t) \leq 0$, then we know that $V(t)$ is a decreasing or constant function of time (i.e., the energy is being dissipated or

[1] For formal and detailed presentations of this theory, the reader is refered to [2, 6].

[2] Since $V(t)$ is related to the energy of the system, it must be a *non-negative* function.

held at a constant level) and will eventually reach a constant; hence, the closed-loop system is considered to be stable in some sense.

The main dilemmas that one has to face in a Lyapunov-based framework are (i) how does one select the energy-related function? and (ii) should the energy-related function simply be composed of the system's kinetic and potential energy or should it include other terms as well? Unfortunately, there do not seem to be straightforward answers to these questions because the selection of the energy-related function is not unique. Indeed, this selection of $V(t)$ is the very fact that makes Lyapunov design so interesting; moreover, it is here that the iterative nature of the control design and the closed-loop stability analysis materializes. Specifically, the selection of $V(t)$ will directly depend on the structure of the closed-loop system and, hence the structure of the controller. It is the authors' belief that a better understanding of this synergy can be obtained from experience and physical insight into the system. In fact, one of the main objectives of this book is to assist the reader in acquiring these skills. We hope to achieve this goal by presenting Lyapunov-based control strategies for a variety of nonlinear control problems that capture the strengths and inherent flexibility of the approach. In particular, we will illustrate how Lyapunov designs can be utilized to address the following issues:

- Model uncertainty
- Unavailability of full-state feedback
- Global stability *vs.* semiglobal stability
- Modular design of control and parameter adaptation laws
- Control of distributed parameter systems.

The discussion of Lyapunov-based designs in this book will be restricted to control problems for mechanical systems. The name "mechanical system" is used loosely to denote a class of physical systems whose dynamic behavior can be derived via well-known methods of analytical dynamics [4] such as Newton's laws, Lagrange's equations of motion, or Hamilton's principle. The focus on mechanical systems can be explained by the observation that many applied nonlinear control researchers, despite their technical background, have often found themselves working on applications involving some kind of mechanical subsystem. A few typical examples of such practice are robotics, magnetic bearings, and spacecraft applications. This emphasis may also be attributed to the fact that many nonlinear control techniques often owe their existence to a particular class of mechanical system. The mechanical systems considered in this book will be subdivided into two classes:

- **Rigid Mechanical Systems** — Systems represented by *ordinary* differential equations (ODE).
- **Flexible Mechanical Systems** — Systems represented by *partial* differential equations (PDE) owing to the independence of the spatial and time variables. (These systems are also often called distributed parameter or infinite dimensional systems.)

In the following two sections, we briefly preview the control problems for these classes of systems that will be presented in the subsequent chapters.

1.2 Rigid Mechanical Systems

Consider the following ODE:

$$\ddot{x}(t) + f\left(x(t), \dot{x}(t), \theta\right) = u, \tag{1.1}$$

where $x(t)$, $\dot{x}(t)$, $\ddot{x}(t) \in \mathbb{R}$ denote the position, velocity, and acceleration, respectively, of the system; $f\left(x, \dot{x}, \theta\right) \in \mathbb{R}$ is some nonlinear function; θ is a constant system parameter; and $u(t) \in \mathbb{R}$ is the control input. The main control objective for the system of (1.1) can be stated as follows: given a sufficiently smooth, desired trajectory for $x(t)$, denoted by $x_d(t)$, design a control law $u(t)$ such that $x(t) \to x_d(t)$ as $t \to \infty$. The solution to this control design problem is based on the paradigm of feedforward and feedback control. To illustrate this paradigm, let us examine a few cases that are dependent on the level of information available with regard to the model of (1.1). For the sake of simplicity, we will assume that $x_d(t) = 0$ $\forall t \geq 0$. First, in the standard case, where $x(t)$, $\dot{x}(t)$ are measurable (i.e., *full-state feedback*) and the nonlinearity $f\left(\cdot\right)$ is exactly known, one could design $u(t)$ as follows:

$$u = \underbrace{f\left(x, \dot{x}, \theta\right)}_{\text{feedforward}} - \underbrace{\left(k_d \dot{x} + k_p x\right)}_{\text{feedback}}. \tag{1.2}$$

This control law is commonly known as a feedback linearizing controller since the nonlinearities are canceled out and the closed-loop system becomes the linear system:

$$\ddot{x} + k_d \dot{x} + k_p x = 0. \tag{1.3}$$

A Lyapunov-based stability analysis of (1.3) can be performed by selecting the following non-negative function:

$$V = \underbrace{\frac{1}{2}\dot{x}^2}_{\text{kinetic energy-type term}} + \overbrace{\frac{1}{2}k_p x^2}^{\text{potential energy-type term}} \tag{1.4}$$

and calculating its derivative along (1.3) as follows:

$$\dot{V} = -k_d \dot{x}^2. \tag{1.5}$$

The application of LaSalle's invariance principle [6] to (1.3) and (1.5) can be used to show that $x(t) \to 0$ as $t \to \infty$ for all $x(0)$, $\dot{x}(0) \in \mathbb{R}$.

Consider now the case, where θ is unknown. Based on the form of (1.2), a good choice for the control law would be the certainty equivalence controller given by

$$u = f(x, \dot{x}, \hat{\theta}) - k_d \dot{x} - k_p x, \tag{1.6}$$

where $\hat{\theta}(t)$ represents an estimate of θ. Note that now the feedforward term of (1.6) will not directly cancel the nonlinearity in (1.1) owing to the mismatch between θ and $\hat{\theta}(t)$; hence, the linear closed-loop system given by (1.3) will not be achieved. The question then becomes how to design $\hat{\theta}(t)$ and select $V(t)$ such that we can still show that $x(t) \to 0$ as $t \to \infty$, $\forall x(0)$, $\dot{x}(0) \in \mathbb{R}$. In anticipation of what will be seen in the subsequent chapters, the solution to this problem is greatly facilitated if the function $f(\cdot)$ can be *linearly parameterized* in the sense that

$$f(x, \dot{x}, \theta) = Y(x, \dot{x})\, \theta, \tag{1.7}$$

where $Y(x, \dot{x}) \in \mathbb{R}$ is a known function dependent only on $x(t)$ and $\dot{x}(t)$.

Consider now the case, where the velocity $\dot{x}(t)$ is not directly available for feedback owing to noisy measurements or the lack of velocity sensors. Since the only state information available to the controller is the position, this problem is usually referred to as the *output feedback* (OFB) problem. If the parameter θ is known, a natural choice for the control law might be to replace all occurrences of the actual velocity with an estimated velocity signal as follows:

$$u = f(x, \dot{\hat{x}}, \theta) - k_d \dot{\hat{x}} - k_p x, \tag{1.8}$$

where $\dot{\hat{x}}(t)$ denotes the estimated velocity. One must now determine how to properly design a velocity observer that generates a velocity surrogate from position measurements while also ensuring that $x(t) \to 0$ as $t \to \infty$.

Further difficulties arise if the parameter θ is also unknown. This additional uncertainty might require an entirely different control approach, a different method of constructing the velocity surrogate, and a different Lyapunov-like function. Another issue that often arises in the OFB problem concerns the set of initial conditions for which the stability result holds. That is, is it still feasible to achieve a *global* stability result (i.e., one that holds for all $x(0), \dot{x}(0) \in \mathbb{R}$)? For example, an OFB controller may only guarantee a *semiglobal* stability result.[3]

1.3 Flexible Mechanical Systems

The term *flexible mechanical system* used in this book refers to systems modeled by PDEs. The presence of partial derivatives in the dynamic equations can be attributed to the independence of the spatial and time variables. That is, the motion of this type of distributed parameter mechanical system depends not only on time but also on the spatial position. Distributed parameter systems are typically governed by a PDE that must be satisfied over all interior points of the domain and a set of boundary conditions. The boundary conditions can either be static or dynamic in nature and must be satisfied at the points bounding the domain.

Some of the more popular methods used to control distributed parameter systems are based on (i) applying traditional linear control techniques to a discretized system model obtained through modal analyses, (ii) applying a distributed control by using smart sensors and actuators, or (iii) using *active boundary* control strategies. The primary disadvantage of designing a controller based on a discretized model is that the controller could potentially excite unmodeled, high-order vibration modes that were neglected during the discretization process (i.e., spillover effects). The primary disadvantage of distributed control is that the sensing and actuation can be costly to implement. In contrast, boundary control techniques utilize the distributed parameter model for control design purposes (hence, spillover effects are avoided) and require relatively few sensors and actuators. For these reasons, we will limit the discussion of control techniques for distributed parameter systems to the boundary control philosophy.

[3] To guarantee a semiglobal stability result, a control gain often has to be adjusted according to the "size" of the initial conditions.

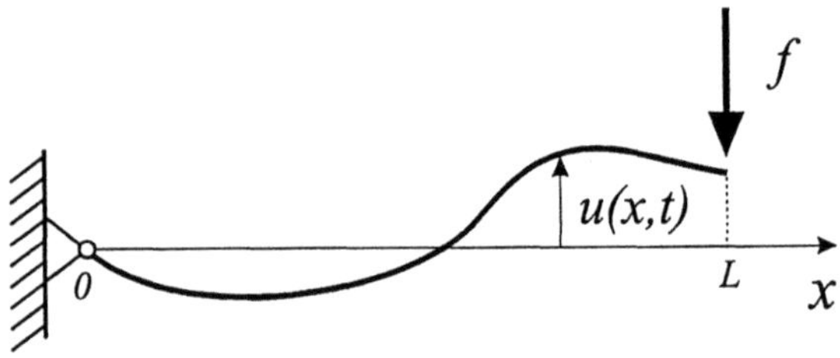

FIGURE 1.1. Typical flexible mechanical system.

To examine the boundary control problem for flexible mechanical systems, let us consider the following illustrative field equation:

$$u_{tt}(x,t) - u_{xx}(x,t) = 0 \tag{1.9}$$

along with the boundary conditions

$$u(0,t) = 0 \qquad u_x(L,t) = f(t), \tag{1.10}$$

where $x \in [0, L]$ denotes the independent position variable; t denotes the independent time variable; $u(x,t) \in \mathbb{R}$ denotes the displacement at position x for time t; the subscripts x, t represent partial derivatives with respect to x, t, respectively; and $f(t) \in \mathbb{R}$ is a control input applied at the boundary position $x = L$. The simple string-like system shown in Figure 1.1 can be seen as a graphical representation of the flexible system described by (1.9) and (1.10).

The control objective for the above system is to design a control law for $f(t)$, a function only of boundary variables, such that vibrations throughout the whole system domain are eliminated (i.e., drive $u(x,t) \to 0\ \forall x \in [0, L]$ as $t \to \infty$). Similar to the concept of feedforward/feedback control outlined in Section 1.2, the solution to this control problem is founded on the paradigm that $f(t)$ should act as an active *damper-like* force that sucks the energy out of the system. The level of sophistication of this damper-like force is directly related to system model. For example, as will be shown below, the linear distributed parameter model of (1.9) and (1.10) requires only a simple damper in the form of a negative boundary velocity feedback term to force $u(x,t)$ to zero $\forall x \in [0, L]$. On the other hand, as will be seen in Chapters 5 to 7, a more complicated flexible, mechanical system model mandates the use of a more complex boundary control law.

To illustrate the boundary control design and analysis procedure, let us consider the following boundary control law for the system described by (1.9) and (1.10):

$$f = -ku_t(L,t), \tag{1.11}$$

where k is a positive scalar control gain. Note that the above controller is only dependent on the measurement of the signal $u_t(x,t)$ at the boundary position $x = L$. The form of (1.11) is based on the concept that negative velocity feedback increases the damping in the system. To illustrate this fact via a Lyapunov-like analysis, one could use the function

$$V(t) = \underbrace{\frac{1}{2}\int_0^L u_t^2(\sigma,t)d\sigma}_{\text{kinetic energy}} + \overbrace{\frac{1}{2}\int_0^L u_\sigma^2(\sigma,t)\,d\sigma}^{\text{potential energy}} + 2\beta\int_0^L \sigma u_t(\sigma,t)\,u_\sigma(\sigma,t)\,d\sigma, \tag{1.12}$$

where β is a small, positive weighting constant that is used to ensure that $V(t)$ is non-negative. After some algebraic manipulation and some integration by parts, the evaluation of the time derivative of (1.12) along (1.9), (1.10), and (1.11) produces

$$\dot{V}(t) \le -\beta\int_0^L \left(u_t^2(\sigma,t) + u_\sigma^2(\sigma,t)\right)d\sigma \tag{1.13}$$

for sufficiently small β. The application of some standard integral inequalities to (1.12) and (1.13) can now be used to show that $u(x,t) \to 0$ $\forall x \in [0,L]$ as $t \to \infty$. It is important to mention that the third term in the definition of (1.12), although seeming to lack a physical interpretation, is crucial in obtaining (1.13).

In contrast to the simple example above, the boundary control problems attacked in this book concentrate on distributed parameter models that are more complex and that are often used to describe a specific engineering application. As a result, the system models will contain the following complicating features: (i) nonlinear field equation; (ii) linear or nonlinear, dynamic boundary condition; (iii) unknown, constant parameters (e.g., tension, mass, inertia, bending stiffness, etc.); and/or (iv) PDE-ODE coupling (i.e., a hybrid, flexible-rigid mechanical system).

1.4 Real-Time Control Implementation

The development of any control algorithm is of limited interest if it cannot be implemented in real-time. The implementation issue for nonlinear controllers is particularly important because the algorithms tend to have a complex structure in comparison to standard linear controllers. In fact, this issue often raises doubts about the feasibility of advanced controllers.

To complement the theoretical control developments presented in the subsequent chapters of this book, we now briefly describe the real-time control environments used in their experimental validations.

In 1993, *WinMotor*, a Windows 3.11-based graphical user interface (GUI) and programming environment, was developed at Clemson University to compile and execute C programs in real-time to control various electromechanical systems [3]. The *WinMotor* environment used a TMS320C30 digital signal processor (DSP) board to execute the control program, and a host Intel-486 personal computer (PC) to perform user interface tasks such as on-line graphing and tuning of control parameters. Data storage was provided in a MATLAB-ready format, thereby allowing the user to analyze the data further. The data acquisition interface consisted of a DS2 encoder interface board from Integrated Motions. As PC CPU technology advanced and real-time operating systems for PCs became widely available, we developed the *Qmotor* control environment [1] to execute both the GUI and the control program on the same CPU, thereby eliminating the need for special purpose DSPs. *Qmotor* is a QNX-based [5] (micro-kernel, real-time operating system) GUI and programming environment that runs on an Intel Pentium-II PC. In addition to the user interface features provided in *WinMotor*, some enhancements were introduced in *Qmotor*, such as the ability to display additional plot variables during a control run and an off-line graphing capability to compare the data stored from previous runs.

References

[1] N. Costescu, D. Dawson, and M. Loffler, Qmotor 2.0 - A Real-Time PC Based Control Environment, *IEEE Control Systems Magazine*, Vol. 19, No. 3, pp. 68 – 76, June 1999.

[2] H. K. Khalil, *Nonlinear Systems*, Upper Saddle River, NJ: Prentice Hall, 1996.

[3] S. Lim, D. Dawson, and P. Vedagarbha, Advanced Motion Control of Mechatronic Systems via High-Speed DSP and Parallel Processing Transputer Network, *Mechatronics - An International Journal*, Vol. 6, No. 1, pp. 101–122, 1996.

[4] L. Meirovitch, *Methods of Analytical Dynamics*, New York, NY: McGraw-Hill, 1970.

[5] http://www.qnx.com.

[6] J. J. Slotine and W. Li, *Applied Nonlinear Control,* Englewood Cliffs, NJ: Prentice Hall, 1991.

2

Control Techniques for Friction Compensation

2.1 Introduction

Friction is a natural phenomenon that affects almost all motion. It has been the subject of extensive studies for centuries, with the main objectives being the design of effective lubricating processes and the understanding of the mechanisms of wear. Whereas friction effects at moderate velocities are somewhat predictable, it is the effect of friction at low velocities that is very difficult to model. The facts that friction changes sign with velocity, is asymmetric about the velocity axis, has evolutionary characteristics, and exhibits the stick-slip phenomenon, etc., aggravates the problem. Although friction effects have been well understood qualitatively, researchers have often relied on experimental data to formulate various mathematical models. A heuristic model for friction was first proposed by Leonardo da Vinci [11] in 1519; however, the model failed to capture the low-velocity friction effects such as the Stribeck effect, presliding displacement, rising static friction, etc., which play a major role in high-precision position/velocity tracking applications. In recent years, several dynamic models have been introduced to describe this highly nonlinear behavior exhibited by friction. For example, Dahl [12] proposed a dynamic model to capture the spring-like behavior during stiction. Canudas et al. [9] proposed a dynamic state-variable model to capture friction effects such as the Stribeck effect, hysteresis, spring-like behavior of stiction, and varying breakaway force.

Precise control of mechanical systems in the presence of friction-related effects is a very challenging task. The coefficients of the various friction-related effects are usually very difficult to measure. In addition, the friction-related coefficients may exhibit evolutionary characteristics; therefore, effective compensation for friction effects via adaptive control seems well motivated.

In the first part of this chapter (Sections 2.2 and 2.3), we consider a reduced-order dynamic representation of the nonlinear friction model proposed in [9], which incorporates viscous, Coulomb, static, and Stribeck friction-related effects. A second-order single-input, single-output (SISO) mechanical system augmented with this reduced-order friction model is used as the basis for the design of two adaptive tracking controllers and one adaptive setpoint controller. For benchmark purposes, the first adaptive tracking controller is based on a traditional Lyapunov design framework. That is, the controller and the parameter update law are designed in conjunction via a single Lyapunov function to ensure global asymptotic tracking. This standard adaptive controller is reminiscent of the SISO version of the well-known, adaptive controller of Slotine and Li [30].

The use of a single Lyapunov function often restricts the design of the parameter update law to a standard, gradient-type adaptation driven by the position and velocity tracking errors. Since gradient-type update laws often exhibit slow parameter convergence, the need for an adaptive control methodology that provides flexibility in design of the parameter estimation update law is well reasoned. Passivity-based adaptive controllers provide for some flexibility in the parameter identifier design; however, the designed update law must satisfy a passive mapping condition [20, 25]. Although mainly applied to linear systems, the *estimation-based* approach to adaptive control allows for further flexibility in the construction of parameter update laws (e.g., prediction error[1]-based gradient or normalized/unnormalized least-squares update laws can be used in lieu of a standard, gradient-type update law). This flexibility is achieved as a result of a modular design of the controller and the update law. Recent work by Krstić and Kokotović [19] has successfully extended previous linear, estimation-based techniques to a class of parametric-strict-feedback, nonlinear systems by cleverly exploiting the nonlinear damping tool (see Lemma A.10 in Appendix A for details). Motivated by the above facts, the second adaptive tracking controller is based on the modular design approach proposed in [19]. Specifically, the

[1] The prediction error is defined as the difference between an estimated, filtered version of the the mechanical system dynamics and a filtered version of the control input [30].

controller is designed to ensure *input-to-state stability* with respect to the parameter estimation error while the parameter update law is used to compensate for the unknown, constant parameters that appear linearly in the model. The design of the controller and the update law is justified by means of a composite stability analysis that illustrates global asymptotic position tracking.

Whereas the effects owing to viscous and Coulomb friction can be represented by a linearly parameterized model, the Stribeck effect is often modeled as a nonlinear velocity-dependent function with an unknown, constant parameter appearing nonlinearly in the model. To make this problem more tractable, several control designs resort to using approximate linearly parameterizable models to represent the Stribeck effect. With the goal of avoiding this approximation, we then address the position setpoint control problem for mechanical systems with a nonlinearly parameterized Stribeck model. Specifically, we utilize the technique discussed in [14, 24] to design an adaptive controller that compensates for parametric uncertainties throughout the entire mechanical system, including the Stribeck friction constant even though it does not appear linearly in the model.

In the latter part of this chapter (Sections 2.4 and 2.5), we consider for control design purposes the second-order SISO mechanical system along with the *full-order* friction model proposed in [9]. A detrimental aspect of the full-order friction model is the existence of an *unmeasurable*, dynamic friction state that mandates the inclusion of a friction state estimator in the control scheme. To this end, we demonstrate how the mechanical system/friction dynamics can be exploited to design different observers or filters that lead to the development of a class of *partial-state feedback*, position tracking controllers. First, by assuming exact knowledge of the system parameters, we present two model-based observer/controllers that share a common control law structure but differ in the observer construction. Specifically, while the first observer only partially exploits the system model and provides asymptotic position tracking, the second observer has a more model-based structure and produces exponential position tracking. Next, we consider uncertainty in selected parameters of the mechanical system/friction dynamics and present two adaptive controllers that achieve asymptotic position tracking. The first adaptive controller utilizes a nonlinear observer/filter structure to compensate for uncertainty in all of the system parameters except for those associated with the Stribeck effect. The second adaptive controller utilizes a different nonlinear observer/filter to compensate for only a single parameter associated with normal force variation in the Stribeck effect.

2.2 Reduced-Order Friction Model

The mathematical model for a SISO mechanical system in the presence of friction is assumed to be of the form

$$M\ddot{q} + B\dot{q} + \left(F_c + F_s \exp\left(-F_\tau \dot{q}^2\right)\right)\operatorname{sat}(\dot{q}) = \tau, \tag{2.1}$$

where $q(t), \dot{q}(t), \ddot{q}(t) \in \mathbb{R}$ represent the position, velocity, and acceleration, respectively; M denotes the constant inertia of the system; B denotes the constant viscous friction coefficient; F_c denotes the Coulomb friction-related constant; F_s denotes the static friction-related constant; F_τ is a positive constant corresponding to the Stribeck effect; and $\tau(t) \in \mathbb{R}$ is the control input. The model of (2.1) is a simplified version of the seven-parameter model proposed in [1]. Specifically, this model ignores frictional effects such as presliding-displacement, frictional memory, and rising-static friction.

Remark 2.1 *The saturation function used in (2.1) is defined such that* $\operatorname{sat}(0) = 0$, $\dot{q}\operatorname{sat}(\dot{q}) \geq 0$, *and* $\operatorname{sat}(\dot{q}) \in \mathcal{L}_\infty$. *For the modular adaptive controller presented in Section 2.3.2, the* $\operatorname{sat}(\cdot)$ *function will be defined as the standard, discontinuous signum function. In contrast, the setpoint controller presented in the Section 2.3.3 will require the* $\operatorname{sat}(\cdot)$ *function to be first-order differentiable (i.e.,* $\frac{\partial \operatorname{sat}(\dot{q})}{\partial \dot{q}} \in \mathcal{L}_\infty \ \forall \dot{q} \in \mathcal{L}_\infty$*), or a non-multivalued, discontinuous function as shown below:*

$$\operatorname{sat}(\cdot) = \begin{cases} 1 & \text{for } \dot{q} > 0 \\ 0 & \text{for } \dot{q} = 0 \\ -1 & \text{for } \dot{q} < 0 \end{cases}. \tag{2.2}$$

An example of a first-order, differentiable saturation function is $\tan^{-1}(\kappa\dot{q})$, *where* κ *is some large positive constant. Note that*

$$\kappa\dot{q}\tan^{-1}(\kappa\dot{q}) \geq 0, \ \tan^{-1}(\kappa\dot{q}) \in \mathcal{L}_\infty, \ \frac{\partial \tan^{-1}(\kappa\dot{q})}{\partial \dot{q}} = \frac{\kappa}{1+\kappa^2\dot{q}^2} \in \mathcal{L}_\infty. \tag{2.3}$$

The mechanical system given by (2.1) can be rewritten in the following concise form:

$$M\ddot{q} + W(\dot{q})\theta = \tau, \tag{2.4}$$

where $W(\dot{q}) \in \mathbb{R}^{1\times 3}$ is the regression vector given by

$$W(\dot{q}) = \left[\begin{array}{ccc} \dot{q} & \operatorname{sat}(\dot{q}) & \operatorname{sat}(\dot{q})\exp\left(-F_\tau \dot{q}^2\right) \end{array}\right], \tag{2.5}$$

and $\theta \in \mathbb{R}^3$ is the vector of constant parameters defined as

$$\theta = \begin{bmatrix} B & F_c & F_s \end{bmatrix}^T . \tag{2.6}$$

Since we are interested in compensating for friction, we will assume in the subsequent control designs that the coefficients B, F_c, and F_s are unknown constants; hence, θ will be an unknown parameter vector. However, we will assume that the rotor inertia M and the Stribeck parameter F_τ are known; hence, $W(\dot{q})$ will be known.

Remark 2.2 *Since the friction parameter F_τ does not appear linearly in (2.1), it is not obvious how to compensate for uncertainty associated with F_τ for the case of position tracking applications. Also, note that the subsequent control strategies can easily be modified to cope with uncertainty associated with the parameter M; however, this minor extension has been ignored because it is deemed uninteresting.*

2.3 Control Designs for Reduced-Order Model

In this section, we present the design of three adaptive controllers for the dynamics given by (2.4). The control objective is to achieve position/velocity tracking or regulation despite the uncertainty associated with the parameter vector θ of (2.6). To this end, we define the position and velocity tracking error signals as follows:

$$e(t) = q_d(t) - q(t) \qquad \dot{e}(t) = \dot{q}_d(t) - \dot{q}(t), \tag{2.7}$$

where $q_d(t)$ is the desired position trajectory that must be constructed to ensure that $q_d(t), \dot{q}_d(t), \ddot{q}_d(t) \in \mathcal{L}_\infty$. Since the control objective is to be met under the constraint of parametric uncertainty, the controllers of this section will contain an adaptation law to estimate the unknown parameters. The difference between the actual and estimated parameters is defined by

$$\tilde{\theta}(t) = \theta - \hat{\theta}(t), \tag{2.8}$$

where $\tilde{\theta}(t) \in \mathbb{R}^3$ denotes the parameter estimation error vector, and $\hat{\theta}(t) \in \mathbb{R}^3$ denotes the dynamic estimate of θ.

In addition, we define the filtered tracking error as

$$r(t) = \dot{e}(t) + \alpha e(t), \tag{2.9}$$

where α is a positive control gain. Note that the Laplace transform of (2.9) yields the following strictly proper, exponentially stable, transfer function between $r(t)$ and $e(t)$:

$$\frac{E(s)}{R(s)} = \frac{1}{s+\alpha}, \tag{2.10}$$

where s denotes the Laplace transform variable. In addition, the definition of (2.9) will allow us to consider the second-order dynamic equation of (2.4) as a first-order equation, thereby simplifying the control design. To see this, we rewrite the dynamics of (2.4) in terms of the definition given in (2.9) as described below. First, we differentiate (2.9) with respect to time and multiply both sides by M to yield

$$M\dot{r} = M(\ddot{q}_d + \alpha\dot{e}) - M\ddot{q}, \tag{2.11}$$

where we have used the fact developed from (2.7) that $\ddot{e} = \ddot{q}_d - \ddot{q}$. After utilizing the system dynamics of (2.4) to substitute for $M\ddot{q}$ into (2.11), we have

$$M\dot{r} = M(\ddot{q}_d + \alpha\dot{e}) + W(\dot{q})\theta - \tau. \tag{2.12}$$

The above first-order, nonlinear, ordinary differential equation represents the open-loop dynamics of $r(t)$ and will be used as the foundation for the design of the first two adaptive controllers of this section.

2.3.1 Standard Adaptive Control

To formulate the standard adaptive controller, we use the structure of the open-loop dynamics of (2.12) to design the control input $\tau(t)$ as follows:

$$\tau = M(\ddot{q}_d + \alpha\dot{e}) + W(\dot{q})\hat{\theta} + kr, \tag{2.13}$$

where k is a positive control gain and $\hat{\theta}(t)$ was defined in (2.8). Motivated by the subsequent Lyapunov-type stability analysis, the parameter estimate vector $\hat{\theta}(t)$ is updated using the adaptation algorithm,

$$\dot{\hat{\theta}} = \Gamma W^T(\dot{q})r, \tag{2.14}$$

where $\Gamma \in \mathbb{R}^{3\times 3}$ is a diagonal, positive-definite, adaptation gain matrix. After substituting the control input of (2.13) into (2.12), we obtain the following closed-loop dynamics for $r(t)$:

$$M\dot{r} = W(\dot{q})\tilde{\theta} - kr, \tag{2.15}$$

where $\tilde{\theta}(t)$ was defined in (2.8). In addition, by differentiating (2.8) with respect to time, we can use (2.14) to form the following closed-loop dynamics for the parameter estimation error:

$$\dot{\tilde{\theta}} = -\Gamma W^T(\dot{q})r. \tag{2.16}$$

To analyze the stability of the closed-loop dynamics given by (2.15) and (2.16), we define the non-negative function

$$V(t) = \frac{1}{2}Mr^2 + \frac{1}{2}\tilde{\theta}^T\Gamma^{-1}\tilde{\theta}. \tag{2.17}$$

After taking the time derivative of (2.17), and then substituting for $M\dot{r}$ from (2.15), we obtain

$$\begin{aligned} \dot{V} &= M\dot{r}r + \tilde{\theta}^T\Gamma^{-1}\dot{\tilde{\theta}} \\ &= -kr^2 + \tilde{\theta}^T\left(W^T(\dot{q})r + \Gamma^{-1}\dot{\tilde{\theta}}\right). \end{aligned} \tag{2.18}$$

After substituting (2.16) into (2.18) for $\dot{\tilde{\theta}}(t)$, we get

$$\dot{V} = -kr^2. \tag{2.19}$$

From (2.19) it is clear that $\dot{V}(t)$ is a nonpositive function; hence, we know that $V(t)$ is either decreasing or constant. Since $V(t)$ of (2.17) is a non-negative function (i.e., $V(t)$ is lower-bounded by zero), we can conclude that $V(t) \in \mathcal{L}_\infty$; hence, $r(t) \in \mathcal{L}_\infty$ and $\tilde{\theta}(t) \in \mathcal{L}_\infty$. Since $r(t) \in \mathcal{L}_\infty$, we can use (2.10) and Lemma A.8 in Appendix A to illustrate that $e(t), \dot{e}(t) \in \mathcal{L}_\infty$; hence, owing to the boundedness of $q_d(t)$ and $\dot{q}_d(t)$, we can use (2.7) to conclude that $q(t), \dot{q}(t) \in \mathcal{L}_\infty$. Since $\tilde{\theta}(t) \in \mathcal{L}_\infty$ and θ is a constant vector, (2.8) can be used to show that $\hat{\theta}(t) \in \mathcal{L}_\infty$. From the above boundedness statements and the fact that $\ddot{q}_d(t)$ is assumed bounded, the definition of (2.5) can be used to state that $W(\dot{q}) \in \mathcal{L}_\infty$. It is now easy to see from (2.13) that the control input $\tau(t) \in \mathcal{L}_\infty$. The above information can be applied to (2.4) and (2.15) to illustrate that $\ddot{q}(t), \dot{r}(t) \in \mathcal{L}_\infty$. Thus, we have illustrated that all signals in the adaptive controller and the system remain bounded during closed-loop operation.

We have already proved that $r(t), \dot{r}(t) \in \mathcal{L}_\infty$. If we can now show that $r(t) \in \mathcal{L}_2$, then Lemma A.3 in Appendix A can be exploited to prove that $r(t)$ goes to zero. To this end, we integrate with respect to time, both sides of (2.19) to produce

$$k\int_0^\infty \|r(t)\|^2 dt \leq -\int_0^\infty \dot{V}(t)dt. \tag{2.20}$$

The evaluation of the integral on the right-hand side of (2.20) yields

$$k\int_0^\infty \|r(t)\|^2 dt \leq V(0) - V(\infty) \leq V(0) < \infty, \tag{2.21}$$

where we have used the fact that $V(0) \geq V(\infty) \geq 0$ because $\dot{V} \leq 0$. The above inequality can be rewritten as follows:

$$\sqrt{\int_0^\infty \|r(t)\|^2 \, dt} \leq \sqrt{\frac{V(0)}{k}} < \infty, \tag{2.22}$$

which indicates from Definition A.1 in Appendix A that $r(t) \in \mathcal{L}_2$. We can now invoke Lemma A.3 to conclude that

$$\lim_{t\to\infty} r(t) = 0. \tag{2.23}$$

Finally, we can use (2.10) and (2.23) along with Lemma A.8 in Appendix A to prove the global asymptotic stability of the position and velocity tracking errors in the sense that

$$\lim_{t\to\infty} e(t), \dot{e}(t) = 0. \tag{2.24}$$

2.3.2 Modular Adaptive Control

We now develop an adaptive position tracking controller that achieves controller/update law modularity. Specifically, we first design a control law to achieve input-to-state stability (ISS) with the parameter estimation error $\tilde{\theta}(t)$ treated as a disturbance input to the filtered tracking error dynamics; hence, the controller guarantees closed-loop stability for any adaptive update law design, which ensures that the parameter estimates remain bounded. We then illustrate how the proposed control law yields global asymptotic position tracking for any parameter update law design, which also ensures that the prediction error is square integrable.

In order to facilitate the modular control design, we first define the *filtered* regression vector as follows:[2] [30]

$$\dot{W}_f(\dot{q}) + \beta W_f(\dot{q}) = \beta W(\dot{q}) \qquad W_f(\dot{q}(0)) = 0, \tag{2.25}$$

where β is a positive, constant design parameter. As in [19], the open-loop dynamics of (2.12) motivates us to design the control input $\tau(t)$ as follows:

$$\begin{aligned} \tau = \; & M(\ddot{q}_d + \alpha\dot{e}) + W(\dot{q})\hat{\theta} + kr \\ & + k_n \|W(\dot{q})\|^2 r + \frac{1}{\beta} W_f(\dot{q})\,\dot{\hat{\theta}} + k_n \left(\frac{1}{\beta} W_f(\dot{q})\,\dot{\hat{\theta}}\right)^2 r, \end{aligned} \tag{2.26}$$

[2] See [21] for details on how to compute $W_f(\dot{q})$ without the need for acceleration measurements.

where k is a positive control gain, and k_n is a positive, nonlinear damping gain. To give some insight into the structure of the control input, we note that the terms on the first line of (2.26) are standard feedforward/feedback terms, while the terms on the second line are nonlinear damping functions injected to achieve the ISS property of the closed-loop system with respect to $\tilde{\theta}(t)$. After substituting the control input of (2.26) into (2.12), we obtain the following closed-loop system:

$$M\dot{r} = -kr + W(\dot{q})\tilde{\theta} - k_n \left\| W(\dot{q}) \right\|^2 r - \frac{1}{\beta} W_f(\dot{q})\,\dot{\hat{\theta}} - k_n \left(\frac{1}{\beta} W_f(\dot{q})\,\dot{\hat{\theta}} \right)^2 r. \tag{2.27}$$

In order to show that the controller achieves boundedness of $r(t)$, we state the following theorem.

Theorem 2.1 *Given the closed-loop dynamics of (2.27), if* $\tilde{\theta}(t) \in \mathcal{L}_\infty$, *then*

$$|r(t)|^2 \leq |r(0)|^2 \exp\left(-\frac{2k}{M} t \right) + \frac{1}{kk_n} \left(\left\| \tilde{\theta} \right\|_\infty^2 + 1 \right), \tag{2.28}$$

where $\left\| \cdot \right\|_\infty$ *denotes the infinity norm; hence,* $r(t) \in \mathcal{L}_\infty$.

Proof. To prove the ISS property of the proposed controller, we define the following non-negative function:

$$V_1 = \frac{1}{2} M r^2. \tag{2.29}$$

After taking the time derivative of (2.29) along the closed-loop expression given by (2.27), we get

$$\begin{aligned} \dot{V}_1 = \; & -kr^2 + \left[W(\dot{q}) r \tilde{\theta} - k_n \left\| W(\dot{q}) \right\|^2 r^2 \right] \\ & + \left[-\frac{1}{\beta} W_f(\dot{q})\,\dot{\hat{\theta}}\, r - k_n \left(\frac{1}{\beta} W_f(\dot{q})\,\dot{\hat{\theta}} \right)^2 r^2 \right]. \end{aligned} \tag{2.30}$$

After applying the nonlinear damping argument given by Lemma A.10 in Appendix A to the bracketed terms of (2.30), we obtain the following upper bound on $\dot{V}_1(t)$:

$$\dot{V}_1 \leq -\frac{2k}{M} V_1 + \frac{\left\| \tilde{\theta} \right\|^2}{k_n} + \frac{1}{k_n}, \tag{2.31}$$

where (2.29) has been used. If $\tilde{\theta}(t) \in \mathcal{L}_\infty$, by application of Lemma A.5 in Appendix A, the above inequality can be solved to yield

$$V_1(t) \leq V_1(0) \exp\left(-\frac{2k}{M} t \right) + \frac{M}{2kk_n} \left(\left\| \tilde{\theta} \right\|_\infty^2 + 1 \right), \tag{2.32}$$

which, by the use of (2.29), can be rewritten as

$$\frac{1}{2}Mr^2(t) \leq \frac{1}{2}Mr^2(0)\exp\left(-\frac{2k}{M}t\right) + \frac{M}{2kk_n}\left(\left\|\tilde{\theta}\right\|_\infty^2 + 1\right). \tag{2.33}$$

It is now easy to see the result of (2.28); thus, we can infer that $r(t) \in \mathcal{L}_\infty$. Note that, since $r(t) \in \mathcal{L}_\infty$, we can use (2.10) and invoke Lemma A.8 in Appendix A to also illustrate that $e(t), \dot{e}(t) \in \mathcal{L}_\infty$ (and hence, $q(t), \dot{q}(t) \in \mathcal{L}_\infty$, owing to the boundedness of $q_d(t)$ and $\dot{q}_d(t)$). □

Parameter Update Laws

Since the ISS property is achieved by the controller, we are left with the design of adaptive update laws that ensure boundedness of the parameter estimates. To this end, we define the *prediction error*

$$\epsilon = \tau_f - \hat{\tau}_f, \tag{2.34}$$

where $\tau_f(t) \in \mathbb{R}$ is the filtered control input [21] defined as follows:

$$\dot{\tau}_f = -\beta\tau_f + \beta\tau \qquad \tau_f(0) = 0, \tag{2.35}$$

and $\hat{\tau}_f(t) \in \mathbb{R}$ is the estimated, filtered control input given by

$$\hat{\tau}_f = \beta M\dot{q} - \zeta + W_f(\dot{q})\hat{\theta}, \tag{2.36}$$

with β being the same constant defined in (2.25), $W_f(\dot{q})$ being defined in (2.25), and $\zeta(t) \in \mathbb{R}$ being an auxiliary filter variable defined as follows:

$$\dot{\zeta} = -\beta\zeta + \beta^2 M\dot{q} \qquad \zeta(0) = 0. \tag{2.37}$$

From (2.36) and (2.37), it is not difficult to see that the prediction error $\epsilon(t)$ is a measurable signal.

Upon substituting (2.4) into (2.35) for $\tau(t)$, we obtain the following expression:

$$\dot{\tau}_f + \beta\tau_f = \beta M\ddot{q} + \beta W(\dot{q})\theta. \tag{2.38}$$

Upon differentiating (2.36) with respect to time, and then utilizing (2.36) and (2.37) to eliminate the variable $\zeta(t)$, we obtain the following expression:

$$\dot{\hat{\tau}}_f + \beta\hat{\tau}_f = \beta M\ddot{q} + \beta W_f(\dot{q})\hat{\theta} + \frac{d}{dt}\left(W_f(\dot{q})\hat{\theta}\right). \tag{2.39}$$

After subtracting (2.39) from (2.38), and utilizing (2.25) and (2.34), we get

$$\dot{\epsilon} + \beta\epsilon = \frac{d}{dt}\left(W_f(\dot{q})\tilde{\theta}\right) + \beta\left(W_f(\dot{q})\tilde{\theta}\right). \tag{2.40}$$

From the structure of (2.40), we can see that an *unmeasurable* representation for the prediction error $\epsilon(t)$ can be defined as follows:

$$\epsilon = W_f(\dot{q})\tilde{\theta}. \tag{2.41}$$

We are now in a position to design the adaptive update laws to estimate θ. Specifically, we employ the following gradient and least-squares-type update laws [19]:

$$\begin{array}{ll} \text{Gradient update law:} & \dot{\hat{\theta}} = \Gamma W_f^T(\dot{q})\epsilon, \quad \text{where} \quad \Gamma = \Gamma^T > 0. \\ & \\ \text{Least-squares update law:} & \dot{\hat{\theta}} = \Gamma(t) W_f^T(\dot{q})\epsilon, \\ & \text{where} \quad \dot{\Gamma} = -\Gamma W_f^T(\dot{q}) W_f(\dot{q})\Gamma \\ & \qquad\qquad \Gamma(0) = \Gamma^T(0) > 0. \end{array} \tag{2.42}$$

The adaptive update laws given by (2.42) not only ensure boundedness of the parameter estimation error but also guarantee that $\epsilon(t) \in \mathcal{L}_2$. In order to delineate this fact, we state the following theorem.

Theorem 2.2 *The update laws given by (2.42) guarantee that $\tilde{\theta}(t) \in \mathcal{L}_\infty$; furthermore, the proposed update laws guarantee that $\epsilon(t) \in \mathcal{L}_2$.*

Proof. To prove this theorem, we utilize the following function:

$$V_2 = \tilde{\theta}^T \Gamma^{-1} \tilde{\theta}. \tag{2.43}$$

After taking the time derivative of $V_2(t)$ along (2.42), we obtain

$$\dot{V}_2 = \begin{cases} -2\tilde{\theta}^T \Gamma^{-1} \dot{\hat{\theta}} & \text{Gradient update law} \\ & \\ -2\tilde{\theta}^T \Gamma^{-1} \dot{\hat{\theta}} - \tilde{\theta}^T \Gamma^{-1} \dot{\Gamma} \Gamma^{-1} \tilde{\theta} & \text{Least-squares update law.} \end{cases} \tag{2.44}$$

After substituting for $\dot{\hat{\theta}}(t)$ from (2.42), utilizing (2.41) to substitute for ϵ, and then simplifying the resulting expression, we obtain the following upper bound on $\dot{V}_2(t)$:

$$\dot{V}_2 \leq \begin{cases} -2\epsilon^2 & \text{Gradient update law} \\ & \\ -\epsilon^2 & \text{Least-squares update law.} \end{cases} \tag{2.45}$$

For the gradient update law, Γ is a constant, positive-definite matrix; thus, $V_2(t)$ of (2.43) is a non-negative function. Since from (2.45) we know

$\dot{V}_2(t)$ is nonpositive, we can infer that $V_2(t)$ is either decreasing or constant; hence, $V_2(t) \in \mathcal{L}_\infty$, which implies that $\tilde{\theta}(t) \in \mathcal{L}_\infty$. After integrating both sides of the first inequality in (2.45), and following a similar procedure as in (2.20)–(2.22), we can easily show that $\epsilon(t) \in \mathcal{L}_2$.

For the least-squares update law, we will first show that $\Gamma^{-1}(t)$ in (2.43) is bounded. To this end, after applying the identity given by

$$\dot{\Gamma}^{-1} = -\Gamma^{-1}\dot{\Gamma}\Gamma^{-1} \tag{2.46}$$

to (2.42), we can show that the solution for $\Gamma^{-1}(t)$ is given by

$$\Gamma^{-1}(t) = \Gamma^{-1}(0) + \int_0^t W_f^T(\dot{q}(\sigma))\, W_f(\dot{q}(\sigma))\, d\sigma. \tag{2.47}$$

It is easy to see that $W_f^T(\dot{q})\, W_f(\dot{q})$ is positive semidefinite, which implies that

$$\int_0^t W_f^T(\dot{q}(\sigma))\, W_f(\dot{q}(\sigma))\, d\sigma > 0; \tag{2.48}$$

thus, it follows from (2.47) that $\Gamma^{-1}(t) \leq \Gamma^{-1}(0)$. If $\Gamma(0)$ is selected to be positive-definite and symmetric according to (2.42), then $\Gamma^{-1}(0)$ is also positive-definite and symmetric. Therefore, it follows that both $\Gamma^{-1}(t)$ and $\Gamma(t)$ are positive-definite and symmetric. From (2.42), we know that $\dot{\Gamma}(t)$ is negative semidefinite; therefore, $\Gamma(t)$ is always constant or decreasing, and it follows that $\Gamma(t) \in \mathcal{L}_\infty$. Since $\Gamma(t) \in \mathcal{L}_\infty$, $V_2(t)$ is non-negative, and $\dot{V}_2(t)$ is nonpositive, we can infer that $V_2(t) \in \mathcal{L}_\infty$, which implies that $\tilde{\theta}(t) \in \mathcal{L}_\infty$. Finally, after integrating both sides of the second inequality in (2.45), and again following a similar procedure as in (2.20)–(2.22), we can show that $\epsilon(t) \in \mathcal{L}_2$. □

Composite Stability Analysis

We are now in a position to perform a composite stability analysis for the closed-loop system given by (2.27) with the adaptive update laws given by (2.42).

Theorem 2.3 *The adaptive controller of (2.26) along with the update laws given by (2.42) ensures that all signals remain bounded during closed-loop operation; furthermore, it ensures global asymptotic tracking as illustrated by*

$$\lim_{t\to\infty} e(t), \dot{e}(t) = 0. \tag{2.49}$$

Proof. From the proof of Theorem 2.2, we know that $\tilde{\theta}(t), \hat{\theta}(t) \in \mathcal{L}_\infty$. Since $\hat{\theta}(t) \in \mathcal{L}_\infty$, we can apply Theorem 2.1 to show that $r(t) \in \mathcal{L}_\infty$.

Since $r(t) \in \mathcal{L}_\infty$, we can invoke Lemma A.8 in Appendix A to show that $e(t), \dot{e}(t) \in \mathcal{L}_\infty$; therefore, it follows from (2.7) that $q(t), \dot{q}(t) \in \mathcal{L}_\infty$. Since $\dot{q}(t) \in \mathcal{L}_\infty$, we can show that $W_f(\dot{q}) \in \mathcal{L}_\infty$, and since $\tilde{\theta}(t) \in \mathcal{L}_\infty$, we know from (2.41) that $\epsilon(t) \in \mathcal{L}_\infty$. It is now easy to see from (2.42) that $\dot{\hat{\theta}}(t) \in \mathcal{L}_\infty$. From the above boundedness statements, we use (2.27) to prove that $\dot{r}(t) \in \mathcal{L}_\infty$. Therefore, we have asserted that all signals remain bounded during closed-loop operation.

In order to prove that $\lim_{t\to\infty} r(t) = 0$, we consider the following variable transformation:

$$x = Mr - \frac{1}{\beta}\epsilon. \tag{2.50}$$

After taking the time derivative of $x(t)$ along the closed-loop expression given by (2.27), substituting for $\dot{\epsilon}$ from (2.41), and then simplifying the resulting expression by the use of (2.25), we obtain

$$\dot{x} = -kr - k_n \left\| W(\dot{q}) \right\|^2 r - k_n \left(\frac{1}{\beta} W_f(\dot{q})\, \dot{\hat{\theta}} \right)^2 r + \epsilon \triangleq -A(\dot{q})r + \epsilon, \tag{2.51}$$

where $A(\dot{q})$ is a positive function. Using the relationship given by (2.50), we can rewrite (2.51) in terms of only $x(t)$ and $\epsilon(t)$ as follows:

$$\dot{x} = -\frac{A(\dot{q})}{M}x + C(\dot{q})\epsilon, \tag{2.52}$$

where $C(\dot{q})$ is defined by

$$C(\dot{q}) = \left(1 - \frac{A(\dot{q})}{M\beta}\right). \tag{2.53}$$

We now differentiate the non-negative expression $\frac{1}{2}x^2$ along (2.52) to obtain

$$\begin{aligned} \frac{d}{dt}\left(\frac{1}{2}x^2\right) &= -\frac{A(\dot{q})}{M}x^2 + C(\dot{q})x\epsilon \\ &\leq -\frac{A(\dot{q})}{M}x^2 + \left\| C(\dot{q}) \right\|_\infty \left\| \epsilon \right\| \left\| x \right\|. \end{aligned} \tag{2.54}$$

We now let

$$y = \sqrt{\frac{1}{2}x^2}, \tag{2.55}$$

and rewrite (2.54) as

$$\frac{d}{dt}\left(y^2\right) = 2y\dot{y} \leq -\frac{A(\dot{q})}{M}y^2 + \sqrt{2}\left\| C(\dot{q}) \right\|_\infty \left\| \epsilon \right\| y. \tag{2.56}$$

Since $y(t) \geq 0$, we can use (2.56) to produce the following upper bound for $\dot{y}(t)$:

$$\dot{y} \leq -\zeta_1 y + \zeta_2 \|\epsilon\|, \tag{2.57}$$

where the positive scalar constants[3] ζ_1 and ζ_2 are defined as

$$\zeta_1 = \inf_t \{A(\dot{q})\} \frac{1}{M} \quad \text{and} \quad \zeta_2 = \frac{\sqrt{2}}{2} \|C(\dot{q})\|_\infty. \tag{2.58}$$

The solution of the differential inequality of (2.57) is given by

$$y(t) \leq y(0)e^{-\zeta_1 t} + \zeta_2 \int_0^t e^{-\zeta_1(t-\xi)} \|\epsilon(\xi)\| \, d\xi. \tag{2.59}$$

After utilizing Holder's inequality [33], we can rewrite (2.59) as follows:

$$\begin{aligned} |y(t)| &\leq |y(0)| e^{-\zeta_1 t} + \zeta_2 \sqrt{\int_0^t e^{-\zeta_1(t-\xi)} d\xi} \sqrt{\int_0^t e^{-\zeta_1(t-\xi)} \|\epsilon(\xi)\|^2 d\xi} \\ &\leq |y(0)| e^{-\zeta_1 t} + \frac{\zeta_2}{\sqrt{\zeta_1}} \sqrt{\int_0^t e^{-\zeta_1(t-\xi)} \|\epsilon(\xi)\|^2 d\xi}. \end{aligned} \tag{2.60}$$

After squaring both sides of (2.60) and integrating the resulting expression, we obtain

$$\int_0^t |y(\sigma)|^2 d\sigma \leq \frac{|y(0)|^2}{\zeta_1} + \frac{2\zeta_2^2}{\zeta_1} \int_0^t \left(\int_0^\sigma e^{-\zeta_1(\sigma-\xi)} \|\epsilon(\xi)\|^2 d\xi \right) d\sigma. \tag{2.61}$$

If we reverse the order of integration in (2.61), we have

$$\begin{aligned} \int_0^t |y(\sigma)|^2 d\sigma &\leq \frac{|y(0)|^2}{\zeta_1} + \frac{2\zeta_2^2}{\zeta_1} \int_0^t e^{\zeta_1 \xi} \|\epsilon(\xi)\|^2 \left(\int_\xi^t e^{-\zeta_1 \sigma} d\sigma \right) d\xi \\ &\leq \frac{|y(0)|^2}{\zeta_1} + \frac{2\zeta_2^2}{\zeta_1} \int_0^t \|\epsilon(\xi)\|^2 \frac{1}{\zeta_1} d\xi. \end{aligned} \tag{2.62}$$

After utilizing the fact that $\varepsilon(t) \in \mathcal{L}_2$, we can use (2.62) to state that

$$\|y\|_2 \leq \frac{|y(0)|^2}{\zeta_1} + \frac{2\zeta_2^2}{\zeta_1^2} \|\epsilon\|_2 < \infty, \tag{2.63}$$

where $\|\cdot\|_2$ denotes the $\mathcal{L}_2$ norm of a signal. From Definition A.1 in Appendix A, we know that the result given by (2.63) illustrates that $y(t) \in \mathcal{L}_2$.

[3] Since we know all signals are bounded from Theorem 2.3, the constants ζ_1 and ζ_2 exist.

Hence, from the definition presented in (2.55), we can see that $x(t) \in \mathcal{L}_2$. Since $x(t), \epsilon(t) \in \mathcal{L}_2$, we can state from (2.50) that $r(t) \in \mathcal{L}_2$. Since $r(t), \dot{r}(t) \in \mathcal{L}_\infty$ and $r(t) \in \mathcal{L}_2$, we can invoke Lemma A.3 in Appendix A to state that $\lim_{t\to\infty} r(t) = 0$; hence, we can use (2.10) and Lemma A.8 in Appendix A to show that $\lim_{t\to\infty} e(t), \dot{e}(t) = 0$. □

2.3.3 Adaptive Setpoint Control

Recall that the adaptive tracking controllers of Sections 2.3.1 and 2.3.2 assumed perfect knowledge of the Stribeck parameter F_τ, which does not appear linearly in the model (see Remark 2.2). We will now design an adaptive position *setpoint* controller (i.e., q_d is now set to some constant) to compensate for uncertainty associated with all of the parameters in the friction model, including the Stribeck parameter F_τ. To this end, we begin by rewriting the mechanical system dynamics given by (2.1) as follows:

$$M\ddot{q} = W_s(\dot{q})\theta_s - F_s \operatorname{sat}(\dot{q}) \exp\left(-F_\tau \dot{q}^2\right) + \tau, \tag{2.64}$$

where

$$W_s(\dot{q}) = \begin{bmatrix} -\dot{q} & -\operatorname{sat}(\dot{q}) \end{bmatrix} \in \mathbb{R}^{1\times 2} \tag{2.65}$$

and

$$\theta_s = \begin{bmatrix} B & F_c \end{bmatrix}^T \in \mathbb{R}^2. \tag{2.66}$$

The structure of (2.64) and the subsequent stability analysis motivate us to design the control input as follows:

$$\tau = -W_s\hat{\theta}_s - k_d\dot{q} + k_p(q_d - q) + \hat{F}_s \operatorname{sat}(\dot{q}) \exp\left(-\hat{F}_\tau \dot{q}^2\right), \tag{2.67}$$

where k_d, k_p are positive control gains, and $\hat{\theta}_s(t) \in \mathbb{R}^2$ and $\hat{F}_s(t), \hat{F}_\tau(t) \in \mathbb{R}$ denote the estimates of ϕ, F_s, and F_τ, respectively, which are updated according to the following adaptation laws:

$$\begin{gathered} \dot{\hat{\theta}}_s = \Gamma_s W_s^T \dot{q} \qquad \dot{\hat{F}}_s = -\gamma_0 \dot{q} \operatorname{sat}(\dot{q}) \exp(-\hat{F}_\tau \dot{q}^2) \\ \dot{\hat{F}}_\tau = \gamma_1 \dot{q}^3 \operatorname{sat}(\dot{q}) \exp(-\hat{F}_\tau \dot{q}^2), \end{gathered} \tag{2.68}$$

where $\Gamma_s \in \mathbb{R}^{2\times 2}$ is a diagonal, positive-definite, adaptation gain matrix and γ_0, γ_1 are positive adaptation gains. After substituting (2.67) into the open-loop system of (2.64), adding and subtracting the term $F_s \operatorname{sat}(\dot{q}) \exp\left(-\hat{F}_\tau \dot{q}^2\right)$, and then simplifying the resulting expression, we obtain the following closed-loop system:

$$\begin{aligned} M\ddot{q} &= W_s(\dot{q})\tilde{\theta}_s - k_d\dot{q} + k_p(q_d - q) - \tilde{F}_s \operatorname{sat}(\dot{q}) \exp\left(-\hat{F}_\tau \dot{q}^2\right) \\ &\quad - F_s \operatorname{sat}(\dot{q}) \left[\exp\left(-F_\tau \dot{q}^2\right) - \exp\left(-\hat{F}_\tau \dot{q}^2\right)\right], \end{aligned} \tag{2.69}$$

where $\tilde{\theta}_s(t) = \theta_s - \hat{\theta}_s(t)$ and $\tilde{F}_s(t) = F_s - \hat{F}_s(t)$ denote the parameter estimation error terms.

In order to examine the stability of the closed-loop system given by (2.69) along with the adaptive update laws of (2.68), we perform the following analysis. We define the non-negative function

$$V = \frac{1}{2}M\dot{q}^2 + \frac{1}{2}k_p(q_d - q)^2 + \frac{1}{2}\tilde{\theta}_s^T \Gamma_s^{-1} \tilde{\theta}_s + \frac{1}{2\gamma_0}\tilde{F}_s^2 + \frac{1}{2}\frac{F_s}{\gamma_1}\tilde{F}_\tau^2, \tag{2.70}$$

where $\tilde{F}_\tau(t) = F_\tau - \hat{F}_\tau(t)$. After taking the time derivative of $V(t)$ along (2.69) and (2.68), we obtain, after some simplifications

$$\dot{V} = -k_d\dot{q}^2 - F_s\dot{q} \operatorname{sat}(\dot{q}) \left[\exp(-F_\tau\dot{q}^2) - \exp(-\hat{F}_\tau\dot{q}^2) + \tilde{F}_\tau\dot{q}^2 \exp(-\hat{F}_\tau\dot{q}^2)\right], \tag{2.71}$$

which can be rewritten as follows:

$$\dot{V} = -k_d\dot{q}^2 - F_s\dot{q} \operatorname{sat}(\dot{q}) \exp(-\hat{F}_\tau\dot{q}^2) \left[\exp(-\tilde{F}_\tau\dot{q}^2) - 1 + \tilde{F}_\tau\dot{q}^2\right]. \tag{2.72}$$

After defining $\varsigma = 1 - \tilde{F}_\tau\dot{q}^2$ in (2.72), we have

$$\dot{V} = -k_d\dot{q}^2 - F_s\dot{q} \operatorname{sat}(\dot{q}) \exp(-\hat{F}_\tau\dot{q}^2) \left[\exp(\varsigma - 1) - \varsigma\right]. \tag{2.73}$$

Since $\operatorname{sat}(\dot{q})\dot{q} \geq 0 \ \forall \dot{q}(t)$ and $\exp(\varsigma - 1) \geq \varsigma \ \forall \varsigma(t)$, $\dot{V}(t)$ can be upper bounded as follows:

$$\dot{V} \leq -k_d\dot{q}^2. \tag{2.74}$$

From (2.70) and (2.74), we can follow similar arguments as in Sections 2.3.1 and 2.3.2 to show that $q(t), \dot{q}(t), \tilde{F}_s(t), \tilde{F}_\tau(t) \in \mathcal{L}_\infty$ and $\tilde{\theta}_s(t) \in \mathcal{L}_\infty$. Since B, F_c, F_s, and F_τ are constant parameters, $\hat{F}_s(t), \hat{F}_\tau(t) \in \mathcal{L}_\infty$ and $\hat{\theta}_s(t) \in \mathcal{L}_\infty$. Since the right-hand side of the equations in (2.67) and (2.68) contain only bounded quantities, we can assert that $\tau(t), \dot{\hat{F}}_s(t), \dot{\hat{F}}_\tau(t) \in \mathcal{L}_\infty$ and $\dot{\hat{\theta}}_s(t) \in \mathcal{L}_\infty$. Similarly, since the right-hand side of (2.64) contains only bounded quantities, we can say that $\ddot{q}(t) \in \mathcal{L}_\infty$. Thus, we can infer that all the signals remain bounded during closed-loop operation.

After integrating both sides of the expression given by (2.74), and then proceeding as in (2.20)–(2.22), we have

$$\sqrt{\int_0^\infty \dot{q}^2(\sigma)d\sigma} \leq \sqrt{\frac{V(0)}{k_d}} < \infty, \tag{2.75}$$

which, according to Definition A.1 in Appendix A, indicates that $\dot{q}(t) \in \mathcal{L}_2$. Since, $\dot{q}(t), \ddot{q}(t) \in \mathcal{L}_\infty$, and $\dot{q}(t) \in \mathcal{L}_2$, we can invoke Lemma A.3 in Appendix A to show that $\lim_{t\to\infty} \dot{q}(t) = 0$. Now, note that $\dot{q}(t)$ can also be rewritten as

$$\dot{q}(t) = \int_0^t \frac{d\dot{q}(\sigma)}{d\sigma} d\sigma + C, \tag{2.76}$$

where C is some constant. Since we have already shown that $\lim_{t\to\infty} \dot{q}(t) = 0$, we can use (2.76) to show that

$$\lim_{t\to\infty} \int_0^t \frac{d\dot{q}(\sigma)}{d\sigma} d\sigma \quad \text{exists and is finite.} \tag{2.77}$$

After differentiating the expression given by (2.69), we obtain

$$\begin{aligned} M \dddot{q} \;=\; & \dot{W}_s \tilde{\theta}_s + W_s \dot{\tilde{\theta}}_s - k_d \ddot{q} - k_p \dot{q} \\ & + \dot{\hat{F}}_s \operatorname{sat}(\dot{q}) \exp(-\hat{F}_\tau \dot{q}^2) - \tilde{F}_s \frac{\partial \operatorname{sat}(\dot{q})}{\partial \dot{q}} \ddot{q} \exp(-\hat{F}_\tau \dot{q}^2) \\ & - \hat{F}_s \operatorname{sat}(\dot{q}) \left(- \dot{\hat{F}}_\tau \dot{q}^2 - 2\hat{F}_\tau \dot{q}\ddot{q} \right) \exp(-\hat{F}_\tau \dot{q}^2) \\ & - F_s \frac{\partial \operatorname{sat}(\dot{q})}{\partial \dot{q}} \ddot{q} \left[\exp(-F_\tau \dot{q}^2) - \exp(-\hat{F}_\tau \dot{q}^2) \right] \\ & - F_s \operatorname{sat}(\dot{q}) \left[-2F_\tau \dot{q}\ddot{q} \exp(-F_\tau \dot{q}^2) \right. \\ & \left. + (\dot{\hat{F}}_\tau \dot{q}^2 + 2\hat{F}_\tau \dot{q}\ddot{q}) \exp(-\hat{F}_\tau \dot{q}^2) \right], \end{aligned} \tag{2.78}$$

where

$$\dot{W}_s = \begin{bmatrix} -\ddot{q} & -\frac{\partial \operatorname{sat}(\dot{q})}{\partial \dot{q}} \ddot{q} \end{bmatrix} \in \mathcal{L}_\infty.$$

Since the right-hand side of (2.78) contains only bounded quantities, we can infer that $\dddot{q}(t) \in \mathcal{L}_\infty$. Since $\dddot{q}(t) \in \mathcal{L}_\infty$, we can state that $\ddot{q}(t)$ is uniformly continuous. Since $\ddot{q}(t)$ is uniformly continuous, we can use (2.77) and Lemma A.2 in Appendix A to show that

$$\lim_{t\to\infty} \ddot{q}(t) = 0. \tag{2.79}$$

Since $\lim_{t\to\infty} \dot{q}(t), \ddot{q}(t) = 0$, we can now see from (2.69) and (2.65) that $\lim_{t\to\infty} q(t) = q_d$. □

Remark 2.3 *Note that LaSalle's invariance principle [30] can be employed to illustrate asymptotic stability if the* sat(·) *function is defined to be the non-multivalued, discontinuous function defined in (2.2). That is, we can find the largest invariant set of (2.68) and (2.69) contained in the set*

$$\Omega = \left\{ (q, \dot{q}, \tilde{\phi}, \tilde{F}_s, \tilde{F}_\tau) \in \mathbb{R}^5 \mid \dot{q} = 0 \right\}. \tag{2.80}$$

Specifically, from (2.74), $\dot{q} = 0$*; hence, from (2.2), the regression vector of (2.65) reduces to*

$$W(\dot{q})|_{\dot{q}=0} = \left[\begin{array}{cc} 0 & 0 \end{array} \right]. \tag{2.81}$$

Since $\dot{q} = 0$*, the acceleration is also equal to zero (i.e.,* $\ddot{q} = 0$*). Therefore, we can use (2.69), (2.2), and (2.81) to show that*

$$k_p(q_d - q) = 0, \tag{2.82}$$

which means

$$q = q_d. \tag{2.83}$$

We can now invoke LaSalle's principle to show that $\lim_{t \to \infty} \dot{q}(t), e(t) = 0$.

2.3.4 Experimental Evaluation

Experimental Setup

A schematic diagram of the experimental setup used to implement the controllers is shown in Figure 2.1. The setup consisted of a 88 N-m switched reluctance motor (NSK Corp., model RS-0810) operating in torque-controlled mode. A metal disk weighing 11 lbs was attached to the rotor concentrically. Friction was introduced via a block of *Teflon* held against the disk, as shown in Figure 2.1. The amount of friction could be varied by adjusting a vice arrangement. An integrally mounted resolver provided rotor position measurement. The resolver signal was converted to quadrature format by a resolver-to-digital converter mounted inside the driver unit. The resolution of the position sensor was 0.00234 deg. The rotor velocity signal was obtained by applying a standard backwards difference algorithm to the position signal with the resulting signal being filtered by a second-order digital filter. A Pentium 266 MHz PC running *Qmotor* provided the environment to write the control algorithm in the C programming language.

The various parameters associated with the friction model were calculated using the following procedure. First, a sinusoidal torque with the amplitude selected to be more than the required static friction torque was applied to the motor. The inertia of the disk along with the rotor inertia were determined using standard procedures. The rotor acceleration was

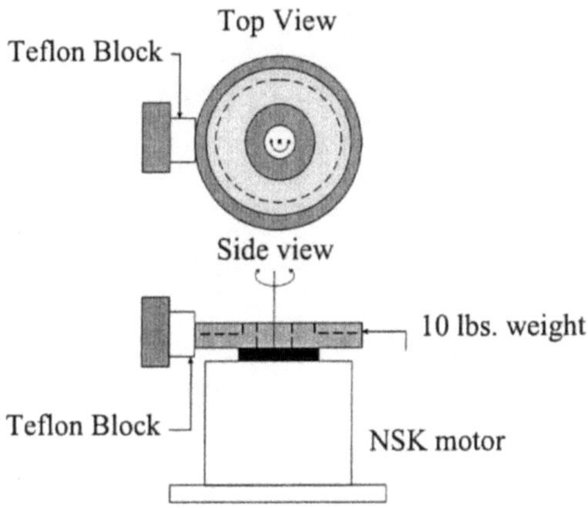

FIGURE 2.1. Schematic diagram of the experimental setup.

computed by applying a backwards difference algorithm to the velocity signal and then filtering the resulting signal using a second-order digital filter. The frictional torque was obtained by subtracting the inertial torque from the applied torque. The parameters of the friction model were then adjusted to fit the experimentally obtained profile of the friction torque. For the model given by (2.1), the numerical values of the parameters were determined to be

$$M = 0.125 \text{ kg-m}^2, \quad B = 1.5 \text{ N-m-sec/rad}, \quad F_c = 3.50 \text{ N-m},$$
$$F_s = 4.90 \text{ N-m}, \quad F_\tau = 0.1890 \text{ sec}^2/\text{rad}^2.$$

The saturation function in (2.1) was defined to be the function given in (2.2).

Experimental Results

Experiment 1: For comparison purposes, the standard, gradient update law-based adaptive controller given by (2.13) and (2.14) was implemented. The desired position trajectory was selected to be (see Figure 2.2)

$$q_d(t) = \tan^{-1}(4\sin(0.5t))\left(1 - \exp(-0.01t^3)\right) \text{ rad.} \tag{2.84}$$

The parameter estimates were initialized to 50% of their nominal measured values (i.e., $\hat{\theta}(0) = 0.5\theta$). The best tracking performance was achieved with the following control and adaptation gains:

$$\alpha = 110, \qquad k_s = 14, \qquad \Gamma = \text{diag}\{1.2, 0.8, 1.5\}.$$

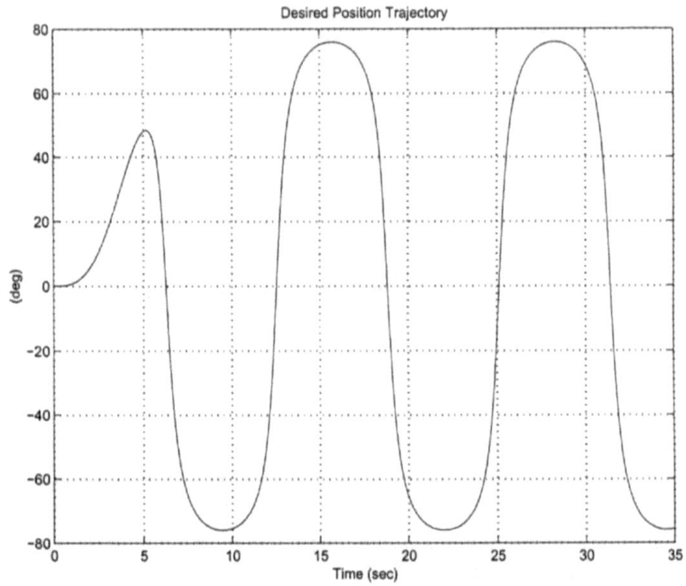

FIGURE 2.2. Desired position trajectory.

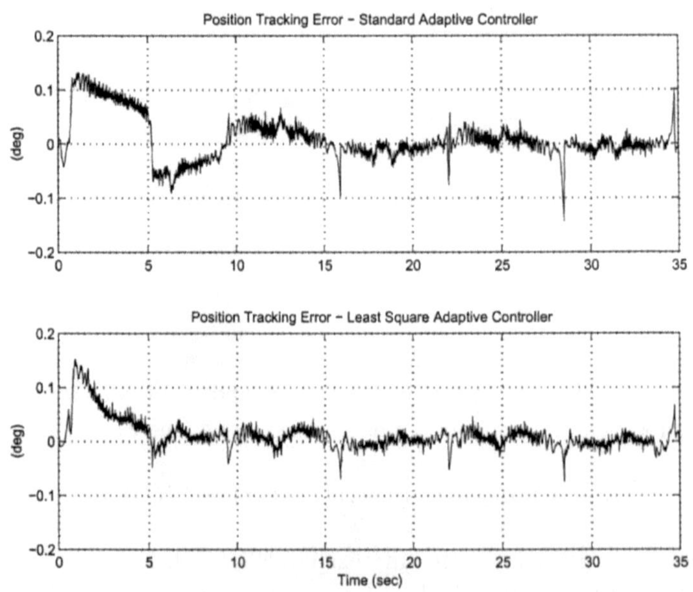

FIGURE 2.3. Position tracking errors.

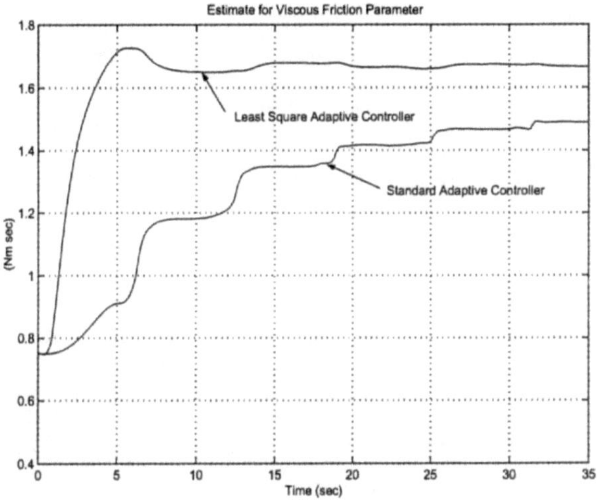

FIGURE 2.4. Estimate of the viscous friction coefficient.

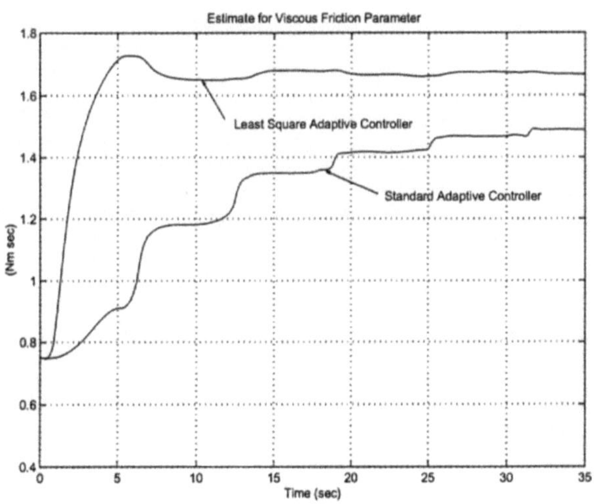

FIGURE 2.5. Estimate of the Coulomb friction coefficient.

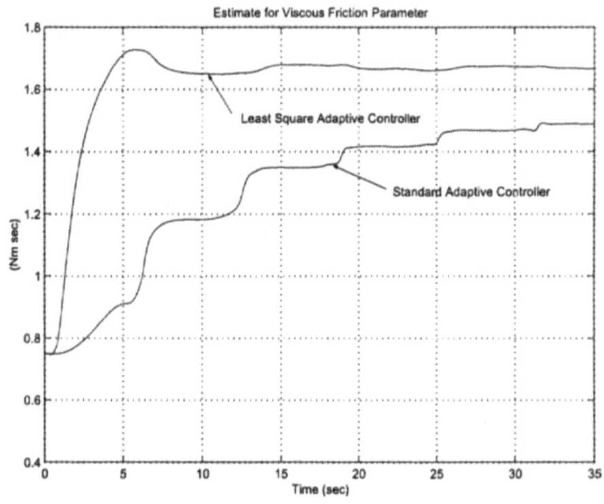

FIGURE 2.6. Estimate of the static friction coefficient.

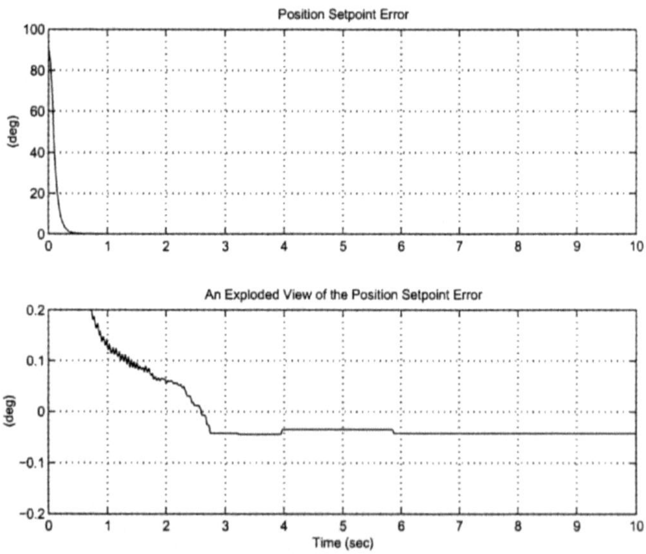

FIGURE 2.7. Position setpoint error.

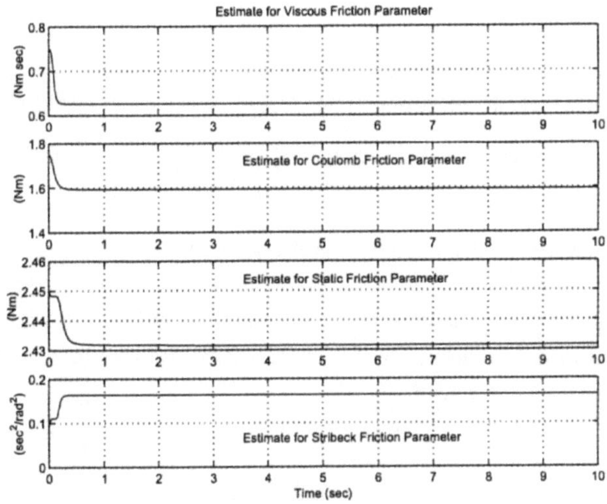

FIGURE 2.8. Parameter estimates for setpoint controller.

The performance of the controller as quantified by the position tracking error is shown in Figure 2.3. The parameter estimates are shown in Figures 2.4 to 2.6.

Next, the modular adaptive controller of (2.26) with the least-squares update law given in (2.42) was implemented with $\hat{\theta}(0) = 0.5\theta$. The desired position trajectory was selected as in (2.84). The best tracking performance was achieved using the following control gains:

$$\alpha = 40, \qquad k_s = 30, \qquad k_N = 1, \qquad \beta = 1,$$

with the adaptation gain matrix initialized to

$$\Gamma(0) = \begin{bmatrix} 0.8 & 1.0 & 1.0 \\ 1.0 & 1.8 & 1.0 \\ 1.0 & 1.0 & 3.0 \end{bmatrix}.$$

The position tracking error is shown Figure 2.3. The parameter estimates are shown in Figures 2.4–2.6.

The experimental results shown in Figures 2.4–2.6 indicate that the least-squares-type estimation rule promotes better transient performance than the standard gradient update law. The improved transient performance of the parameter update laws results in faster decay of the position tracking error for the least-squares based controller, as illustrated by Figure 2.3.

Experiment 2: The setpoint controller of (2.67) along with the update laws given by (2.68) was implemented with $\hat{\theta}(0) = 0.5\theta$. The desired setpoint was selected to be

$$q_d = 90 \text{ deg}.$$

The best regulation performance was achieved using the following gains:

$$k_p = 800, \qquad \Gamma_s = \text{diag}\{0.01, 0.1\}, \qquad \gamma_1 = 0.08,$$
$$k_d = 55, \qquad \gamma_0 = 0.1.$$

The position error and parameter estimates are shown in Figures 2.7, 2.8, respectively. The position setpoint error shown in Figure 2.7 clearly indicates asymptotic regulation with a very small steady-state error. The parameter estimates, as shown by Figure 2.8, settle down to constant values after a brief transient period.

2.4 Full-Order Friction Model

A different friction model from the one presented in Section 2.2 will now be considered. Specifically, we will consider a second-order, SISO mechanical system with nonlinear load dynamics and a nonlinear, dynamic friction model. The mathematical model for this system is assumed to be of the form [9]

$$M\ddot{q} + B\dot{q} + T_L(q, \dot{q}) + \chi(\dot{q})z = \tau \tag{2.85}$$

$$\dot{z} = \dot{q} - f(\dot{q})z, \tag{2.86}$$

where the auxiliary functions $\chi(\dot{q})$ and $f(\dot{q})$ are defined as follows:

$$\chi(\dot{q}) = \theta_0 - f(\dot{q})\theta_1, \tag{2.87}$$

$$f(\dot{q}) = \frac{|\dot{q}|}{g(\dot{q})}, \tag{2.88}$$

where $q(t), \dot{q}(t), \ddot{q}(t)$ represent the position, velocity, and acceleration, respectively; $z(t)$ denotes the *unmeasurable* internal friction state; M denotes the constant inertia of the system; B denotes the constant viscous friction coefficient; $T_L(q, \dot{q})$ denotes a nonlinear load function dependent on the position and velocity; θ_0, θ_1 are positive constant parameters used to weight how the friction dynamics enters the mechanical system; the function $g(\dot{q})$ is used to describe the Stribeck effect; and $\tau(t)$ is the control input.

The model given above accounts for stick-slip friction effects via a so-called *bristle* model [16]. The bristle model was conceived to capture the stick-slip friction effects at the microscopic level. According to this model, the friction between the two interacting surfaces is assumed to be caused by a large number of bristles in the narrow interstices of the rubbing surfaces.

The state $z(t)$ in (2.85) and (2.86) represents the average deflection of thousands of such bristles. Understandably, the state $z(t)$ does not offer itself for direct measurement.

The following assumptions are made regarding the model given in (2.85)–(2.88).

Assumption 1: The load function $T_L(q, \dot{q}) \in \mathcal{L}_\infty$ if $q(t), \dot{q}(t) \in \mathcal{L}_\infty$.

Assumption 2: The Stribeck function $g(\dot{q})$ is constructed such that (i) it ensures that $f(\dot{q}) \in \mathcal{L}_\infty$ if $q(t), \dot{q}(t) \in \mathcal{L}_\infty$ and (ii) $0 < g(\dot{q}) < \delta_o$ where δ_o is some positive constant [9]. It should be noted that since $0 < g(\dot{q}) < \delta_o$, it can easily be shown that if $|z(0)| \leq \delta_o$, then $|z(t)| \leq \delta_o \ \forall t \geq 0$ [9] (i.e., $z(t) \in \mathcal{L}_\infty$).

Remark 2.4 *An example of a function that has been proposed to describe the Stribeck effect [2, 31] is*

$$g(\dot{q}) = \beta_0 + \beta_1 \exp\left(-\left(\frac{\dot{q}}{\beta_2}\right)^2\right), \tag{2.89}$$

where β_0, β_1, and β_2 are positive constant parameters.

2.5 Control Designs for Full-Order Model

In this section, we present a class of partial-state feedback (PSFB), position-tracking controllers for the friction model described in Section 2.4. The PSFB condition stems from the fact that the controllers will only require direct measurement of $q\,(t)$ and $\dot{q}\,(t)$, while a filter and/or observer will generate an estimate for the unmeasurable friction state $z\,(t)$.

To facilitate the subsequent control designs, we redefine the filtered tracking error $r(t)$ given in (2.9) as follows:

$$r = \dot{e} + \mathfrak{L}^{-1}\left\{\frac{1}{s}K_F\,(s)\,E\,(s)\right\}, \tag{2.90}$$

where $e\,(t)$ is the same tracking error defined in (2.7)[4]; $E(s) = \mathfrak{L}\,\{e\,(t)\}$, where $\mathfrak{L}$ denotes the Laplace transform operation; s denotes the Laplace transform variable; and $K_F\,(s)$ is a linear filter that is selected to ensure that the transfer function given by

$$\frac{E(s)}{R(s)} = H(s) \triangleq \frac{s}{s^2 + K_F\,(s)} \tag{2.91}$$

[4] As in Section 2.3, it is assumed that the desired position trajectory $q_d(t)$ is constructed such that $q_d(t), \dot{q}_d(t), \ddot{q}_d(t) \in \mathcal{L}_\infty$.

is strictly proper and exponentially stable. We also define the observation error for the unmeasurable friction state as follows:

$$\tilde{z}(t) = z(t) - \hat{z}(t), \tag{2.92}$$

where $\hat{z}(t)$ represents the estimate of the state $z(t)$.

Remark 2.5 *The filtered tracking error of (2.90) is defined in terms of the function $K_F(s)$ to facilitate the design of standard linear feedback laws for the mechanical dynamics of (2.85). For example, in order to implement a proportional-derivative feedback law, the function $K_F(s)$ can be designed as follows:*

$$K_F(s) = \alpha s \quad \Longrightarrow \quad r = \dot{e} + \alpha e, \tag{2.93}$$

where α is a positive control gain (note that this is the case of (2.9)). Similarly, a proportional-integral-derivative feedback law can be designed by defining $K_F(s)$ as follows:

$$K_F(s) = \alpha s + \beta \quad \Longrightarrow \quad r = \dot{e} + \alpha e + \beta \int_0^t e(\tau)\, d\tau, \tag{2.94}$$

where β is a positive control gain.

2.5.1 Model-Based Control: Asymptotic Tracking

First, a model-based observer/controller will be designed by assuming that all the parameters in the mechanical system given by (2.85)–(2.88) are exactly known. To simplify the control development, we rewrite the dynamics given by (2.85) in terms of the filtered tracking error $r(t)$ as follows:

$$M\dot{r} = w(q, \dot{q}, t) + \chi(\dot{q})z - \tau, \tag{2.95}$$

where $w(\cdot)$ is an auxiliary measurable function defined as follows:

$$w(q, \dot{q}, t) = M\left(\ddot{q}_d + \mathcal{L}^{-1}\left\{K_F(s)\, E(s)\right\}\right) + B\dot{q} + T_L(q, \dot{q}). \tag{2.96}$$

The structure of open-loop dynamics of (2.95) motivates the design of the control input $\tau(t)$ as

$$\tau = w(q, \dot{q}, t) + \chi(\dot{q})\hat{z} + kr, \tag{2.97}$$

where k is a positive control gain. After substituting (2.97) into (2.95), we obtain the following closed-loop error system:

$$M\dot{r} = -kr + \chi(\dot{q})\tilde{z}. \tag{2.98}$$

To provide motivation for some of the terms injected during the subsequent observer design, we now conduct a preliminary stability analysis. To this end, we define the following non-negative function:

$$V_c = \frac{1}{2} M r^2. \tag{2.99}$$

After taking the time derivative of (2.99) along (2.98), we obtain the following expression for $\dot{V}_c(t)$:

$$\dot{V}_c = -kr^2 + \chi(\dot{q})\tilde{z}r. \tag{2.100}$$

Note that, whereas the first term on the right-hand side of (2.100) is non-positive, the second term will have to be directly canceled by the injection of appropriate terms during construction of the observer.

Now we turn our attention to the observer design. An obvious method of constructing an observer is to rewrite the dynamic equation governing the unmeasurable state in terms of the estimated state. For example, we can design an observer for $z(t)$ by utilizing the structure of (2.86) and replacing all occurrences of $z(t)$ with the estimate $\hat{z}(t)$ as shown below:

$$\dot{\hat{z}} = \dot{q} - f(\dot{q})\hat{z} + \chi(\dot{q})r. \tag{2.101}$$

Note that the $\chi(\dot{q})r$ term has been injected to compensate for the $\chi(\dot{q})\tilde{z}r$ term in (2.100). To obtain the observation error dynamics, we take the time derivative of (2.92) and then substitute the expressions given by (2.86) and (2.101) as follows:

$$\dot{\tilde{z}} = -f(\dot{q})\tilde{z} - \chi(\dot{q})r. \tag{2.102}$$

The stability of the proposed observer/controller can be demonstrated by defining the following non-negative function:

$$V_A = V_c + \frac{1}{2}\tilde{z}^2, \tag{2.103}$$

where $V_c(t)$ was defined earlier in (2.99). After taking the time derivative of (2.103), substituting the expressions given by (2.100) and (2.102), and then simplifying the resulting expression, we obtain

$$\dot{V}_A = -kr^2 - f(\dot{q})\tilde{z}^2. \tag{2.104}$$

Since $f(\dot{q}) \geq 0$, as illustrated by (2.88) and the assumptions on $g(\dot{q})$ given in Section 2.4, we can upper bound $\dot{V}(t)$ as follows:

$$\dot{V}_A \leq -kr^2. \tag{2.105}$$

From (2.103) and (2.105), we know $V_A(t) \in \mathcal{L}_\infty$, which implies that $r(t), \tilde{z}(t) \in \mathcal{L}_\infty$. From (2.98), it is now easy to show that $\dot{r}(t) \in \mathcal{L}_\infty$. Using the standard signal-chasing arguments shown in Section 2.3, we can prove the boundedness of all other signals during closed-loop operation. Furthermore, the form of (2.105) allows us to show that $r(t) \in \mathcal{L}_2$. With the above information, we can invoke Lemma A.3 in Appendix A to show that

$$\lim_{t\to\infty} r(t) = 0. \tag{2.106}$$

Now we can use (2.106), (2.91), and Lemma A.8 in Appendix A to show global asymptotic position and velocity tracking in the sense that

$$\lim_{t\to\infty} e(t), \dot{e}(t) = 0. \tag{2.107}$$

From (2.104), we can see that if a given position tracking application somehow ensured that $f(\dot{q})$ remained positive for all time, then we could use (2.103) and (2.104) to prove exponential convergence of the filtered tracking error and the observation error. Since exponential convergence is a desirable property, we will illustrate next how the structure of the mechanical model given in (2.85)–(2.88) can be exploited to design an observer which guarantees exponential convergence of $r(t)$ and $\tilde{z}(t)$.

2.5.2 Model-Based Control: Exponential Tracking

An observer that exponentially estimates the state $z(t)$ is given by

$$\hat{z} = p - \frac{M}{\theta_1}\dot{q}, \tag{2.108}$$

where $p(t)$ is an auxiliary variable that is updated according to

$$\dot{p} = \frac{1}{\theta_1}\left[-\theta_0 p + \left(-B + \theta_1 + M\frac{\theta_0}{\theta_1}\right)\dot{q} - T_L(q,\dot{q}) + \tau + \chi(\dot{q})r\right]. \tag{2.109}$$

While the appearance of most of the terms in (2.108) and (2.109) is motivated by the structure of the mechanical model, the $\chi(\dot{q})r$ term has been injected to cancel the $\chi(\dot{q})\tilde{z}r$ term in (2.100). To formulate the observation error dynamics, we first take the time derivative of (2.108), multiply the resulting equation by θ_1, and then substitute (2.109) for $\dot{p}(t)$ to obtain

$$\theta_1 \dot{\hat{z}} = \theta_1\dot{q} + \chi(\dot{q})z - \theta_0\hat{z} + \chi(\dot{q})r. \tag{2.110}$$

After taking the time derivative of (2.92), multiplying the resulting equation by θ_1, substituting (2.86) and (2.110), and simplifying the resulting expression, we finally obtain

$$\theta_1 \dot{\tilde{z}} = -\theta_0\tilde{z} - \chi(\dot{q})r. \tag{2.111}$$

In order to prove the exponential stability of the closed-loop system formed by (2.98) and (2.111), we define the following non-negative function:

$$V_E = V_c + \frac{1}{2}\theta_1 \tilde{z}^2, \tag{2.112}$$

where $V_c(t)$ was defined in (2.99). Note that $V_E(t)$ can be bounded as follows:

$$\frac{1}{2}\lambda_1 \|x\|^2 \leq V_E \leq \frac{1}{2}\lambda_2 \|x\|^2, \tag{2.113}$$

where $x(t) \in \mathbb{R}^2$ and λ_1, λ_2 are positive bounding constants defined as

$$x = \begin{bmatrix} r & \tilde{z} \end{bmatrix}^T, \quad \lambda_1 = \min\{M, \theta_1\}, \quad \lambda_2 = \max\{M, \theta_1\}. \tag{2.114}$$

After taking the time derivative of (2.112) along (2.100) and (2.111), we get

$$\dot{V}_E = -kr^2 - \theta_0 \tilde{z}^2. \tag{2.115}$$

Utilizing (2.115) and (2.113), an upper bound can be placed on $\dot{V}_E(t)$ as shown below:

$$\dot{V}_E \leq -\min\{k, \theta_0\} \|x\|^2 \leq -\frac{2\min\{k, \theta_0\}}{\lambda_2} V_E. \tag{2.116}$$

We can now apply Lemma A.4 in Appendix A to (2.116) to show that

$$V_E(t) \leq V_E(0) \exp\left(-\frac{2\min\{k, \theta_0\}}{\lambda_2} t\right). \tag{2.117}$$

After applying (2.113) to (2.117), we obtain the following upper bound for $x(t)$:

$$\|x(t)\| \leq \sqrt{\frac{\lambda_2}{\lambda_1}} \|x(0)\| \exp\left(-\frac{\min\{k, \theta_0\}}{\lambda_2} t\right), \tag{2.118}$$

which, according to the definition of (2.114), illustrates the global exponential convergence of $r(t)$ and $\tilde{z}(t)$. Since $r(t)$ is exponentially stable, we can invoke Lemma A.7 in Appendix A to prove the global exponential convergence of $e(t)$ and $\dot{e}(t)$. Note that it can also be easily shown that all other closed-loop signals remain bounded.

Remark 2.6 *The two model-based observer/controllers just presented have the drawback of assuming exact knowledge of all the parameters in the system given by (2.85)–(2.88). In the following two sections, we present two adaptive PSFB observer/controllers that compensate for uncertainty in selected parameters of the system model.*

2.5.3 Adaptive Control: Case 1

We design the first adaptive controller based on the assumption that all of the parameters, except those associated with the Stribeck function $g(\dot{q})$ introduced in (2.88), are unknown. Specifically, we first parameterize the dynamics given by (2.95) and (2.96) as follows:

$$M\dot{r} = W_a(q,\dot{q})\theta_a + (\theta_0 - f(\dot{q})\theta_1)\, z - \tau, \tag{2.119}$$

where the explicit definition of $\chi(\dot{q})$ defined in (2.87) has been utilized, $W_a(q,\dot{q}) \in \mathbb{R}^{1\times p}$ denotes a known regression matrix, and $\theta_a \in \mathbb{R}^p$ is a unknown, constant vector containing the inertia, viscous friction, and load parameters.[5]

The structure of (2.119) and the subsequent stability analysis motivates us to design the control input $\tau(t)$ as follows:

$$\tau = W_a(q,\dot{q})\hat{\theta}_a + \left(\hat{\theta}_0 - f(\dot{q})\hat{\theta}_1\right)\hat{z} + \left(\hat{\theta}_0\zeta_0 - f(\dot{q})\hat{\theta}_1\zeta_1\right) + kr, \tag{2.120}$$

where k is a positive control gain, $\zeta_0(t), \zeta_1(t)$ are filter states defined below, and $\hat{\theta}_a(t)$, $\hat{\theta}_0(t)$, and $\hat{\theta}_1(t)$ are the dynamic estimates of the unknown parameters that are updated using the following adaptation laws:

$$\dot{\hat{\theta}}_a = \Gamma W^T(q,\dot{q})r, \qquad \dot{\hat{\theta}}_0 = \gamma_0\left(\hat{z}+\zeta_0\right)r, \qquad \dot{\hat{\theta}}_1 = -\gamma_1\left(\hat{z}+\zeta_1\right)f(\dot{q})r, \tag{2.121}$$

with $\Gamma \in \mathbb{R}^{p\times p}$ being a diagonal, positive-definite, adaptation gain matrix, and γ_0, γ_1 being positive adaptation gains. The unmeasurable state $z(t)$ is now estimated by the following observer:[6]

$$\dot{\hat{z}} = \dot{q} - f(\dot{q})\hat{z}, \tag{2.122}$$

while the filter states $\zeta_0(t), \zeta_1(t)$ are updated according to the following dynamic equations:

$$\dot{\zeta}_0 = -f(\dot{q})\zeta_0 + r, \qquad \dot{\zeta}_1 = -f(\dot{q})\zeta_1 - f(\dot{q})r. \tag{2.123}$$

To obtain the observation error dynamics, we take the time derivative of (2.92), and then substitute the expressions given by (2.86) and (2.122) to produce

$$\dot{\tilde{z}} = -f(\dot{q})\tilde{z}. \tag{2.124}$$

---

[5] In formulating the $W_a(q,\dot{q})\theta_a$ term, we have assumed that the load function $T_L(q,\dot{q})$ introduced in (2.85) is linearly parameterizable.

[6] Note that the structure of this observer is similar to the previous observer of (2.101), but without the term $\chi(\dot{q})r$.

After substituting the control input given by (2.120) into the open-loop system of (2.119), we get the following closed-loop filtered tracking error system:

$$\begin{aligned} M\dot{r} = & -kr + \left[W_a(q,\dot{q})\tilde{\theta}_a + (\hat{z}+\zeta_0)\,\tilde{\theta}_0 - (\hat{z}+\zeta_1)\,f(\dot{q})\tilde{\theta}_1\right] \\ & +\theta_0\,(\tilde{z}-\zeta_0) - \theta_1\,(\tilde{z}-\zeta_1)\,f(\dot{q}), \end{aligned} \tag{2.125}$$

where the parameter estimation errors are defined as follows:

$$\tilde{\theta}_m(t) = \theta_m - \hat{\theta}_m(t), \qquad \tilde{\theta}_0(t) = \theta_0 - \hat{\theta}_0(t), \qquad \tilde{\theta}_1(t) = \theta_1 - \hat{\theta}_1(t). \tag{2.126}$$

The following observations which motivate the design of the adaptive update laws and the auxiliary filters can be made with regard to the form of the closed-loop equation given by (2.125). As will be seen in the subsequent stability analysis: (i) the update laws given in (2.121) compensate for the bracketed term in the first line of (2.125), and (ii) the filters defined in (2.123) compensate for the two terms on the second line of (2.125).

We will now perform a composite stability analysis for the closed-loop observer/controller system. Specifically, we define the following non-negative function:

$$\begin{aligned} V_{A1} = & \frac{1}{2}Mr^2 + \frac{1}{2}\tilde{z}^2 + \frac{1}{2}\theta_0\,(\tilde{z}-\zeta_0)^2 + \frac{1}{2}\theta_1\,(\tilde{z}-\zeta_1)^2 \\ & +\frac{1}{2}\tilde{\theta}_a\Gamma^{-1}\tilde{\theta}_a + \frac{1}{2}\frac{1}{\gamma_0}\tilde{\theta}_0^2 + \frac{1}{2}\frac{1}{\gamma_1}\tilde{\theta}_1^2. \end{aligned} \tag{2.127}$$

After taking the time derivative of (2.127) along (2.121), (2.123), (2.124), (2.125), and (2.126), and then simplifying the resulting expression, we obtain

$$\dot{V}_{A1} = -kr^2 - f(\dot{q})\tilde{z}^2 - \theta_0 f(\dot{q})\,(\tilde{z}-\zeta_0)^2 - \theta_1 f(\dot{q})\,(\tilde{z}-\zeta_1)^2. \tag{2.128}$$

Since $f(\dot{q}) \geq 0$, as illustrated by (2.88), and according to the assumptions on $g(\dot{q})$ given in Assumption 1 in Section 2.4, we can upper bound $\dot{V}_{A1}(t)$ as follows:

$$\dot{V}_{A1} \leq -kr^2. \tag{2.129}$$

Based on the form of (2.127) and (2.129), we can follow the same arguments described in Section 2.3 to show the boundedness of all closed-loop signals and that $r(t) \in \mathcal{L}_2$. Since we know $r(t), \dot{r}(t) \in \mathcal{L}_\infty$ and $r(t) \in \mathcal{L}_2$, we can use Lemma A.3 in Appendix A to state that $\lim_{t\to\infty} r(t) = 0$. Finally, we can apply Lemma A.8 in Appendix A to show that $\lim_{t\to\infty} e(t), \dot{e}(t) = 0$.

2.5.4 Adaptive Control: Case 2

We now synthesize an adaptive controller that compensates for parametric uncertainty associated only with the normal force variation in the Stribeck effect. That is, we will assume that $f(\dot{q})$ can be linearly parameterized and expressed as follows [10]:

$$f(\dot{q}) = \kappa f_{\kappa}(\dot{q}), \tag{2.130}$$

where $f_{\kappa}(\dot{q})$ denotes a known function, and κ is an unknown, positive, constant parameter.

We begin by parameterizing the dynamics given by (2.95) as follows:

$$M\dot{r} = w(q, \dot{q}, t) + (\theta_0 - \kappa f_{\kappa}(\dot{q})\theta_1)\, z - \tau, \tag{2.131}$$

where (2.87) and (2.130) have been utilized. The structure of (2.131) and the subsequent stability analysis motivates the design of the control input $\tau(t)$ as follows:

$$\tau = w + \theta_0 \hat{z} - \hat{\kappa} f_{\kappa}(\dot{q})\theta_1 (\hat{z} - \zeta_{\kappa}) + kr, \tag{2.132}$$

where $\zeta_{\kappa}(t)$ is a filter variable that is defined below, and $\hat{\kappa}(t)$ represents the dynamic estimate of the unknown parameter κ, which is updated using the following rule:

$$\dot{\hat{\kappa}} = -\gamma_{\kappa} (\hat{z} + \zeta_{\kappa})\, \theta_1 f_{\kappa}(\dot{q}) r, \tag{2.133}$$

with γ_{κ} being a positive adaptation gain. The filter state $\zeta_{\kappa}(t)$ is generated according to the following dynamic equation:

$$\dot{\zeta}_{\kappa} = -\frac{\theta_0}{\theta_1}\zeta_{\kappa} - \frac{\theta_0}{\theta_1} r - \theta_1 f_{\kappa}(\dot{q}) r. \tag{2.134}$$

The unmeasurable state $z(t)$ is now estimated by the following observer:[7]

$$\hat{z} = p - \frac{M}{\theta_1}\dot{q}, \tag{2.135}$$

where $p(t)$ is an auxiliary variable having the following dynamics:

$$\dot{p} = \frac{1}{\theta_1}\left[-\theta_0 p + \left(-B + \theta_1 + M\frac{\theta_0}{\theta_1}\right)\dot{q} - T_L(q, \dot{q}) + \tau + \theta_0 r\right]. \tag{2.136}$$

The observation error dynamics can be formulated from (2.92), (2.135), and (2.136) using a procedure similar to that in Section 2.5.2 to yield

$$\theta_1 \dot{\tilde{z}} = -\theta_0 \tilde{z} - \theta_0 r. \tag{2.137}$$

---

[7] Note that the structure of this observer is similar to the previous observer of (2.108) and (2.109), except that it does utilize the function $f(\dot{q})$.

To formulate the closed-loop expression for the filtered tracking error dynamics, we substitute the control input of (2.132) into (2.131) to obtain

$$M\dot{r} = -kr + \theta_0\tilde{z} - (\hat{z} + \zeta_\kappa)\,\theta_1 f_\kappa(\dot{q})\tilde{\kappa} - \kappa\,(\tilde{z} - \zeta_\kappa)\,\theta_1 f_\kappa(\dot{q}), \tag{2.138}$$

where the parameter estimation error signal $\tilde{\kappa}\,(t)$ is defined as follows:

$$\tilde{\kappa}(t) = \kappa - \hat{\kappa}(t). \tag{2.139}$$

To analyze the stability of the closed-loop systems, we define the non-negative function

$$V_{A2} = \frac{1}{2}Mr^2 + \frac{1}{2}\theta_1\tilde{z}^2 + \frac{1}{2}\kappa\,(\tilde{z} - \zeta_\kappa)^2 + \frac{1}{2}\frac{1}{\gamma_\kappa}\tilde{\kappa}^2. \tag{2.140}$$

After taking the time derivative of $V_{A2}(t)$ along (2.139), (2.138), (2.137), (2.133), and (2.134), we have

$$\begin{aligned}\dot{V}_{A2} &= -kr^2 - \theta_0\tilde{z}^2 - \kappa\frac{\theta_0}{\theta_1}\,(\tilde{z} - \zeta_\kappa)^2 \\ &\leq -kr^2 - \theta_0\tilde{z}^2 \leq -\min\{k, \theta_0\}\,\|x\|^2,\end{aligned} \tag{2.141}$$

where $x(t)$ was previously defined in (2.114). From (2.140), (2.141), and Lemma A.6 in Appendix A, it is clear that all system signals are bounded and $\lim_{t\to\infty} \|x(t)\| = 0$; hence, $\lim_{t\to\infty} r(t), \tilde{z}(t) = 0$. We can now use Lemma A.8 in Appendix A to show that $\lim_{t\to\infty} e(t), \dot{e}(t) = 0$.

2.5.5 Experimental Evaluation

The same experimental setup discussed in Section 2.3.4 (see Figure 2.1) was used to evaluate the controllers presented in this section. The various parameters associated with the friction model were calculated using a procedure similar to that discussed in Section 2.3.4. For the model given by (2.85)–(2.88), the numerical values of the parameters were determined to be

$$\begin{array}{c}M = 0.125\text{ kg-m}^2, \quad B = 1.42\text{ N-m-sec/rad}, \quad \theta_0 = 12\text{ N-m/rad}, \\ \theta_1 = 0.1\text{ N-m-sec/rad}, \quad \beta_0 = 3.24, \quad \beta_1 = 5.21, \quad \beta_2 = 3.00.\end{array} \tag{2.142}$$

The desired position trajectory $q_d(t)$ was selected as in (2.84). Since there was no external load torque acting on the motor other than the friction torque, $T_L(q, \dot{q})$ in (2.85) was equal to zero. For the sake of brevity, only the experimental results for the model-based controller of Section 2.5.1 and the adaptive controller of Section 2.5.3 are presented.

Experiment 1: First, the model-based observer/controller represented by (2.97), (2.93), (2.108), and (2.109) was implemented. The best tracking performance was achieved with the control gains set to $\alpha = 80$ and $k = 4.0$. The performance of the controller, as quantified by the position tracking error, is shown in Figure 2.9.

Experiment 2: The adaptive controller of (2.120) and (2.93) was implemented with the parameter estimates initialized to 50% of the nominal values given in (2.142) with the exception of the inertia parameter M which was assumed to be exactly known. The best tracking performance was achieved using the following control gains:

$$\alpha = 340, \quad k_c = 22, \quad \Gamma = 1.5, \quad \gamma_0 = 5, \quad \gamma_1 = 0.02.$$

The position tracking error is shown Figure 2.10, while the parameter estimates are shown in Figure 2.11.

Experiment 3: For comparison purposes, we also evaluated a controller based on the reduced-order friction model described in Section 2.2 and given by (2.1). Specifically, we implemented the following model-based control law:

$$\tau = M\left(\ddot{q}_d + \alpha \dot{e}\right) + B\dot{q} + \left(F_c + F_s \exp\left(-F_\tau \dot{q}^2\right)\right) \operatorname{sgn}(\dot{q}) + kr, \qquad (2.143)$$

where $r(t)$ is defined by (2.93) and F_c, F_s were determined experimentally to be 3.24 N-m and 5.21 N-m, respectively. The best tracking performance was achieved using the following control gains, $\alpha = 110$ and $k = 14.0$. The resulting position tracking error is shown Figure 2.12 (for comparison purposes, the tracking error of Experiment 1 has been replotted on the same scale).

It is interesting to note from Figure 2.12 that the full-order, model-based controller of Experiment 1 outperformed the reduced-order, model-based controller of Experiment 3. Also, as indicated by Figure 2.10, the adaptive controller of Experiment 2 achieved about the same level of tracking performance as the other two controllers, despite the uncertainty associated with the parameters B, θ_0, and θ_1.

2.6 Notes

To capture low-velocity friction effects, several researchers, such as Tustin [31] and Hess and Soom [17], have proposed empirical models between friction and velocity to fit the Stribeck curve. In [7], Canudas de Wit et al.

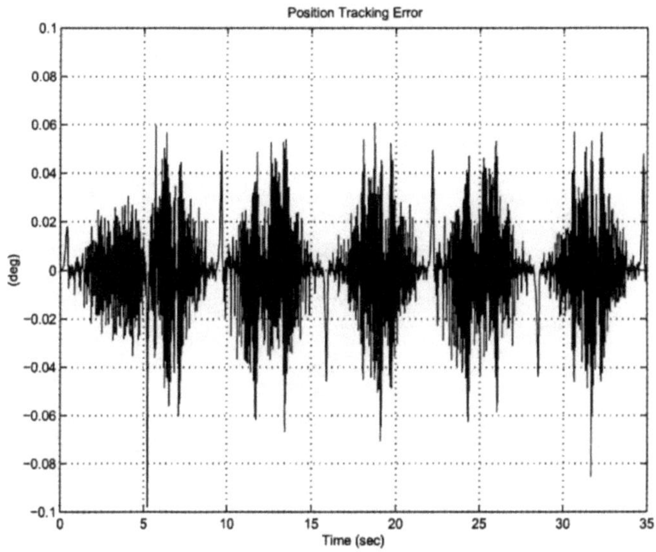

FIGURE 2.9. Position tracking error: full-order, model-based controller.

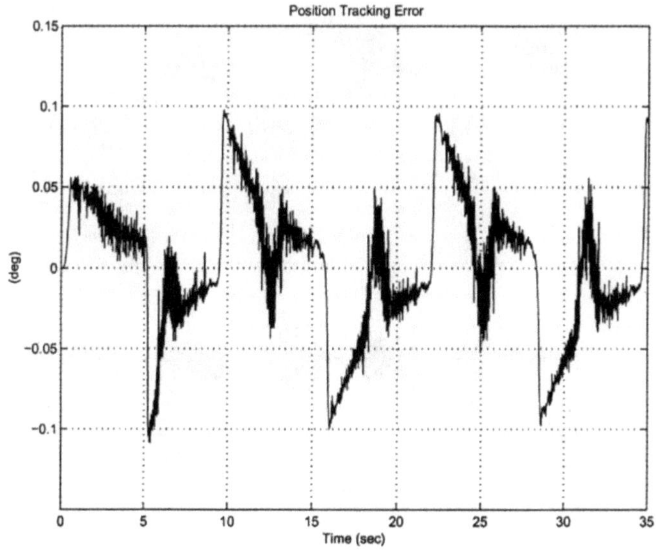

FIGURE 2.10. Position tracking error: adaptive controller.

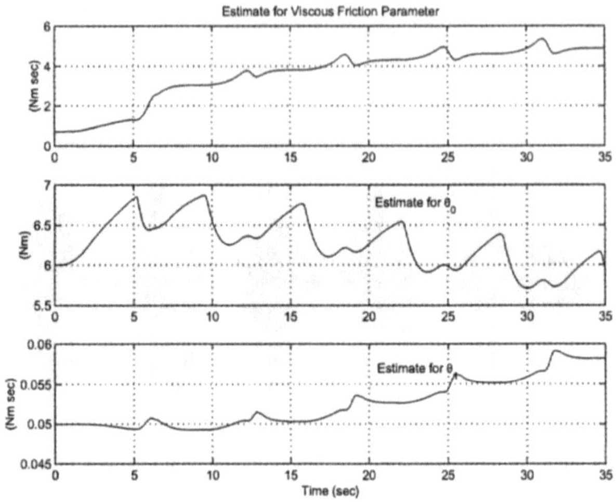

FIGURE 2.11. Parameter estimates: adaptive controller.

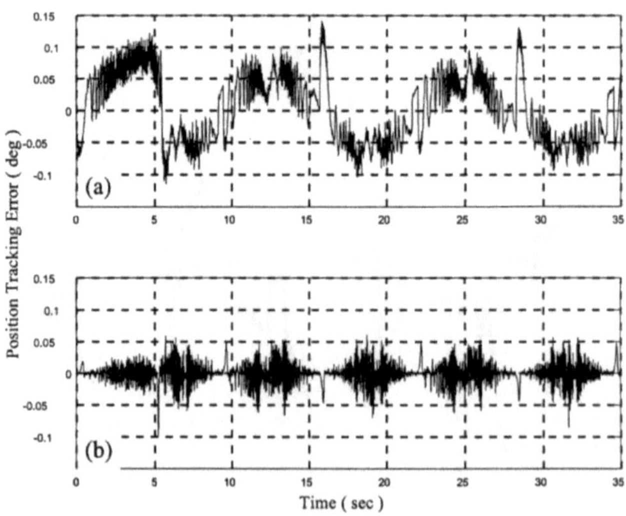

FIGURE 2.12. Position tracking errors: (a) reduced-order, model-based controller and (b) full-order, model-based controller.

approximated the exponential relationship proposed in [31] by a linearly parametrized model. Rabinowicz [26] integrated the stick-slip motion into a friction model that takes into account the temporal phenomena of dwell time and lag time. An alternative approach, known as the *state-variable* model, was proposed in [13, 27, 28] to capture the effects of time delay associated with friction. For further information on similar work, the reader is referred to [5]. Many of these mathematical models were later given physical interpretations. For example, the bristle model utilized for control synthesis in Section 2.4 is the outcome of one such interpretation [16]. For an excellent and comprehensive review on the subject of friction modelling, the reader is referred to [1, 6].

Following the development of empirical mathematical models, several researchers have used these models to design adaptive controllers that compensate for parametric uncertainty. In [2], Armstrong-Hélouvry explored the implications of the Stribeck effect on feedback control to stabilize the performance of the controller at low velocities. Later in [3], Armstrong-Hélouvry applied dimensional and perturbation analysis to solve a nonlinear, low-velocity friction control problem. In [34], Walrath designed an adaptive controller for airborne servo mechanism based on a first-order dynamic friction model. Friedland and Park [15] proposed an adaptive control scheme that entailed the use of an observer to estimate the kinetic friction coefficient.While in [7], Canudas de Wit et al. designed an adaptive controller for DC motor drives utilizing a friction model that was asymmetric in angular velocity, [8] addressed the problem of adaptive friction compensation for robot manipulators operating at low velocities based on the assumption of a linearly parameterized model for the Stribeck effect. Later, Canudas de Wit et al. designed a model-based [9] and an adaptive controller [10] to achieve position/velocity tracking for a second-order linear system in the presence of nonlinear, dynamic friction effects. The control schemes presented in [9, 10] required the feedback portion of the controller to satisfy a strictly positive real (SPR) condition that precluded the use of a PID feedback law. In [32], Vedagarbha et al. exploited the structure of the friction dynamics proposed in [9] to design several different closed-loop observer/control strategies. Other related work can be found in [4, 18, 22, 23, 29, 35].

With respect to the above literature review, we first note that the controllers presented in Section 2.5 were designed to eliminate the SPR condition required in [9, 10]; hence, many different types of feedback laws are possible in the feedback portion of the controller (see Remark 2.5). In addition, relaxation of the SPR condition endows the adaptive control laws of Sections 2.5.3 and 2.5.4 with the ability to compensate for uncertain nonlinear load dynamics in a straightforward manner. Finally, the observer/controller proposed in [10] to deal with parametric uncertainty associated with $f(\dot{q})$ neglected the observer transients during the stability analysis. On the other hand, the adaptive controller of Section 2.5.4 given by (2.132) utilizes the nonlinear observer given by (2.134) actively to compensate for the observer transients.

References

[1] B. Armstrong-Hélouvry, P. Dupont, and C. Canudas de Wit, A Survey of Models, Analysis Tolls and Compensation Methods for the Control of Machines with Friction, *Automatica,* Vol. 30, No. 7, pp. 1083-1138, July 1994.

[2] B. Armstrong-Hélouvry, Stick-slip Arising from Stribeck Friction, *Proceedings of the IEEE International Conference on Robotics and Automation,* pp. 1377–1382, Cincinnati, OH, May 1990.

[3] B. Armstrong-Hélouvry, Stick-Slip and Control in Low-Speed Motion, *IEEE Transactions on Automatic Control,* Vol. 38, No. 10, pp. 1483–1496, Oct. 1993.

[4] C. G. Batil and P. O. Gutman, Performance Related Adaptive Friction Compensation for Uncertain Systems, *Proceedings of the European Control Conference*, Grenoble, France, 1991.

[5] P. A. Bliman and M. Sorine, A System-Theoretic Approach to Systems with Hysteresis. Application to Friction Modelling and Compensation, *Proceedings of the European Control Conference*, Groningen, The Netherlands, 1993.

[6] P. A. Bliman and M. Sorine, Easy-to-use Realistic Dry Friction Models for Automatic Control, *Proceedings of the European Control Conference,* Rome, Italy, 1995.

[7] C. Canudas de Wit, K. J. Aström, and K. Braun, Adaptive Friction Compensation in DC-Motor Drives, *IEEE Journal of Robotics and Automation,* Vol. RA-3, No. 6, pp. 681–685, Dec. 1987.

[8] C. Canudas de Wit, P. Noel, A. Aubin, and B. Brogliato, Adaptive Friction Compensation in Robot Manipulators: Low-velocities, *Inter-*

national Journal of Robotics Research, Vol. 10, No. 3, pp. 189–199, June 1991.

[9] C. Canudas de Wit, H. Olsson, K. J. Aström, and P. Lischinsky, A New Model for Control of Systems with Friction, *IEEE Transactions on Automatic Control,* Vol. 40, No. 3, pp. 419–425, Mar. 1995.

[10] C. Canudas de Wit and P. Lischinsky, Adaptive Friction Compensation with Dynamic Friction Model, *IFAC World Congress,* pp. 197–202, San Francisco, CA, June 1996.

[11] L. da Vinci, *The Notebooks,* New York, NY: Dover.

[12] P. R. Dahl, A Solid Friction Model,TOR-158(3107-18), The Aerospace Corporation, El Sugundo, CA.

[13] P. E. Dupont and E. P. Dunlap, Friction Modeling and Control in Boundary Lubrication, *Proceedings of the American Control Conference*, pp. 1910–1914, San Francisco, CA, 1993.

[14] V. Fomin, A. Fradkov, and V. Yakubovich, *Adaptive Control of Dynamical Systems*, Nauka, Moscow, 1981 (in Russian).

[15] B. Friedland and Y. J Park., On Adaptive Friction Compensation, *IEEE Transactions on Automatic Control,* Vol. 37, No. 10, pp. 1609-1612, Oct. 1992.

[16] D. A. Haessig and B. Friedland, On the Modeling and Simulation of Friction, *ASME Journal of Dynamic Systems, Measurement and Control*, Vol. 113, No. 3, pp. 354-362, 1991.

[17] D. P. Hess and A. Soom, Friction at a Lubricated Line Contact Operating at Oscillating Sliding Velocities, *Journal of Tribology,* Vol. 112, pp. 147–152, Jan. 1990.

[18] S. Jain, F. Khorrami, N. Ahmad, and S. Sankaranarayanan, Friction Compensation for Drives with and without Transmission Compliance, *Proceedings of the American Control Conference*, pp. 2925-2929, Albuquerque, NM, June 1997.

[19] M. Krstić and P. Kokotović, Adaptive Nonlinear Design with Controller-Identifier Separation and Swapping, *IEEE Transactions on Automatic Control*, Vol. 40, No. 3, pp. 426-440, Mar. 1995.

[20] M. Krstić, I. Kanellakopoulos, and P. Kokotović, *Nonlinear and Adaptive Control Design*, New York, NY: Wiley Interscience, 1995.

[21] F. L. Lewis, C. T. Abdallah, and D. M. Dawson, *Control of Robot Manipulators,* New York, NY: Macmillan Publishing Co., 1993.

[22] B. Maqueira and M. K. Masten, Adaptive Friction Compensation for Line-of-Sight Pointing and Stabilization, *Proceedings of the American Control Conference,* pp. 1942–1945, San Francisco, CA, June 1993.

[23] H. Olsson and K. Aström, Observer-Based Friction Compensation, *Proceedings of the Conference on Decision and Control,* pp. 4345–4350, Kobe, Japan, Dec. 1996.

[24] R. Ortega, Some Remarks on Adaptive Neuro-Fuzzy Systems, *International Journal of Adaptive Control and Signal Processing*, Vol. 10, No. 1, pp. 79–83, Jan. 1996.

[25] R. Ortega, A. Loria, P. J. Nicklasson, and H. Sira-Ramirez, *Passivity-based Control of Euler-Lagrange Systems*, London: Springer-Verlag, 1998.

[26] E. Rabinowicz, The Intrinsic Variables Affecting the Stick-Slip Process, *Proceedings of Physical Society of London,* Vol. 71, No. 4, pp. 668–675.

[27] J. R. Rice and A. L. Ruina, Stability of Steady Frictional Slipping, *Journal of Applied Mechanics*, Vol. 50, pp. 343–349, 1983.

[28] A. L. Ruina, Friction Laws and Instabilities: A Quasistatic Analysis of Some Dry Frictional Behavior, *Ph.D. Dissertation,* Division of Engineering, Brown University.

[29] U. Schafer and G. Brandenburg, Model Reference Position Control of an Elastic Two-Mass System with Compensation of Coulomb Friction, *Proceedings of the American Control Conference*, pp. 1937–1941, San Francisco, CA, June 1993.

[30] J. J. Slotine and W. Li, *Applied Nonlinear Control,* Englewood Cliffs, NJ: Prentice Hall Co., 1991.

[31] A. Tustin, The Effects of Backlash and of Speed-Dependent Friction on the Stability of Closed-cycle Control Systems, *IEE Journal*, Vol. 94, Part 2A, pp. 143–151, 1947.

[32] P. Vedagarbha, D. M. Dawson, and M. Feemster, Tracking Control of Mechanical Systems in the Presence of Nonlinear Dynamic Friction Effects, *IEEE Transactions on Control Systems Technology*, Vol.7, No.4, pp.446–456, July 1999.

[33] M. Vidyasagar, *Nonlinear Systems Analysis*, Englewood Cliffs, NJ: Prentice Hall Co., 1978.

[34] C. D. Walrath, Adaptive Bearing Friction Compensation Based on Recent Knowledge of Dynamic Friction, *Automatica*, Vol. 20, No. 6, pp. 717–727, June 1984.

[35] A. Yazdizadeh and K. Khorasani, Adaptive Friction Compensation using a Lyapunov-Based Design Scheme, *Proceedings of the Conference on Decision and Control*, pp. 2830–2831, Kobe, Japan, Dec. 1996.

3

Full-State Feedback Tracking Controllers

3.1 Introduction

This chapter will concentrate on the *full-state feedback* (FSFB) (i.e., position and velocity are available for feedback) control problem for general, nonlinear, MIMO, rigid mechanical systems with *parametric uncertainty.* Specifically, we will emphasize the adaptive control solution to this problem. First (as in Chapter 2), we will present, owing to its popularity and relative simplicity, the standard adaptive controller of Slotine and Li [27] in its original MIMO version. The idea is for this adaptive controller to serve as a benchmark for the subsequent MIMO, FSFB controllers of this chapter and the output feedback controllers of Chapter 4.

In the standard adaptive controller of [27], the regression matrix depends on the *actual* position and velocity; hence, the regression matrix must be calculated on-line for use in the feedforward portion of the control input signal. Since the regression matrix often contains many nonlinear terms, its on-line computation substantially increases the burden on the real-time implementation of the adaptive control scheme. To reduce on-line computation, Sadegh and Horowitz [26] proposed the so-called *desired compensation adaptation law* (DCAL). The DCAL controller is composed of (i) a nonlinear feedforward term consisting of a *desired* regression matrix and the parameter update law, (ii) a linear feedback term, and (iii) a nonlinear feedback term that is used to compensate for the difference between the actual and desired regression matrices. In this scheme, the nonlinear functions appearing in the regression matrix are computed using the desired position, velocity, and acceleration trajectories; hence, the feedforward control term can usually be precalculated off-line (i.e., provided the desired trajectory is known a priori). In this chapter, we reexamine the original DCAL controller proposed in [26]. By employing a slightly different structure for the nonlinear/linear feedback terms and making use of the nonlinear damping design tool (see Lemma A.10 in Appendix A), we formulate a class of DCAL-type control laws that includes adaptive controllers as well as some nonadaptive extensions.

The DCAL and standard adaptive controllers are founded on a traditional Lyapunov design approach that does not offer much freedom in the selection of the parameter update laws. As shown in Chapter 2, this problem for nonlinear, SISO, rigid mechanical systems was overcome by employing the estimation-based design of [10], which provides a certain degree of modularity in the design of the controller and update law. Unfortunately, the modular adaptive control framework of [10] cannot be directly applied to nonlinear, MIMO mechanical systems owing to the inertia matrix-related couplings in the system dynamics. However, the work presented in [10] does provide some theoretical ammunition that can be used to achieve controller/update law modularity for MIMO mechanical systems, as will be illustrated in Section 3.6.

3.2 System Model

The dynamic model of an n degrees-of-freedom, rigid mechanical system is assumed to be given by the following nonlinear, ordinary differential equation:

$$M(q)\ddot{q} + V_m(q,\dot{q})\dot{q} + G(q) + F_d\dot{q} = \tau, \tag{3.1}$$

where $q(t), \dot{q}(t), \ddot{q}(t) \in \mathbb{R}^n$ denote the position, velocity, and acceleration vectors, respectively; $M(q) \in \mathbb{R}^{n\times n}$ represents the system's inertia matrix; $V_m(q,\dot{q}) \in \mathbb{R}^{n\times n}$ represents the centripetal-Coriolis matrix; $G(q) \in \mathbb{R}^n$ denotes the gravity effects; $F_d \in \mathbb{R}^{n\times n}$ is the constant, diagonal, positive-definite, viscous friction coefficient matrix; and $\tau(t) \in \mathbb{R}^n$ represents the control input vector. We will assume that the left-hand side of (3.1) is first-order differentiable with respect to time and is bounded provided $q(t)$, $\dot{q}(t)$, and $\ddot{q}(t)$ are bounded.

The dynamic model of (3.1) is assumed to possess the following standard properties [11, 16] that will be used in the control design and stability analysis.

Property 3.1 The inertia matrix $M(q)$ is symmetric and positive-definite, and satisfies the following inequalities:

$$m_1 \|\xi\|^2 \leq \xi^T M(q)\xi \leq m_2 \|\xi\|^2 = \|M(q)\|_{i2} \|\xi\|^2 \qquad \forall \xi \in \mathbb{R}^n, \tag{3.2}$$

where m_1, m_2 are known, positive bounding constants, $\|\cdot\|$ denotes the standard Euclidean norm, and $\|\cdot\|_{i2}$ represents the matrix-induced two norm [30].

Property 3.2 The inertia and centripetal-Coriolis matrices satisfy the following skew-symmetric relationship:

$$\xi^T \left(\frac{1}{2} \dot{M}(q) - V_m(q, \dot{q}) \right) \xi = 0 \qquad \forall \xi \in \mathbb{R}^n, \tag{3.3}$$

where $\dot{M}(q)$ denotes the time derivative of the inertia matrix.

Property 3.3 The left-hand side of the dynamic equation of (3.1) can be linearly parameterized as

$$M(q)\ddot{q} + V_m(q, \dot{q})\dot{q} + G(q) + F_d\dot{q} = Y(q, \dot{q}, \ddot{q})\theta, \tag{3.4}$$

where $\theta \in \mathbb{R}^p$ contains the constant system parameters, and $Y(\cdot) \in \mathbb{R}^{n \times p}$ is the regression matrix, which contains known functions of the signals $q(t)$, $\dot{q}(t)$, and $\ddot{q}(t)$.

3.3 Problem Statement

The primary control objective is to design the control input $\tau(t)$ such that $q(t) \to q_d(t)$ as $t \to \infty$ with the assumption of FSFB. The closeness of this control objective is quantified by the position tracking error $e(t) \in \mathbb{R}^n$ defined as follows:

$$e(t) = q_d(t) - q(t). \tag{3.5}$$

Hence, from the definition of (3.5), the control objective is to make $e(t) \to 0$ as $t \to \infty$. We will assume the desired motion trajectory is specified such that $q_d(t)$ and its first two time derivatives are all bounded functions of time. Since the control objective is to be met under the constraint of parametric uncertainty, the controllers of this chapter will contain an adaptation law to estimate the unknown parameters. The difference between the actual and estimated parameters is defined by

$$\tilde{\theta}(t) = \theta - \hat{\theta}(t), \tag{3.6}$$

where $\tilde{\theta}(t) \in \mathbb{R}^p$ denotes the parameter estimation error vector, and $\hat{\theta}(t) \in \mathbb{R}^p$ denotes the dynamic estimate of the unknown, constant, parameter vector θ defined in (3.4).

In addition, a filtered tracking error denoted by $r(t) \in \mathbb{R}^n$ is defined as follows:

$$r(t) = \dot{e}(t) + \alpha e(t), \tag{3.7}$$

where $\alpha \in \mathbb{R}^{n \times n}$ is a diagonal, positive-definite, control gain matrix. Note that this n-dimensional, filtered tracking error has the same properties as the one-dimensional, filtered tracking error defined in Section 2.3 in Chapter 2. We now proceed as in Section 2.3 to rewrite the dynamics of (3.1) in terms of the definition given in (3.7) as described below. First, we differentiate (3.7) with respect to time, and multiply both sides by $M(q)$ to yield

$$M(q)\dot{r} = M(q)\left(\ddot{q}_d + \alpha\dot{e}\right) - M(q)\ddot{q}. \tag{3.8}$$

After utilizing the system dynamics of (3.1) to substitute for $M(q)\ddot{q}$ into (3.8), we have

$$\begin{aligned} M(q)\dot{r} &= M(q)\left(\ddot{q}_d + \alpha\dot{e}\right) + V_m(q,\dot{q})\dot{q} + G(q) + F_d\dot{q} - \tau \\ &= M(q)\left(\ddot{q}_d + \alpha\dot{e}\right) + V_m(q,\dot{q})(\dot{q}_d + \alpha e) + G(q) \\ &\quad + F_d\dot{q} - V_m(q,\dot{q})r - \tau, \end{aligned} \tag{3.9}$$

where (3.5) and (3.7) were used to rewrite the velocity as $\dot{q} = \dot{q}_d + \alpha e - r$ in the term $V_m(q,\dot{q})\dot{q}$. Property 3.4 can now be exploited to rewrite (3.9) as

$$M(q)\dot{r} = -V_m(q,\dot{q})r + Y_s(q,\dot{q},t)\theta - \tau, \tag{3.10}$$

where the linear parametrization $Y_s(\cdot)\theta$ is defined as

$$Y_s(\cdot)\theta = M(q)\left(\ddot{q}_d + \alpha\dot{e}\right) + V_m(q,\dot{q})(\dot{q}_d + \alpha e) + G(q) + F_d\dot{q}. \tag{3.11}$$

The above first-order, nonlinear, ordinary differential equation represents the open-loop dynamics of $r(t)$, and will be used as the foundation for the design of the adaptive controllers throughout this chapter.

3.4 Standard Adaptive Control

In this section, for the purpose of clarity, we review the well-known, model-based, MIMO adaptive controller of Slotine and Li [27] in the context of Lyapunov-type design and analysis techniques used throughout the book.

3.4.1 Controller Formulation

Based on the form of the open-loop dynamics of (3.10), the control input $\tau(t)$ is designed as follows: [27]

$$\tau = Y_s(\cdot)\hat{\theta} + Kr, \tag{3.12}$$

where $K \in \mathbb{R}^{n \times n}$ is a diagonal, positive-definite, control gain matrix, and $\hat{\theta}(t)$ was defined in (3.6). Based on the subsequent Lyapunov-type stability analysis, the parameter estimate vector $\hat{\theta}(t)$ is updated using the adaptation algorithm

$$\dot{\hat{\theta}} = \Gamma Y_s^T(\cdot)r, \tag{3.13}$$

where $\Gamma \in \mathbb{R}^{p \times p}$ is a diagonal, positive-definite, adaptation gain matrix. Substituting (3.12) into (3.10) produces the closed-loop dynamics for $r(t)$ as shown below:

$$M(q)\dot{r} = -V_m(q,\dot{q})r - Kr + Y_s(\cdot)\tilde{\theta}, \tag{3.14}$$

where the definition of (3.6) was used. In addition, by differentiating (3.6) with respect to time, we can use (3.13) to form the following closed-loop dynamics for the parameter estimation error:

$$\dot{\tilde{\theta}} = -\Gamma Y_s^T(\cdot)r. \tag{3.15}$$

3.4.2 Stability Result

The structure of the error systems of (3.14) and (3.15) yields a stability result for the position tracking error as delineated by the following theorem.

Theorem 3.1 *The control law of (3.12) and the parameter estimation update law of (3.13) ensure the global asymptotic convergence of the position and velocity tracking error as illustrated by*

$$\lim_{t \to \infty} e(t), \dot{e}(t) = 0. \tag{3.16}$$

Proof. To prove the above result, we define the non-negative function

$$V = \frac{1}{2}r^T M(q) r + \frac{1}{2}\tilde{\theta}^T \Gamma^{-1} \tilde{\theta}. \tag{3.17}$$

Differentiating (3.17) with respect to time yields

$$\dot{V} = \frac{1}{2}r^T \dot{M}(q) r + r^T M(q) \dot{r} + \tilde{\theta}^T \Gamma^{-1} \dot{\tilde{\theta}}\,. \tag{3.18}$$

After substituting the closed-loop dynamics of (3.14) into (3.18), we obtain

$$\dot{V} = \left[\frac{1}{2}r^T \dot{M}(q) r - r^T V_m(q,\dot{q}) r\right] - r^T K r + \tilde{\theta}^T \left(Y_s^T(\cdot) r + \Gamma^{-1} \dot{\tilde{\theta}}\right). \tag{3.19}$$

Applying Property 3.2 to the bracketed term and substituting (3.15) into the parenthesized term allows (3.19) to be simplified as follows:

$$\dot{V} = -r^T K r \leq -\lambda_{\min}\{K\} \|r\|^2\,, \tag{3.20}$$

where $\lambda_{\min}\{\cdot\}$ denotes the minimum eigenvalue of a matrix, and Lemma A.9 in Appendix A has been utilized.

According to the form of (3.20), we know that $V(t)$ is either decreasing or constant. Since $V(t)$ of (3.17) is a non-negative function (i.e., $V(t)$ is lower bounded by zero), we can conclude that $V(t) \in \mathcal{L}_\infty$; hence, $r(t) \in \mathcal{L}_\infty$ and $\tilde{\theta}(t) \in \mathcal{L}_\infty$. Since $r(t) \in \mathcal{L}_\infty$, we can utilize Lemma A.8 in Appendix A to show that $e(t), \dot{e}(t) \in \mathcal{L}_\infty$; hence, owing to the boundedness of $q_d(t)$ and $\dot{q}_d(t)$, we can use (3.5) to conclude that $q(t), \dot{q}(t) \in \mathcal{L}_\infty$. Since $\tilde{\theta}(t) \in \mathcal{L}_\infty$ and θ is a constant vector, (3.6) can be used to show that $\hat{\theta}(t) \in \mathcal{L}_\infty$. From the above boundedness statements and the fact that $\ddot{q}_d(t)$ is assumed bounded, the definition of (3.11) can be used to state that $Y_s(\cdot) \in \mathcal{L}_\infty$. It is now easy to see from (3.12) that the control input $\tau(t) \in \mathcal{L}_\infty$. The above information along with the fact that $M^{-1}(q)$ exists and is bounded can be applied to (3.1) and (3.14) to illustrate that $\ddot{q}(t), \dot{r}(t) \in \mathcal{L}_\infty$. Thus, we have illustrated that all signals in the adaptive controller and in the system remain bounded during closed-loop operation. Furthermore, the form of (3.20) allows us to show that $r(t) \in \mathcal{L}_2$ (see (2.20) to (2.22)). With the above information, we can now invoke Lemma A.3 in Appendix A to conclude that

$$\lim_{t \to \infty} r(t) = 0. \tag{3.21}$$

Finally, according to (3.21) and (3.7), Lemma A.8 in Appendix A can be utilized to obtain the result of (3.16). □

3.5 Desired Trajectory-Based Adaptive Control

Recall that the model-based adaptive controller of Section 3.4 formulates the regression matrix in terms of the actual position $q(t)$ and actual velocity $\dot{q}(t)$ (see (3.11) and (3.12)). Thus, it has the following implementation problems: (i) it carries the burden of requiring the on-line computation of an $n \times p$ nonlinear matrix and (ii) a significant portion of the feedforward terms of the control law may be contaminated with sensor noise from $q(t)$ and $\dot{q}(t)$. Fortunately, this implementation issue can be overcome by formulating the regression matrix as a function of the predefined, desired position, velocity, and acceleration (i.e., the DCAL control scheme), hence allowing its off-line calculation. In this section, we reexamine the original DCAL controller of [26] by initially presenting a simplified stability analysis to illustrate asymptotic position and velocity tracking for a DCAL-like controller. The analysis is simplified by employing a slightly different structure for the nonlinear feedback term than that originally proposed in [26], and by making use of the nonlinear damping design tool of Lemma A.10 in Appendix A. By further exploiting the nonlinear damping tool, we also show that a DCAL-like controller with only *linear* feedback can guarantee semiglobal asymptotic tracking. Finally, we show that the reformulated DCAL control structure can easily be utilized to examine some nonadaptive control extensions.

3.5.1 Controller Formulation

Before we present the controller formulation, let us use (3.4) to define a so-called *desired* linear parameterization as shown below:

$$Y_d(q_d, \dot{q}_d, \ddot{q}_d)\theta = M(q_d)\ddot{q}_d + V_m(q_d, \dot{q}_d)\dot{q}_d + G(q_d) + F_d\dot{q}_d, \tag{3.22}$$

where $Y_d(q_d, \dot{q}_d, \ddot{q}_d) \in \mathbb{R}^{n \times p}$ is the desired regression matrix which is a function only of the desired position, velocity, and acceleration. Based on the above definition and the open-loop dynamics of (3.10), we now design the control input $\tau(t)$ as follows:

$$\tau = Y_d(\cdot)\,\hat{\theta} + K_p e + (K_v + K_e)\,r + v_R, \tag{3.23}$$

where $K_p, K_v, K_e \in \mathbb{R}^{n \times n}$ are diagonal, positive-definite, control gain matrices, and $v_R(t)$ is an auxiliary nonlinear term yet to be determined. The auxiliary $v_R(t)$ term will be used to compensate for the mismatch between the desired regression matrix $Y_d(\cdot)$ and the *actual* regression matrix $Y_s(\cdot)$ defined in (3.11). The parameter estimate vector $\hat{\theta}(t)$ is generated by the dynamic update law

$$\dot{\hat{\theta}} = \Gamma Y_d^T(\cdot) r, \tag{3.24}$$

where $\Gamma \in \mathbb{R}^{p \times p}$ is the diagonal, positive-definite, adaptation gain matrix.

The substitution of (3.23) into (3.10) yields the closed-loop dynamics for $r(t)$ as follows:

$$M(q)\dot{r} = -V_m(q,\dot{q})r - (K_v + K_e)\,r - K_p e + Y_d\tilde{\theta} + \tilde{Y}\,(q,\dot{q},t) - v_R, \tag{3.25}$$

where the variable $\tilde{Y}\,(\cdot) \in \mathbb{R}^n$ is introduced to quantify the difference between the following linear parameterizations:

$$\tilde{Y} = Y_s(\cdot)\theta - Y_d\,(\cdot)\,\theta. \tag{3.26}$$

As in the previous section, we can use the time derivative of (3.6) along with (3.24) to form the closed-loop dynamics for the parameter estimation error as follows:

$$\dot{\tilde{\theta}} = -\Gamma Y_d^T(\cdot) r. \tag{3.27}$$

Remark 3.1 *It can be shown that the norm of the variable $\tilde{Y}\,(\cdot)$, defined in (3.26), can be upper bounded as (see [11, 26] for proof)*

$$\left\|\tilde{Y}\right\| \leq \rho(x)\left\|x\right\|, \tag{3.28}$$

where $x(t) \in \mathbb{R}^{2n}$ is defined as

$$x = \begin{bmatrix} r^T & e^T \end{bmatrix}^T, \tag{3.29}$$

and $\rho(x)$ is a positive function defined by

$$\rho(x) = \zeta_1 + \zeta_2 \left\|x\right\|. \tag{3.30}$$

In (3.30), ζ_1, ζ_2 denote positive bounding constants that depend only on the norm of the control gain matrix α of (3.7), the bounding constant m_2 defined in (3.2), and the bounds on the desired trajectory. The above bound will be exploited to obtain the stability results delineated in the following.

3.5.2 Stability Results

We now describe two stability results for the position tracking performance based on different choices of the control gain matrix K_e and the nonlinear term v_R in the control input of (3.23).

Theorem 3.2 *The control law of (3.23) and the parameter estimation update law of (3.24) ensure the global asymptotic convergence of the position and velocity tracking error, as illustrated by*

$$\lim_{t\to\infty} e(t), \dot{e}(t) = 0. \tag{3.31}$$

This result holds provided that K_e and $v_R(t)$ in (3.23) are selected as follows:

$$K_e = 0, \qquad v_R = k_n \rho^2(x) r, \tag{3.32}$$

where $\rho(x)$ was defined in (3.30), and $k_n \in \mathbb{R}$ is a nonlinear damping gain selected according to

$$k_n > \frac{1}{\lambda_3}, \tag{3.33}$$

with λ_3 being a positive constant defined as

$$\lambda_3 = \min\{\lambda_{\min}\{K_p\alpha\}, \lambda_{\min}\{K_v\}\}. \tag{3.34}$$

Proof. We begin by defining the following non-negative function:

$$V = \frac{1}{2} r^T M(q) r + \frac{1}{2} e^T K_p e + \frac{1}{2} \tilde{\theta}^T \Gamma^{-1} \tilde{\theta}. \tag{3.35}$$

Note that in comparison to (3.17), an additional term, $\frac{1}{2} e^T K_p e$, has been included in (3.35). The need for this extra term will become clear later in the analysis. From the above definition, it is not difficult to see that $V(t)$ can be bounded as

$$\lambda_1 \|x\|^2 \le \lambda_1 \|z\|^2 \le V \le \lambda_2 \|z\|^2, \tag{3.36}$$

where $x(t)$ was defined in (3.29), $z(t) \in \mathbb{R}^{2n+p}$ is defined as

$$z = \begin{bmatrix} r^T & e^T & \tilde{\theta}^T \end{bmatrix}^T, \tag{3.37}$$

and the positive bounding constants λ_1, λ_2 are defined by

$$\begin{aligned} \lambda_1 &= \frac{1}{2} \min\{m_1, \lambda_{\min}\{K_p\}, \lambda_{\min}\{\Gamma^{-1}\}\}, \\ \lambda_2 &= \frac{1}{2} \max\{m_2, \lambda_{\max}\{K_p\}, \lambda_{\max}\{\Gamma^{-1}\}\}, \end{aligned} \tag{3.38}$$

with m_1, m_2 being the bounding constants given in (3.2). Note that $\lambda_{\max}\{\cdot\}$ denotes the maximum eigenvalue of a matrix.

The differentiation of (3.35) with respect to time results in

$$\dot{V} = \frac{1}{2} r^T \dot{M}(q) r + r^T M(q) \dot{r} + e^T K_p \dot{e} + \tilde{\theta}^T \Gamma^{-1} \dot{\tilde{\theta}}. \tag{3.39}$$

Following the same procedure as in the proof of Theorem 3.1, we can substitute the closed-loop dynamics of (3.25) and (3.27) into (3.39) to get

$$\dot{V} = -r^T \left(K_v + K_e\right) r - r^T K_p e + r^T \tilde{Y} - r^T v_R + e^T K_p \dot{e}, \tag{3.40}$$

where Property 3.2 has been used. In order to substitute for $\dot{e}(t)$ in the last term in (3.40), we will use the fact, derived from (3.7), that

$$\dot{e} = -\alpha e + r. \tag{3.41}$$

After substituting (3.41) into (3.40), and making use of Lemma A.9 in Appendix A, we can place an upper bound on $\dot{V}(t)$ as follows:

$$\begin{aligned} \dot{V} \leq\ & -\lambda_{\min}\{K_v\} \left\|r\right\|^2 - \lambda_{\min}\{K_p \alpha\} \left\|e\right\|^2 \\ & -\lambda_{\min}\{K_e\} \left\|r\right\|^2 + \left\|r\right\| \left\|\tilde{Y}\right\| - r^T v_R. \end{aligned} \tag{3.42}$$

Using the definitions of (3.37) and (3.34), we see that the right-hand side of (3.42) can be upper bounded as

$$\dot{V} \leq -\lambda_3 \left\|x\right\|^2 - \lambda_{\min}\{K_e\} \left\|r\right\|^2 + \left\|r\right\| \left\|\tilde{Y}\right\| - r^T v_R. \tag{3.43}$$

After substituting (3.28) and (3.32) into (3.43), we have

$$\dot{V} \leq -\lambda_3 \left\|x\right\|^2 + \left[\rho \left\|x\right\| \left\|r\right\| - k_n \rho^2 \left\|r\right\|^2\right]. \tag{3.44}$$

Since the bracketed term in (3.44) forms a nonlinear damping pair, we can apply Lemma A.10 in Appendix A to further upper bound $\dot{V}(t)$ as

$$\dot{V} \leq -\left(\lambda_3 - \frac{1}{k_n}\right) \left\|x\right\|^2. \tag{3.45}$$

If the nonlinear damping gain k_n is selected such that $k_n > \frac{1}{\lambda_3}$, (3.45) can be rewritten as

$$\dot{V} \leq -\beta \left\|x\right\|^2 \leq -\beta \left\|r\right\|^2, \tag{3.46}$$

where (3.29) has been used, and β is some positive constant that approaches λ_3 as k_n is increased. The proof of Theorem 3.1 can now be followed to illustrate the boundedness of all closed-loop signals and the result of (3.31).
□

Remark 3.2 *The introduction of the term $\frac{1}{2} e^T K_p e$ in the definition of $V(t)$ of (3.35) is motivated by the need to obtain a negative $\left\|x\right\|^2$ term in $\dot{V}(t)$ (i.e., $-\lambda_3 \left\|x\right\|^2$ in (3.44)). This term is then used to dominate the positive $\left\|x\right\|^2$ term that arises from the application of the nonlinear damping argument to the bracketed term of (3.44) (i.e., the $\frac{1}{k_n} \left\|x\right\|^2$ term in (3.45)).*

Theorem 3.3 *The control law of (3.23) and the parameter estimation update law of (3.24) ensure the semiglobal asymptotic convergence of the position and velocity tracking errors as illustrated by*

$$\lim_{t\to\infty} e(t), \dot{e}(t) = 0. \tag{3.47}$$

This result is valid provided that K_e and $v_R(t)$ are now chosen as follows:

$$\lambda_{\min}\{K_e\} = k_n\left(\zeta_1^2 + \zeta_2^2\right), \qquad v_R = 0, \tag{3.48}$$

where ζ_1, ζ_2 were defined in (3.30), and k_n is selected according to

$$k_n > \frac{1}{\lambda_3}\left(\frac{\lambda_2}{\lambda_1}\|z(0)\|^2 + 1\right), \tag{3.49}$$

with λ_3, $z(t)$, and λ_1, λ_2 being defined in (3.34), (3.37), and (3.38), respectively.

Proof. The proof of this theorem makes use of the same non-negative function given by (3.35). The proofs, however, slightly diverge as described below.

After substituting (3.28), (3.30), and (3.48) into (3.43), we obtain

$$\dot{V} \le -\lambda_3\|x\|^2 + \left[\zeta_1\|x\|\|r\| - k_n\zeta_1^2\|r\|^2\right] + \left[\zeta_2\|x\|^2\|r\| - k_n\zeta_2^2\|r\|^2\right]. \tag{3.50}$$

The application of Lemma A.10 in Appendix A to the bracketed terms will produce

$$\dot{V} \le -\lambda_3\|x\|^2 + \frac{\|x\|^2}{k_n} + \frac{\|x\|^4}{k_n}, \tag{3.51}$$

which can be rewritten as

$$\dot{V} \le -\left[\lambda_3 - \frac{1}{k_n}\left(1 + \|x\|^2\right)\right]\|x\|^2. \tag{3.52}$$

The sign of the upper bound on $\dot{V}(t)$ is determined by the bracketed term of (3.52), which must be positive for one to conclude the negative semidefiniteness of $\dot{V}(t)$. That is, we must have

$$1 - \frac{1}{k_n}\left(1 + \|x(t)\|^2\right) > 0 \tag{3.53}$$

for $\dot{V}(t)$ to be negative semidefinite. From (3.36), a sufficient condition for (3.53) can be derived as follows:

$$1 - \frac{1}{k_n}\left(1 + \frac{V(t)}{\lambda_1}\right) > 0; \tag{3.54}$$

hence, the analysis to this point can be summarized as

$$\dot{V} \leq -\beta \|x\|^2 \qquad \text{for} \qquad k_n > \frac{V(t)}{\lambda_1} + 1, \tag{3.55}$$

where β is some positive constant. From (3.55), we know that since $\dot{V}(t) \leq 0$ $\forall t \in [0, \infty)$, $V(t)$ is decreasing or constant $\forall t \in [0, \infty)$. Therefore, a sufficient condition for (3.55) is given by

$$\dot{V} \leq -\beta \|x\|^2 \qquad \text{for} \qquad k_n > \frac{V(0)}{\lambda_1} + 1; \tag{3.56}$$

or by (3.36), we have a sufficient condition for (3.56), as follows:

$$\dot{V} \leq -\beta \|x\|^2 \leq -\beta \|r\|^2 \qquad \text{for} \qquad k_n > \frac{\lambda_2}{\lambda_1} \|z(0)\|^2 + 1, \tag{3.57}$$

where (3.29) has been utilized. The arguments used in the proof of Theorem 3.1 can now be followed to illustrate the boundedness of all closed-loop signals and the result of (3.47). □

Remark 3.3 *From (3.49), we can see that the control gain k_n can be increased to cover any set of initial conditions; therefore, the stability result of Theorem 3.3 is referred to as semiglobal.*

Remark 3.4 *It is interesting to see how different choices for part of the feedback portion of the control law (3.23) (i.e., the last two terms on the right-hand side of (3.23)) can produce slightly different stability results. Notice from (3.32) that the control law of Theorem 3.2 is a DCAL-like controller with nonlinear feedback that achieves the global convergence of the tracking errors. On the other hand, the control law that results from the selection of (3.32) in Theorem 3.3 is computationally simpler in the sense that it is a DCAL-like controller with linear feedback; however, this controller produces only a semiglobal convergence.*

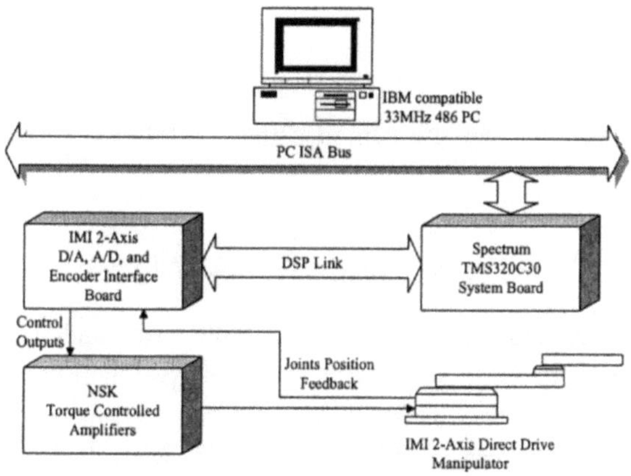

FIGURE 3.1. Block diagram of the experimental setup.

3.5.3 Experimental Results

The two DCAL–like control schemes described above were implemented on a two-link, direct-drive, rigid planar robot manipulator built by Integrated Motion, Inc. (IMI) [7]. This robot is a torque-controlled system that utilizes special-purpose electronic hardware to enable the implementation of torque control input algorithms. Specifically, the robot features links that are directly actuated by switched-reluctance motors that are controlled through Nippon Seiko K. K. (NSK) drives [14]. The *WinMotor* control environment was used to write the control algorithm in the C programming language. See Appendix D for a description of the C code used to implement the DCAL control algorithms. A functional diagram of the experimental setup is shown in Figure 3.1. As provided by the manufacturer, the two-link manipulator has the following dynamic model [7]:

$$\begin{aligned}\begin{bmatrix}\tau_1\\ \tau_2\end{bmatrix} = & \begin{bmatrix} p_1+2p_3c_2 & p_2+p_3c_2 \\ p_2+p_3c_2 & p_2\end{bmatrix}\begin{bmatrix}\ddot{q}_1\\ \ddot{q}_2\end{bmatrix} \\ & +\begin{bmatrix} -p_3s_2\dot{q}_2 & -p_3s_2(\dot{q}_1+\dot{q}_2) \\ p_3s_2\dot{q}_1 & 0\end{bmatrix}\begin{bmatrix}\dot{q}_1\\ \dot{q}_2\end{bmatrix} \\ & +\begin{bmatrix} f_{d1} & 0 \\ 0 & f_{d2}\end{bmatrix}\begin{bmatrix}\dot{q}_1\\ \dot{q}_2\end{bmatrix},\end{aligned} \tag{3.58}$$

where $p_1 = 3.473$ kg-m^2, $p_2 = 0.193$ kg-m^2, $p_3 = 0.242$ kg-m^2, $f_{d1} = 5.3$ N-m-sec, $f_{d2} = 1.1$ N-m-sec, c_2 denotes $\cos(q_2)$, and s_2 denotes $\sin(q_2)$. Based on (3.4) and (3.58), the unknown parameter vector θ can be constructed as

$$\theta = \begin{bmatrix} p_1 & p_2 & p_3 & f_{d1} & f_{d2} \end{bmatrix}^T. \tag{3.59}$$

The experiment was performed using the following desired position trajectories:

$$q_d(t) = \begin{bmatrix} 40.11\sin(t)\left(1-\exp\left(-0.3t^3\right)\right) \\ 68.75\sin(t)\left(1-\exp\left(-0.3t^3\right)\right) \end{bmatrix} \text{ deg.} \tag{3.60}$$

Note that the exponential term in (3.60) was included to ensure that $q_d(0) = \dot{q}_d(0) = \ddot{q}_d(0) = 0$. After several experimental runs, it was found that the nonlinear feedback DCAL controller given by (3.23), (3.24), and (3.32) achieved its best tracking performance with the following set of gains:

$$\begin{array}{c} \alpha = \operatorname{diag}\{180, 70\}, \quad K_v = \operatorname{diag}\{70, 50\}, \quad K_p = \operatorname{diag}\{1, 1\}, \\ k_n = 10^{-4}, \quad \zeta_1 = 70, \quad \zeta_2 = 60, \quad \Gamma = \operatorname{diag}\{7, 0.12, 0.2, 30, 5\}. \end{array}$$

For the linear feedback DCAL controller given by (3.23), (3.24), and (3.48), the set of gains that gave the best tracking performance was

$$\begin{array}{c} \alpha = \operatorname{diag}\{480, 90\}, \quad K_v = \operatorname{diag}\{35, 25\}, \quad K_e = \operatorname{diag}\{35, 25\}, \\ K_p = \operatorname{diag}\{1, 1\}, \quad \Gamma = \operatorname{diag}\{4, 0.1, 0.1, 20, 15\}. \end{array}$$

For both controllers, the initial values of the parameter update law of (3.24) were set to zero (i.e., $\hat{\theta}(0) = 0$) while a trapezoidal algorithm was used to compute the integral. A standard backwards difference algorithm applied to the position measurement $q(t)$ was utilized to generate the required velocity signal $\dot{q}(t)$. The sampling time for all experiments was 200 μsec.

The experimental results are shown in Figures 3.2–3.7. Figure 3.2 through 3.4 illustrate the link position tracking errors, torque control inputs, and parameter estimates for the nonlinear feedback DCAL controller. In Figures 3.5–3.7, the same plots are shown for the linear feedback DCAL controller. As can be seen from the experimental results, the linear feedback DCAL produced tracking errors of about the same magnitude as the nonlinear feedback scheme while maintaining the same level of torque inputs. Therefore, one could postulate that the nonlinear feedback term is not required for DCAL-like control implementations. This fact may be specially attractive in applications where it is deemed that the feedback portion of the control law should be executed as quickly as possible.

3.5.4 Nonadaptive Extensions

Instead of utilizing the dynamic update law of (3.24) to estimate the system parameters on-line, one can use a constant parameter estimate. Depending

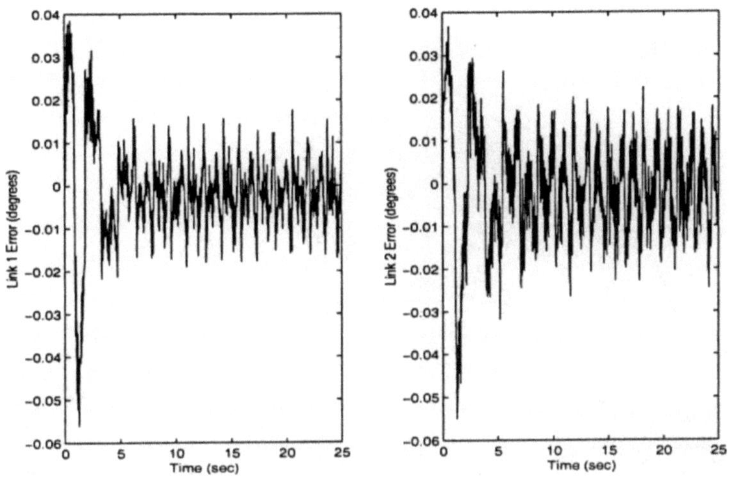

FIGURE 3.2. Link position tracking errors for the nonlinear feedback DCAL.

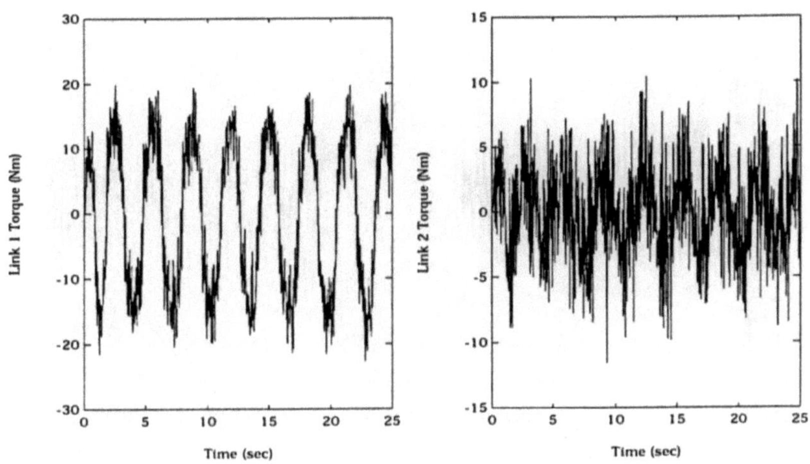

FIGURE 3.3. Torque inputs for the nonlinear feedback DCAL.

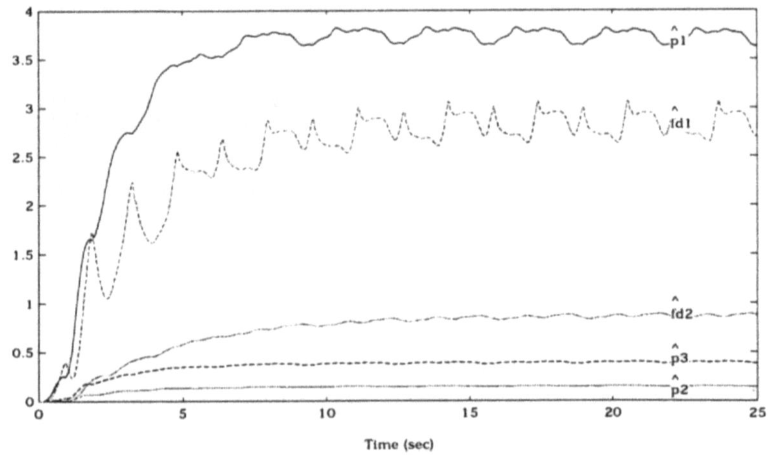

FIGURE 3.4. Parameter estimates for the nonlinear feedback DCAL.

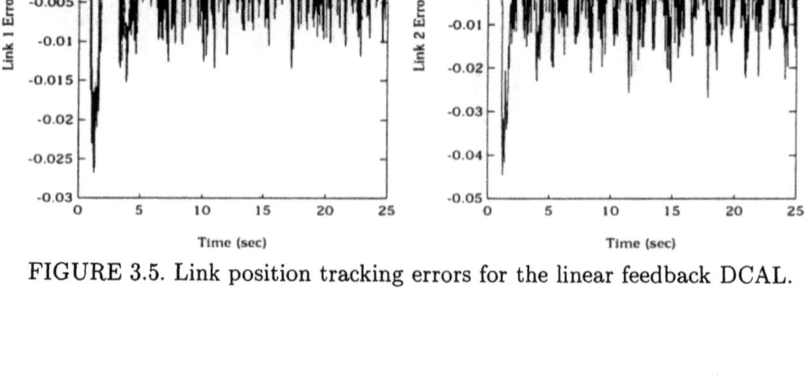

FIGURE 3.5. Link position tracking errors for the linear feedback DCAL.

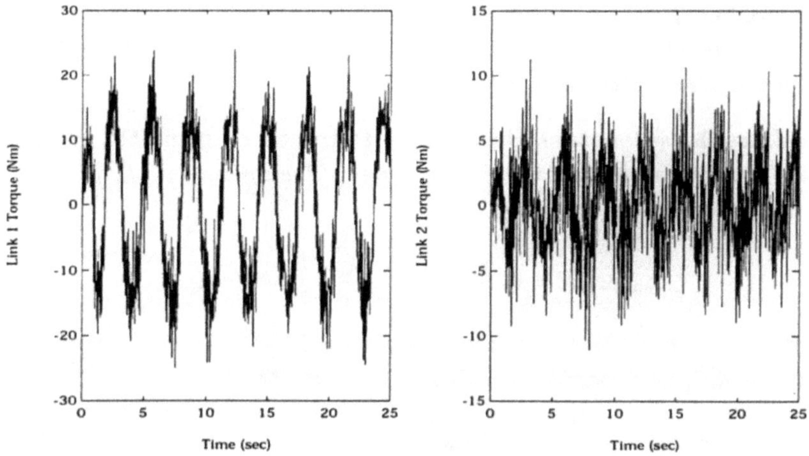

FIGURE 3.6. Torque inputs for the linear feedback DCAL.

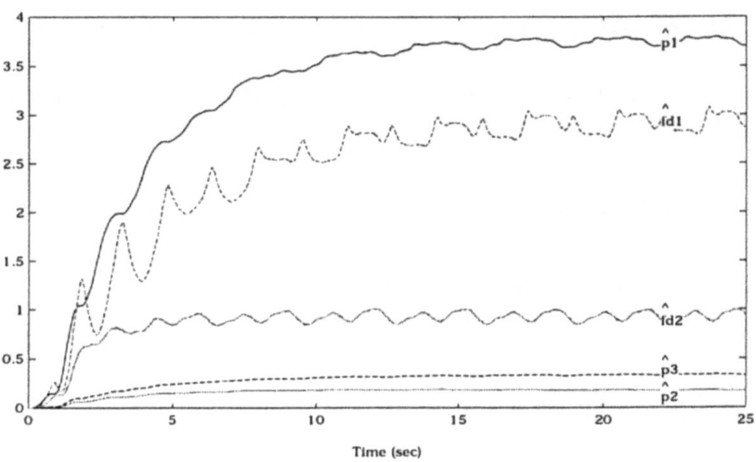

FIGURE 3.7. Parameter estimates for the linear feedback DCAL.

on the choice of the constant parameter estimate, different position tracking stability results will be achieved. The following two corollaries describe three cases for the selection of the fixed parameter estimate vector: (i) constant, best guess estimates of the actual parameters; (ii) the parameter estimates are set to zero; and (iii) the constant parameter estimates exactly match the actual parameter values.

Corollary 3.1 (Constant Best Guess Parameter Estimate) *Let $\hat{\theta}$ in (3.23) contain constant, best guess estimates of the system parameters, and K_e and v_R be chosen such that*

$$\lambda_{\min}\{K_e\} = k_n\left(\zeta_1^2 + \zeta_2^2\right) + \frac{\zeta_3^2}{\epsilon}, \qquad v_R = 0, \tag{3.61}$$

where ϵ, k_n are positive nonlinear damping gains, ζ_1, ζ_2 were defined in (3.30), and ζ_3 is a positive bounding constant defined as [1]

$$\left\| Y_d \tilde{\theta} \right\| \leq \zeta_3. \tag{3.62}$$

If k_n is selected according to

$$k_n > \frac{1}{\lambda_3}\left(\frac{\lambda_2}{\lambda_1}\|x(0)\|^2 + \frac{\lambda_2 \epsilon}{\lambda_1 \beta} + 1\right), \tag{3.63}$$

where $x(t), \lambda_3$ were defined in (3.29) and (3.34), respectively, β is some positive constant, and λ_1, λ_2 are defined as

$$\lambda_1 = \frac{1}{2}\min\{m_1, \lambda_{\min}\{K_p\}\}, \qquad \lambda_2 = \frac{1}{2}\max\{m_2, \lambda_{\max}\{K_p\}\}, \tag{3.64}$$

then the position tracking error is semiglobal uniformly ultimately bounded (SGUUB) in the sense that

$$\|e(t)\| \leq \sqrt{\frac{\lambda_2}{\lambda_1}\|x(0)\|^2 \exp\left(-\frac{\beta}{\lambda_2}t\right) + \frac{\lambda_2 \epsilon}{\lambda_1 \beta}\left(1 - \exp\left(-\frac{\beta}{\lambda_2}t\right)\right)}. \tag{3.65}$$

Proof. We define a non-negative function of the form

$$V = \frac{1}{2} r^T M(q) r + \frac{1}{2} e^T k_p e, \tag{3.66}$$

which can be bounded as

$$\lambda_1 \|x\|^2 \leq V \leq \lambda_2 \|x\|^2, \tag{3.67}$$

---

[1]The bounding constant ζ_3 will depend only on the bounds on the desired trajectory and on the magnitude of the difference between the constant, best guess parameter estimates and the actual parameters.

where x was defined in (3.29) and λ_1, λ_2 were defined in (3.64). The proof of Theorem 3.2 can be followed to form an upper bound on time derivative of $V(t)$ as

$$\dot{V} \leq -\lambda_3 \|x\|^2 - \lambda_{\min}\{K_e\} \|r\|^2 + \|r\| \left\|\tilde{Y}\right\| + \|r\| \left\|Y_d\tilde{\theta}\right\|. \tag{3.68}$$

After substituting (3.61), (3.28), and (3.62) into (3.68), we obtain

$$\begin{aligned} \dot{V} \leq\ & -\lambda_3 \|x\|^2 + \left[\zeta_1 \|x\| \|r\| - k_n\zeta_1^2 \|r\|^2\right] \\ & + \left[\zeta_2 \|x\|^2 \|r\| - k_n\zeta_2^2 \|r\|^2\right] + \left[\zeta_3 \|r\| - \frac{\zeta_3^2}{\epsilon} \|r\|^2\right]. \end{aligned} \tag{3.69}$$

Applying Lemma A.10 in Appendix A to the bracketed terms produces

$$\dot{V} \leq -\lambda_3 \|x\|^2 + \frac{\|x\|^2}{k_n} + \frac{\|x\|^4}{k_n} + \epsilon, \tag{3.70}$$

which can be rewritten as

$$\dot{V} \leq -\left(\lambda_3 - \frac{1}{k_n}\left(1 + \|x\|^2\right)\right) \|x\|^2 + \epsilon. \tag{3.71}$$

From (3.71), it is clear that

$$\dot{V} \leq -\beta \|x\|^2 + \epsilon \quad \text{for} \quad k_n > \frac{1}{\lambda_3}\left(1 + \|x(t)\|^2\right), \tag{3.72}$$

where β is some positive constant that approaches λ_3 as k_n is increased. After utilizing (3.67), a new upper bound can be placed on $\dot{V}(t)$ of (3.72) as follows:

$$\dot{V} \leq -\frac{\beta}{\lambda_2} V + \epsilon \quad \text{for} \quad k_n > \frac{1}{\lambda_3}\left(1 + \|x(t)\|^2\right). \tag{3.73}$$

We can now invoke Lemma A.5 in Appendix A along with (3.67) and (3.29) to conclude that

$$\begin{aligned} \|e(t)\| \leq \|x(t)\| \leq & \sqrt{\frac{\lambda_2}{\lambda_1} \|x(0)\|^2 \exp\left(-\frac{\beta}{\lambda_2}\right) t + \frac{\lambda_2\epsilon}{\lambda_1\beta}\left(1 - \exp\left(-\frac{\beta}{\lambda_2} t\right)\right)} \\ & \text{for} \quad k_n > \frac{1}{\lambda_3}\left(\frac{\lambda_2}{\lambda_1} \|x(0)\|^2 + \frac{\lambda_2\epsilon}{\lambda_1\beta} + 1\right) \end{aligned} \tag{3.74}$$

from which the result of (3.65) follows directly. □

Remark 3.5 *The SGUUB tracking result given by (3.74) can be illuminated by writing the steady state bound on* $\|x(t)\|$ *as*

$$\lim_{t\to\infty}\|x(t)\| = \sqrt{\frac{\lambda_2\epsilon}{\lambda_1\beta}}. \tag{3.75}$$

From (3.75), it is clear that the steady-state value of $x(t)$*; and hence, the steady-state bound for the position tracking error* $e(t)$*, can be made arbitrarily small by decreasing the value of the gain* ϵ*.*

Remark 3.6 *Corollary 3.1 includes the case where* $\hat{\theta} = 0$*. That is, even if there is no feedforward compensation in the control law, the SGUUB tracking result of (3.65) remains valid (note that now (3.62) must be written as* $\|Y_d\theta\| \leq \zeta_3$*). With* $\hat{\theta} = 0$*, the control law is a simple linear controller as proposed in [4, 18, 19, 20].*

Corollary 3.2 (Exact Parameter Knowledge) *Provided that* $\hat{\theta} = \theta$ *and* K_e*,* $v_R(t)$*, and* k_n *are selected as in Theorem 3.2, the position tracking error is semiglobal exponentially stable in the sense that* [2]

$$\|e(t)\| \leq \sqrt{\frac{\lambda_2}{\lambda_1}}\|x(0)\|\exp\left(-\frac{\beta}{2\lambda_2}t\right), \tag{3.76}$$

where β *is some positive constant.*

Proof. The proof follows directly from the proof of Corollary 3.1 by removing the last bracketed term in (3.69). □

Remark 3.7 *It is easy to show that the controllers of Corollaries 3.1 and 3.2 can be redesigned with* v_R *of (3.32) to yield global position tracking (see [5] for related work).*

3.6 Control/Adaptation Law Modularity

In this section, the modular adaptive control design of Section 2.3.2 in Chapter 2 is extended to MIMO mechanical systems. As in Chapter 2, we first design the control law to ensure the ISS property with respect to the parameter estimation error, hence guaranteeing closed-loop stability for any update law design that ensures the boundedness of the parameter estimates. We then show that the control law produces global asymptotic

[2] This stability result is similar to that of references [8, 31].

position tracking for any adaptive update law design that also satisfies the following sufficient conditions: (i) the prediction error is square integrable, and (ii) the estimated inertia matrix is positive-definite.

3.6.1 Input-to-State Stability Result

Before we initiate the controller formulation, let us first define the filtered regression matrix [15] $Y_f(q,\dot{q}) \in \mathbb{R}^{n\times p}$ as[3]

$$\dot{Y}_f(q,\dot{q}) + \beta Y_f(q,\dot{q}) = \beta Y(q,\dot{q},\ddot{q}) \qquad Y_f(q(0),\dot{q}(0)) = 0, \tag{3.77}$$

where $Y(\cdot)$ is the regression matrix defined in (3.4), and β is a positive, constant design parameter. We also define the measurable prediction error $\varepsilon(t) \in \mathbb{R}^n$ as follows:

$$\varepsilon = Y_f\tilde{\theta} = Y_f\theta - Y_f\hat{\theta} = \tau_f - Y_f\hat{\theta}, \tag{3.78}$$

where the filtered torque $\tau_f(t) \in \mathbb{R}^n$ is defined by

$$\dot{\tau}_f + \beta\tau_f = \beta\tau, \qquad \tau_f(0) = 0, \tag{3.79}$$

with $\tau(t)$ being defined in (3.1).

One of the main objectives of the control design is to achieve ISS with respect to the *disturbance input* $\tilde{\theta}(t)$ [10]. Based on this objective, the form of (3.10), and the subsequent analysis, we will utilize the following control law:

$$\begin{aligned}\tau = \ & Y_s(\cdot)\hat{\theta} + Kr + \left[\frac{1}{\beta}Y_f\dot{\hat{\theta}} + \left(\dot{\hat{M}} - \hat{V}_m\right)r\right] + k_n\|Y_s\|^2 r \\ & + k_n\left\|\frac{1}{\beta}Y_f\dot{\hat{\theta}}\right\|^2 r + k_n\left\|\left(\dot{\hat{M}} - \hat{V}_m\right)r\right\|^2 r,\end{aligned} \tag{3.80}$$

where $K \in \mathbb{R}^{n\times n}$ is a diagonal, positive-definite, control gain matrix; k_n is a positive nonlinear damping gain; $\hat{M}(q), \hat{V}_m(q,\dot{q})$ represent the estimates of the inertia and centripetal-Coriolis matrices, respectively; and $Y_s(\cdot), Y_f(\cdot)$ were defined in (3.11) and (3.77), respectively. To clarify the somewhat unusual structure of the control law (3.80), we note that the terms inside the brackets have been introduced to cancel similar terms in the subsequent stability analysis, while the norm-squared terms are nonlinear damping functions injected to achieve the ISS property of the closed-loop system

[3] See [11] for details on how to compute $Y_f(\cdot)$ without the need for acceleration measurements.

with respect to $\tilde{\theta}(t)$. By substituting (3.80) into (3.10), we can form the closed-loop error system for $r(t)$ as follows:

$$\begin{aligned} M(q)\dot{r} = \;& -Kr + Y_s\tilde{\theta} - k_n \left\|Y_s\right\|^2 r - \frac{1}{\beta}Y_f\,\dot{\hat{\theta}} - k_n \left\|\frac{1}{\beta}Y_f\,\dot{\hat{\theta}}\right\|^2 r \\ & - \left(\dot{\hat{M}} - \hat{V}_m\right) r - k_n \left\|\left(\dot{\hat{M}} - \hat{V}_m\right) r\right\|^2 r - V_m(q,\dot{q})r. \end{aligned} \tag{3.81}$$

We now state a theorem regarding the ISS property of (3.81) with respect to $\tilde{\theta}(t)$.

Theorem 3.4 *Given the closed-loop error system of (3.81), if* $\tilde{\theta}(t) \in \mathcal{L}_\infty[0, t_f)$*, then* $r(t) \in \mathcal{L}_\infty[0, t_f)$ *where* t_f *denotes the final time.*

Proof. We start by defining the non-negative function:

$$V = \frac{1}{2}r^T M(q)r. \tag{3.82}$$

After taking the time derivative of (3.82) along (3.81), we can obtain the following upper bound on $\dot{V}(t)$:

$$\begin{aligned} \dot{V} \leq \;& -\lambda_{\min}\{K\}\left\|r\right\|^2 + \left[\left\|Y_s\right\|\left\|r\right\|\left\|\tilde{\theta}\right\| - k_n\left\|Y_s\right\|^2\left\|r\right\|^2\right] \\ & + \left[\left\|\frac{1}{\beta}Y_f\,\dot{\hat{\theta}}\right\|\left\|r\right\| - k_n\left\|\frac{1}{\beta}Y_f\,\dot{\hat{\theta}}\right\|^2\left\|r\right\|^2\right] \\ & + \left[\left\|\left(\dot{\hat{M}} - \hat{V}_m\right) r\right\|\left\|r\right\| - k_n\left\|\left(\dot{\hat{M}} - \hat{V}_m\right) r\right\|^2\left\|r\right\|^2\right]. \end{aligned} \tag{3.83}$$

After applying Lemma A.10 in Appendix A to the bracketed pairs of (3.83), we have

$$\dot{V} \leq -\lambda_{\min}\{K\}\left\|r\right\|^2 + \frac{1}{k_n}\left\|\tilde{\theta}\right\|^2 + \frac{2}{k_n}, \tag{3.84}$$

which can be further upper bounded, by the use of (3.2) and (3.82), as follows:

$$\dot{V}(t) \leq -\lambda V(t) + \frac{1}{k_n}\left\|\tilde{\theta}(t)\right\|^2 + \frac{2}{k_n}, \tag{3.85}$$

where $\lambda = \dfrac{2\lambda_{\min}\{K\}}{m_2}$. After solving the differential inequality of (3.85), we have

$$\begin{aligned} V(t) &\leq V(0)e^{-\lambda t} + e^{-\lambda t}\int_0^t e^{\lambda\xi}\left(\frac{1}{k_n}\left\|\tilde{\theta}(\xi)\right\|^2 + \frac{2}{k_n}\right)d\xi \\ &\leq V(0)e^{-\lambda t} + \frac{1}{\lambda k_n}\left(\left\|\tilde{\theta}(t)\right\|_\infty^2 + 2\right). \end{aligned} \tag{3.86}$$

Hence, from (3.86), (3.2), and (3.17), we have

$$\|r(t)\| \leq \sqrt{\frac{m_2}{m_1}}\,\|r(0)\|\,e^{-\frac{\lambda}{2}t} + \sqrt{\frac{2}{m_1\lambda k_n}\left(\left\|\tilde{\theta}(t)\right\|_\infty^2 + 2\right)} < \infty, \tag{3.87}$$

which proves the theorem. □

Remark 3.8 *Since Theorem 3.4 states that $r(t) \in \mathcal{L}_\infty[0, t_f)$, we can apply Lemma A.8 in Appendix A to (3.7) to show that $e(t), \dot{e}(t) \in \mathcal{L}_\infty[0, t_f)$; hence, $q(t), \dot{q}(t) \in \mathcal{L}_\infty[0, t_f)$ owing to the assumed boundedness on $q_d(t), \dot{q}_d(t)$.*

3.6.2 Position Tracking Result

The control law (3.80) guarantees the boundedness of the system states independent of the adaptation algorithm. In other words, it permits some flexibility in the design of the parameter update law as long as the boundedness condition on $\tilde{\theta}(t)$ is satisfied. We now turn our attention to developing specific technical conditions on the update law that lead to boundedness of all closed-loop signals and position tracking.

Theorem 3.5 *Given the control law (3.80) and any update law, denoted by $\dot{\hat{\theta}}(q, \dot{q}, \varepsilon)$, that ensures $\hat{\theta}(t) \in \mathcal{L}_\infty[0, t_f)$, we can state that (i) all signals are bounded during closed-loop operation on $t \in [0, \infty)$, and (ii) if the update law also ensures that $\varepsilon(t) \in \mathcal{L}_2[0, t_f)$, then*

$$\lim_{t\to\infty} e(t), \dot{e}(t) = 0, \tag{3.88}$$

provided $\hat{M}(q)$ remains positive-definite and K of (3.80) is selected as $K = kI_n$ where k is a positive constant and I_n is the $n \times n$ identity matrix.

Proof. (Part (i)) Since $\hat{\theta}(t) \in \mathcal{L}_\infty[0, t_f)$, we know from Theorem 3.4 and Remark 3.8 that $r(t), q(t), \dot{q}(t) \in \mathcal{L}_\infty[0, t_f)$; hence, $Y_f(q, \dot{q}) \in \mathcal{L}_\infty[0, t_f)$. Since $\hat{\theta}(t) \in \mathcal{L}_\infty[0, t_f)$ and $Y_f(q, \dot{q}) \in \mathcal{L}_\infty[0, t_f)$, we can state from (3.78) that $\varepsilon(t) \in \mathcal{L}_\infty[0, t_f)$. Since $\dot{\hat{\theta}}(t)$ is a function of $q(t)$, $\dot{q}(t)$, and $\varepsilon(t)$, we know that $\dot{\hat{\theta}}(t) \in \mathcal{L}_\infty[0, t_f)$. From the previous boundedness statements, we can see from (3.81) and (3.80) that $\dot{r}(t), \tau(t) \in \mathcal{L}_\infty[0, t_f)$, and therefore, $\ddot{q}(t) \in \mathcal{L}_\infty[0, t_f)$ from (3.1) and (3.2). Thus, we have shown that all closed-loop signals are bounded on $[0, t_f)$. As in [10], the bounds are dependent only on the initial conditions, control gains, and the desired motion trajectory (i.e., not dependent on t_f); hence, this independence of time proves that t_f can be set to ∞. □

Proof. (Part (ii)) We start by noting from (3.4) and (3.11) that

$$Y(q, \dot{q}, \ddot{q})\,\tilde{\theta} = Y_s(q, \dot{q}, t)\,\tilde{\theta} - \tilde{M}(q)\dot{r} - \tilde{V}_m(q, \dot{q})\,r, \tag{3.89}$$

where $\tilde{M}(q) = M(q) - \hat{M}(q)$ and $\tilde{V}_m(q, \dot{q}) = V_m(q, \dot{q}) - \hat{V}_m(q, \dot{q})$. With the intent of writing the term $Y_s\tilde{\theta}$ in terms of $\varepsilon(t)$, we can use (3.78), the time derivative of (3.6), and the relationships given by (3.77) and (3.89) to show that

$$\begin{aligned} \frac{1}{\beta}\dot{\varepsilon} + \varepsilon &= \frac{1}{\beta}\frac{d}{dt}\left(Y_f\tilde{\theta}\right) + Y_f\tilde{\theta} \\ &= \frac{1}{\beta}\dot{Y}_f\tilde{\theta} + Y_f\tilde{\theta} - \frac{1}{\beta}Y_f\,\dot{\hat{\theta}} \\ &= Y\tilde{\theta} - \frac{1}{\beta}Y_f\,\dot{\hat{\theta}} \\ &= Y_s\tilde{\theta} - \tilde{M}(q)\dot{r} - \tilde{V}_m(q, \dot{q})\,r - \frac{1}{\beta}Y_f\,\dot{\hat{\theta}}\,. \end{aligned} \tag{3.90}$$

We can now utilize the last line in (3.90) to obtain the following expression for $Y_s\tilde{\theta}$:

$$Y_s\tilde{\theta} = \frac{1}{\beta}\dot{\varepsilon} + \varepsilon + \tilde{M}(q)\dot{r} + \tilde{V}_m(q, \dot{q})\,r + \frac{1}{\beta}Y_f\,\dot{\hat{\theta}}\,. \tag{3.91}$$

After substituting the right-hand side of (3.91) into (3.81), we have

$$\hat{M}(q)\dot{r} = -A(q, \dot{q}, \hat{\theta}, \dot{\hat{\theta}}, t)r + \frac{1}{\beta}\dot{\varepsilon} + \varepsilon - \dot{\hat{M}}(q)r, \tag{3.92}$$

where the positive function $A(\cdot) \in \mathbb{R}$ is defined as

$$A(q, \dot{q}, \hat{\theta}, \dot{\hat{\theta}}, t) = k + k_n\left\|Y_s\right\|^2 + k_n\left\|\frac{1}{\beta}Y_f\,\dot{\hat{\theta}}\right\|^2 + k_n\left\|\left(\dot{\hat{M}} - \hat{V}_m\right)r\right\|^2. \tag{3.93}$$

Similar to the analysis of the modular adaptive, friction compensation controller of Section 2.3.2, we now define the following variable transformation:

$$x = \hat{M}(q)r - \frac{1}{\beta}\varepsilon. \tag{3.94}$$

After taking the time derivative of (3.94) and substituting (3.92) into the resulting expression, we obtain

$$\dot{x} = -A\left(\cdot\right)r + \varepsilon. \tag{3.95}$$

By utilizing (3.94), the above expression for $\dot{x}\left(t\right)$ can be rewritten as follows:

$$\dot{x} = -A\hat{M}^{-1}x + B\varepsilon, \tag{3.96}$$

where the matrix $B\left(\cdot\right) \in \mathbb{R}^{n\times n}$ is defined as

$$B(q, \dot{q}, \hat{\theta}, \dot{\hat{\theta}}, t) = I_n - \frac{1}{\beta}A\hat{M}^{-1}. \tag{3.97}$$

After differentiating $\frac{1}{2}x^T x$ along (3.96), we obtain[4]

$$\begin{aligned} \frac{d}{dt}\left(\frac{1}{2}x^T x\right) &= -Ax^T\hat{M}^{-1}x + x^T B\varepsilon \\ &\leq -A\lambda_{\min}\left\{\hat{M}^{-1}\right\}\|x\|^2 + \|B\|_{i\infty}\|\varepsilon\|\|x\|, \end{aligned} \tag{3.98}$$

where Lemma A.9 in Appendix A has been utilized.

We now let

$$y = \sqrt{\frac{1}{2}x^T x} \tag{3.99}$$

and rewrite (3.98) as

$$\frac{d}{dt}\left(y^2\right) = 2y\dot{y} \leq -2A\lambda_{\min}\left\{\hat{M}^{-1}\right\}y^2 + \sqrt{2}\|B\|_{i\infty}\|\varepsilon\|y. \tag{3.100}$$

Since $y(t) \geq 0$, we can use (3.100) to produce the following upper bound on $\dot{y}(t)$:

$$\dot{y} \leq -\zeta_1 y + \zeta_2\|\varepsilon\|, \tag{3.101}$$

where the positive constants[5] ζ_1 and ζ_2 are defined as

$$\zeta_1 = \inf_t\{A\}\lambda_{\min}\left\{\hat{M}^{-1}\right\} \qquad \text{and} \qquad \zeta_2 = \frac{\sqrt{2}}{2}\|B\|_{\infty}. \tag{3.102}$$

From this point on, the same derivations described in (2.59) to (2.63) in Chapter 2 can be followed to show (3.88). □

---

[4] Note that if $\hat{M}(q)$ is positive-definite, we know $\hat{M}^{-1}(q)$ is also positive-definite.

[5] Since we know all signals are bounded from the proof of Part (i), the constants ζ_1 and ζ_2 exist.

Remark 3.9 *A typical parameter adaptation algorithm that satisfies the conditions stated in Theorem 3.5 (i.e., $\hat{\theta}(t) \in \mathcal{L}_\infty$ and $\varepsilon(t) \in \mathcal{L}_2$) is the least-squares estimator given by*

$$\dot{\hat{\theta}} = \frac{\Gamma Y_f^T \varepsilon}{1 + \gamma tr\left\{Y_f \Gamma Y_f^T\right\}}, \qquad \dot{\Gamma} = -\frac{\Gamma Y_f^T Y_f \Gamma}{1 + \gamma tr\left\{Y_f \Gamma Y_f^T\right\}}, \tag{3.103}$$

where $\Gamma(t) \in \mathbb{R}^{p \times p}$ is a time-varying, symmetric matrix; γ is a non-negative constant; and $tr\{\cdot\}$ denotes the trace of a matrix. The reader is referred to Section 2.3.2 in Chapter 2 for the proof that illustrates that $\hat{\theta}(t) \in \mathcal{L}_\infty$ and $\varepsilon(t) \in \mathcal{L}_2$. Note that if $\gamma = 0$ in (3.103), we obtain the standard, unnormalized least-squares estimator.

Remark 3.10 *In order to satisfy the condition on the positive-definiteness of $\hat{M}(q)$ stated in Part (ii) of Theorem 3.5, a projection-type algorithm (see [1, 12] for examples) must be incorporated into the design of $\hat{\theta}(t)$.*

3.6.3 Experimental Results

The performance of the control law (3.80) combined with the unnormalized least-squares estimator (i.e., (3.103) with $\gamma = 0$) (hereinafter denoted as *LS* controller) was evaluated on the IMI, two-link, direct-drive, planar robot described in Section 3.5.3. A Pentium 166 MHz PC running *Qmotor* provided the environment to write the control algorithm in the C programming language.

For comparison purposes, the adaptive controller of Section 3.4 given by (3.12) and (3.13), which utilizes a filtered tracking error-based, gradient update law, was also evaluated (hereinafter denoted as *GR* controller). The desired position trajectories were selected as follows:

$$q_d(t) = \begin{bmatrix} 40.11 \sin(2t)\left(1 - \exp\left(-0.3t^3\right)\right) \\ 68.75 \sin(2t)\left(1 - \exp\left(-0.3t^3\right)\right) \end{bmatrix} \text{deg.} \tag{3.104}$$

A standard backwards difference algorithm applied to the position measurements and then passed through a digital low-pass filter was used to generate the velocity signals. The update laws given in (3.103) and (3.13) were initialized to zero, while a trapezoidal algorithm was utilized to compute the integrals. The sampling period for the *LS* controller was 400 μsec while the *GR* controller was run at 150 μsec[6]. All control parameters were

[6] Becuase of its relative simplicity, the *GR* algorithm can be run at a faster sampling period than the *LS* algorithm. Since a faster sampling period will generally improve an algorithm's performance, this experimental comparison is much fairer to the *GR* algorithm.

tuned by trial-and-error until the best position tracking performance was achieved. For the *LS* controller, the control gains and the initial values for the time-varying, adaptation matrix $\Gamma(t)$ that resulted in the best tracking performance were

$$\beta = 23, \quad \alpha = diag\{51.5, 49.0\}, \quad K = diag\{70,\ 26\}, \quad k_n = 1e^{-4},$$

$$\Gamma(0) = \begin{bmatrix} 0.5 & 0.25 & 0.275 & 0.5 & 0.125 \\ 0.0175 & 0.0263 & 0.0175 & 0.0175 & 0.088 \\ 0.042 & 0.045 & 0.036 & 0.03 & 0.012 \\ 0.153 & 0.144 & 0.288 & 0.099 & 0.09 \\ 0.242 & 0.033 & 0.055 & 0.01767 & 0.22 \end{bmatrix} \tag{3.105}$$

For the *GR* controller, the set of control and adaptation gains that resulted in the best tracking performance were

$$\alpha = \text{diag}\{40, 23\}, \quad K = \text{diag}\{75, 33\}, \quad \Gamma = \text{diag}\{8.0, 0.1, 0.2, 7.0, 2.0\}. \tag{3.106}$$

The experimental results are shown in Figures 3.8–3.11. Figures 3.8(a) and 3.8(b) illustrate the link 1 position tracking error for the *LS* and *GR* controllers, respectively, while Figures 3.9(a) and 3.9(b) show the link 2 position tracking error. Figures 3.10(a) and 3.10(b) depict the parameter estimates for the *LS* and *GR* controllers, respectively. In Figures 3.11(a) and 3.11(b), we have included the control inputs for links 1 and 2, respectively, of the *LS* controller, while Figures 3.11(c) and 3.118(d) contain the same signals for the *GR* controller. From these figures, we observe the faster parameter adaptation of the *LS* controller; and hence, the improved transient performance of the tracking error signals. We note that the improved tracking/parameter adaptation performance of the *LS* controller was obtained at the expense of only slightly larger control torques in comparison to those of the *GR* controller.

3.6.4 Discussion of Results

Since the condition on the positive-definiteness of $\hat{M}(q)$ required in Part (ii) of Theorem 3.5 is only a *sufficient* condition for achieving asymptotic position tracking with the *LS* controller, it comes as no surprise that the above experimental results were obtained without utilizing a projection-type algorithm for $\hat{\theta}(t)$, as discussed in Remark 3.10. Furthermore, since the requirement that $K = kI_n$ is also only a *sufficient* condition for achieving asymptotic position tracking with the *LS* controller, it also comes as no surprise that the experimental results were obtained without satisfying this

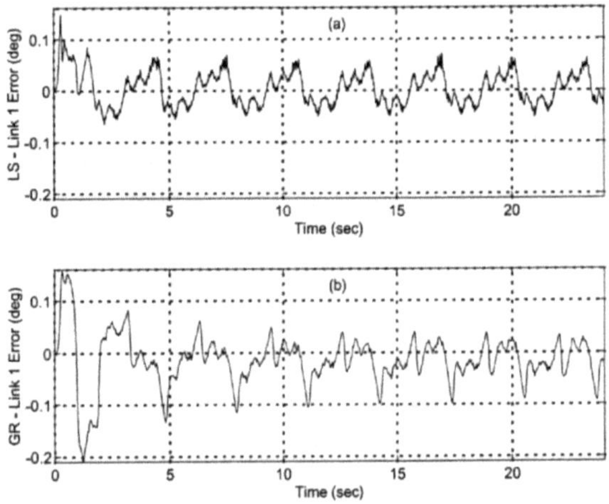

FIGURE 3.8. Tracking error for link 1: (a) *LS* controller and (b) *GR* controller.

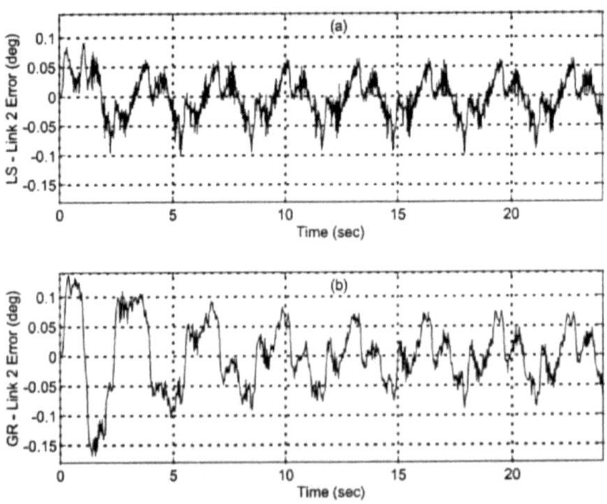

FIGURE 3.9. Tracking error for link 2: (a) *LS* controller and (b) *GR* controller.

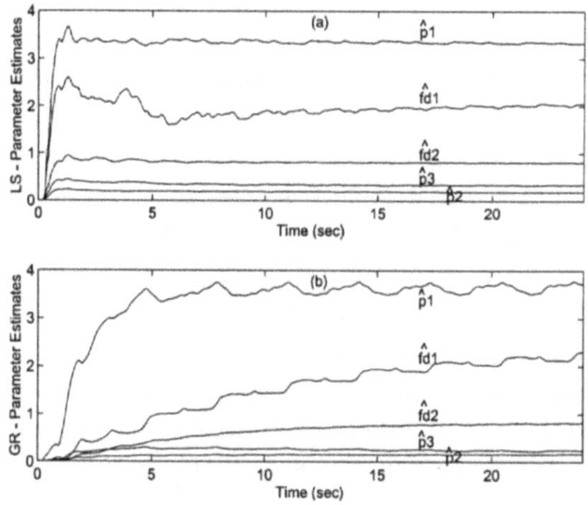

FIGURE 3.10. Parameter estimates: (a) *LS* controller and (b) *GR* controller.

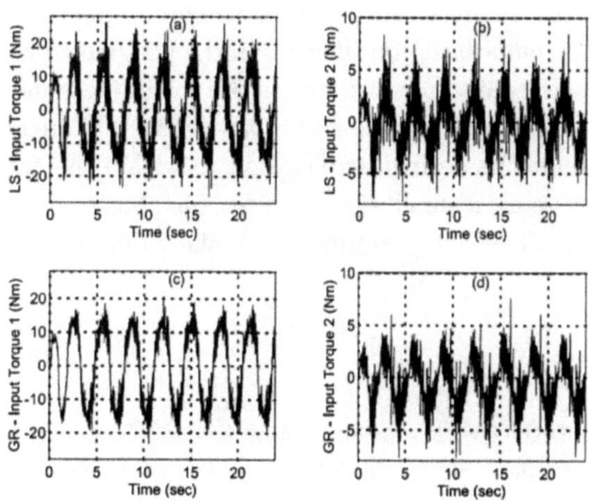

FIGURE 3.11. Torque inputs: (a) τ_1-*LS* controller, (b) τ_2-*LS* controller, (c) τ_1-*GR* controller, and (d) τ_2-*GR* controller.

condition (see (3.105)). We also note that no care was taken during the tuning procedure to ensure that $\Gamma(0)$ of (3.105) was positive-definite or symmetric.

The experimental results indicate that an improvement in the transient response of the tracking performance can be obtained through the use of the *LS* controller. Note that this improvement was also detected in the experimental results of Section 2.3.4. This behavior is due in part to the indirect influence the matrix $\Gamma(0)$ has on the system's transient performance. Specifically, if higher values were utilized for the elements of $\Gamma(0)$, we observed that the parameter estimates greatly overshoot their steady-state values, thus causing the transient tracking errors to be larger. On the other hand, if smaller values were used, the parameter estimates took longer to converge, thus causing the tracking errors to have a larger settling time. We note that owing to the 5×5 dimension and coupled nature of $\Gamma(0)$, it is not easy to tune its elements such that a critically damped behavior is obtained for all the parameter estimates. However, the fact that $\Gamma(0)$ provides some means of indirectly controlling the system's transient performance in the *LS* controller has been shown to constitute an advantage over the *GR* controller.

As stated in the previous section, the control/adaptation gains and $\Gamma(0)$ were selected by trial-and-error to achieve the best tracking performance for the desired trajectory signal given by (3.104). That is, we found that the controller parameters can be tuned to obtain good position tracking performance by monitoring the changes in the position tracking errors and then adjusting the gains in between runs. Although the gains in (3.106) and (3.105) work for different desired trajectory signals that are reasonable (i.e., signals that limit the desired acceleration to within the capability of the motor drive system), a small amount of retuning has to be done to achieve the best position tracking performance for these different trajectories.

3.7 Notes

Some representative work related to adaptive FSFB controllers applicable to MIMO mechanical systems can be found in [2, 3, 6, 9, 17, 21, 24, 28, 32], while a class of DCAL FSFB controllers can be found in [22]. Most of the above mentioned results share the characteristic of utilizing standard, gradient-type, adaptation schemes. With the goal of overcoming the slow parameter convergence of gradient-type update laws, Lozano Leal et al. [13] utilized previous passivity-based adaptive control designs to construct a modified least-squares update law with the position/velocity tracking error as the input. As time goes to infinity, the parameter estimates in this modified least-squares update law converge to the parameter estimates generated by a standard, least-squares update law. A modified least-squares update law based on the position/velocity tracking error was also proposed by Sadegh and Horowitz [25] in the design of an exponentially stable DCAL-type controller. This exponential tracking result was predicated on the assumption that the desired regression matrix satisfies a semipersistency of excitation condition; furthermore, the modified least-squares update law requires the calculation of a matrix dependent on the excitation condition. In [27], Slotine and Li developed a *composite* adaptive controller by constructing the parameter update law as the composite sum of a least-squares update law driven by the prediction error and a modified gradient update law driven by the position/velocity tracking error (i.e., the gradient update law is modified because it uses the same time-varying, adaptation gain as the least-squares update law). As compared to the standard persistency of excitation condition, the composite adaptive controller requires a weaker excitation condition (i.e., the infinite integral condition) for parameter convergence; however, this algorithm still utilizes a very specific choice for the adaptive update law. In [29], Tang and Arteaga developed an adaptive controller that included the standard, gradient update law; the composite adaptation update law; and an averaging gradient update law as special cases.

Modular estimation-based adaptive control approaches were reported by Queiroz et al. [23] and Middleton and Goodwin [15]. Roughly speaking, in [15], the adaptive computed torque controller of Craig et al. [3] was augmented with additional terms that allowed the closed-loop error system to be written as a stable, linear, strictly-proper transfer function with the position tracking error as the output and a prediction error-related term as the input. This input–output relationship facilitated the dual objectives of position tracking and controller/update law modularity in the sense that any parameter update law could be used as long as its design ensured that (i) the parameter estimates remained bounded, (ii) the prediction error was square integrable, and (iii) the estimated inertia matrix was positive-definite (i.e., a projection-type algorithm is required in the parameter update law). We note that in contrast to the design of [15], the modular adaptive controller of Section 3.6 does not need the estimated inertia matrix to be positive-definite to guarantee boundedness of the system signals, but only to prove position tracking (see Theorems 3.4 and 3.5). Moreover, the control law (see (3.80)) does not require the online calculation of the inverse of the estimated inertia matrix, as is the case for the controller of [15].

References

[1] M. M. Bridges, D. M. Dawson, and C. T. Abdallah, Control of Rigid-Link Flexible-Joint Robots: a Survey of Backstepping Approaches, *Journal of Robotic Systems*, Vol. 12, No. 3, pp. 199-216, Mar. 1995.

[2] B. Brogliato, I. Landau, and R. Lozano-Leal, Adaptive Motion Control of Robot Manipulators: A Unified Approach Based on Passivity, *Proceedings of the American Control Conference*, pp. 2259-2264, San Diego, CA, May 1990.

[3] J. J. Craig, P. Hsu, and S. Sastry, Adaptive Control of Mechanical Manipulators, *Proceedings of the IEEE International Conference on Robotics and Automation*, pp. 190-195, San Francisco, CA, Mar. 1986.

[4] D. M. Dawson, Z. Qu, F. L. Lewis, and J. F. Dorsey, Robust Control for the Tracking of Robot Motion, *International Journal of Control*, Vol. 52, pp. 581-595, 1990.

[5] D. M. Dawson and Z. Qu, On Uniform Ultimate Boundedness of a DCAL-like Robot Controller, *IEEE Transactions on Robotics and Automation*, Vol. 8, No. 3, pp. 409-413, June 1992.

[6] P. Hsu, M. Bodson, S. Sastry, and B. Paden, Adaptive Identification and Control for Manipulators without Using Joint Accelerations, *Proceedings of the IEEE International Conference on Robotics and Automation*, pp. 2137-2142, San Antonio, TX, Dec. 1993.

[7] Integrated Motion, Inc., *Direct Drive Manipulator Research and Development Package Operations Manual*, Berkeley, CA, 1992.

[8] R. Kelly and R. Salgado, PD Control with Computed Feedforward of Robot Manipulators: A Design Procedure, *IEEE Transactions on Robotics and Automation*, Vol. 10, No. 4, pp. 566-571, Aug. 1994.

[9] R. Kelly, R. Carelli, and R. Ortega, Adaptive Motion Control Design of Robot Manipulators: An Input-Output Approach, *International Journal of Control*, Vol. 50, No. 6, pp. 2563-2581, June 1989.

[10] M. Krstić and P. V. Kokotović, Adaptive Nonlinear Design with Controller-Identifier Separation and Swapping, *IEEE Transactions on Automatic Control*, Vol. 40, No. 3, pp. 426-440, Mar. 1995.

[11] F. L. Lewis, C. T. Abdallah, and D. M. Dawson, *Control of Robot Manipulators*, New York, NY: Macmillan Publishing Co., 1993.

[12] R. Lozano Leal and B. Brogliato, Adaptive Control of Robot Manipulators with Flexible Joints, *IEEE Transactions on Automatic Control*, Vol. 37, No. 2, pp. 174-181, Feb. 1992.

[13] R. Lozano Leal and C. Canudas de Wit, Passivity Based Adaptive Control for Mechanical Manipulators Using LS-Type Estimation, *IEEE Transactions on Automatic Control*, Vol. 35, No. 12, pp. 1363-1365, Dec. 1990.

[14] *Megatorque Motor System-User's Manual*, Nippon Seiko K.K. Ltd., Tokyo, Japan, 1989.

[15] R. H. Middleton and C. G. Goodwin, Adaptive Computed Torque Control for Rigid Link Manipulators, *Systems & Control Letters*, Vol. 10, pp. 9-16, 1988.

[16] S. Nicosia and P. Tomei, Robot Control by Using Only Joint Position Measurements, *IEEE Transactions on Automatic Control*, Vol. 35, No. 9, pp. 1058-1061, Sept. 1990.

[17] R. Ortega and M. Spong, Adaptive Motion Control of Rigid Robots: A Tutorial, *Automatica*, Vol. 25, No. 6, pp. 877-888, June 1989.

[18] P. Paden and R. Panja, Globally Asymptotic Stable PD+ Controller for Robot Manipulators, *International Journal of Control*, Vol. 47, No. 6, pp. 1696-1712, June 1988.

[19] Z. Qu and J. Dorsey, Robust Tracking Control of Robots by a Linear Feedback Law, *IEEE Transactions on Automatic Control*, Vol. 36, pp. 1081-1084, 1991.

[20] Z. Qu, J. F. Dorsey, and D. M. Dawson, Robust Control of Robots by Computed Torque Law, *Systems & Control Letters*, Vol. 16, pp. 25-32, 1991.

[21] Z. Qu, J. F. Dorsey, and D. M. Dawson, Exponentially Stable Trajectory Following of Robotic Manipulators Under a Class of Adaptive Controls, *Automatica*, Vol. 28, No. 3, pp. 579-586, Mar. 1992.

[22] M. S. de Queiroz, D. Dawson, and T. Burg, Reexamination of the DCAL Controller for Rigid Link Robots, *Robotica*, Vol. 14, pp. 41-49, Jan. 1996.

[23] M. S. de Queiroz, D. M. Dawson, and M. Agarwal, Adaptive Control of Robot Manipulators with Controller/Update Law Modularity, *Automatica*, Vol. 35, No. 8, pp. 1379-1390, Aug. 1999.

[24] J. Reed and P. Ioannou, Instability Analysis and Robust Adaptive Control of Robotic Manipulators, *Proceedings of IEEE Conference on Decision Control*, pp. 1607-1612, Austin, TX, 1988.

[25] N. Sadegh and R. Horowitz, An Exponentially Stable Adaptive Control Law for Robot Manipulators, *IEEE Transactions on Robotics and Automation*, Vol. 6, No. 4, pp. 491-496, Aug. 1990.

[26] N. Sadegh and R. Horowitz, Stability and Robustness Analysis of a Class of Adaptive Controllers for Robotic Manipulators, *International Journal of Robotics Research*, Vol. 9, No. 9, pp. 74-92, June 1990.

[27] J. J. E. Slotine and W. Li, *Applied Nonlinear Control*, Englewood Cliff, NJ: Prentice Hall, Inc., 1991.

[28] M. W. Spong and R. Ortega, On Adaptive Inverse Dynamics Control of Rigid Robots, *IEEE Transactions on Automatic Control*, Vol. 35, No. 1, pp. 92-95, Jan. 1990.

[29] Y. Tang and M.A. Arteaga, Adaptive Control of Robot Manipulators Based on Passivity, *IEEE Transactions on Automatic Control*, Vol. 39, No. 9, pp. 1871-1875, Sept. 1994.

[30] M. Vidyasagar, *Nonlinear Systems Analysis*, Englewood Cliffs, NJ: Prentice Hall, 1978.

[31] J. T. Wen and D. S. Bayard, New Class of Control Laws for Robotic Manipulators: Part 1. Non-Adaptive Case, *International Journal of Control*, Vol. 47, No. 5, pp. 1361-1385, May 1988.

[32] B. Yao and M. Tomizuka, Smooth Robust Adaptive Sliding Mode Control of Manipulators with Guaranteed Transient Performance, *Proceedings of the American Control Conference*, pp. 1176-1180, Baltimore, MD, June 1994.

4
Output Feedback Tracking Controllers

4.1 Introduction

All the position tracking controllers presented in Chapter 3 required full-state feedback (FSFB). That is, the control implementation requires the measurement of the position and velocity of the mechanical system. Since the cost of implementing a FSFB controller for achieving position tracking would typically include the cost of motion sensors, this chapter addresses the problem of position tracking under the constraint of minimizing the sensor count (i.e., elimination of velocity measurements). Hence, we are motivated to investigate means of constructing velocity signal surrogates for use in closed-loop, position tracking control strategies, i.e., output feedback (OFB) controllers. A standard approach for removing velocity measurements is to apply the so-called *backwards difference* algorithm to the position measurements. Even though this method of eliminating velocity may provide reasonable performance, the use of this discrete-time velocity approximation is not satisfying from a theoretical viewpoint since the dynamics of the backwards difference algorithm are normally not included during the closed-loop stability analysis.

An alternate approach illustrated in Chapter 2 for SISO systems that contains a complete mathematical development suggests the use of a nonlinear observer to produce an estimate of the unmeasurable velocity signal. Often this observer contains dynamics that attempt to mimic the behavior of the mechanical system to a certain extent (i.e., model-based observers). Since, in general, the separation principle [19] does not hold for nonlinear systems, a more complex procedure is required to analyze the closed-loop stability of the observer-controller system. To this end, the role of the velocity observer is not merely to provide a velocity estimate but to ensure closed-loop stability. Analogously, the controller also plays a role in the accurate estimation of velocity. That is, the design of the composite observer-controller system is based on a coupled set of equations that describe the observation error dynamics and the position tracking error dynamics.

A drawback of the observer-based approach is that an accurate model of the mechanical system is required to guarantee the stability of the observation error and the position tracking error. As mentioned in Chapter 3, although the structure of the mechanical system dynamic model is fairly well defined, the physical parameters representing the mass, inertia, and friction effects are often not precisely known. Therefore, an alternative, model-independent method is sometimes needed to generate a velocity-related signal from only position measurements. Heuristically, this method, often referred to as a *high-gain observer,* attempts to provide a continuous-time, approximate derivative of the position signal. That is, a linear, high-pass-filter-like structure is used to approximate the behavior of a perfect differentiator over a range of frequencies. A crucial point is that the filter dynamics can be explicitly included in the analysis along with the position tracking error dynamics to deliver the desired closed-loop stability result.

A limitation that exists in most OFB tracking controllers, observer or filter-based, is the *semiglobal* nature of the stability result (i.e., a control gain has to be made sufficiently large to guarantee the desired stability result for a given set of initial conditions). Ideally, it would be desirable to recover the *global* stability results achieved with most of the FSFB designs of Chapter 3, despite having only OFB. A solution to this challenging problem can be accomplished by utilizing a velocity-generating filter with a more complex structure and a nonquadratic Lyapunov function that is *softer*[1] than the standard, quadratic Lyapunov function.

Based on the above discussion, this chapter presents three solutions to the OFB tracking control problem. The first result is a model-based observer-controller that guarantees the semiglobal exponential stabilization of the velocity observation error and the position tracking error. An interesting aspect of this result is that the observer-controller design and analysis is

---

[1]The word *softer* is used to illustrate the fact that if $V(x) = f(x)$, where $f(x)$ is some radially unbounded, non-negative function, then $f(x) \leq x^2$.

presented from an *observed integrator backstepping* [22] perspective. The second result is a linear filter-based, adaptive controller that ensures semi-global asymptotic position tracking while compensating for unknown system parameters. Finally, a nonlinear filter-based, adaptive control is developed that produces global asymptotic position tracking.

4.2 Problem Statement

In this chapter, we will consider the same nonlinear dynamic model described in Section 3.2. However, in addition to the model properties listed in Section 3.2, we will make use in this chapter of the following supplemental properties.

Property 4.1: The centripetal-Coriolis matrix of (3.1) satisfies the following relationship: [27]

$$V_m(q,\xi)\nu = V_m(q,\nu)\xi \qquad \forall \xi, \nu \in \mathbb{R}^n. \tag{4.1}$$

Property 4.2 The norm of the centripetal-Coriolis and gravity terms of (3.1) can be upper bounded as follows [23]:

$$\|V_m(q,\dot{q})\|_{i\infty} \leq \zeta_{c1} \|\dot{q}\| \qquad \|G(q)\|_{i\infty} \leq \zeta_g, \tag{4.2}$$

where ζ_{c1}, ζ_g are positive bounding constants, and $\|\cdot\|_{i\infty}$ denotes the matrix-induced infinity norm [30].

The primary control objective of this chapter is to design the control input $\tau(t)$ such that $q(t) \to q_d(t)$ as $t \to \infty$ with the constraint that only position measurements (i.e., $q(t)$) are available. As in Chapter 3, we define the position tracking error $e(t) \in \mathbb{R}^n$ as

$$e(t) = q_d(t) - q(t). \tag{4.3}$$

We will assume the desired motion trajectory is specified such that $q_d(t)$ and its first three time derivatives are all bounded functions of time.[2] In particular, we assume that

$$\|\dot{q}_d(t)\| \leq \zeta_{d2}, \tag{4.4}$$

where ζ_{d2} is a known, positive bounding constant.

[2]The controller presented in Section 4.3 will however *not* require the boundedness of $\dddot{q}_d(t)$.

4.3 Model-Based Observer/Control

In this section, the OFB control problem is tackled by designing a model-based velocity observer/control algorithm. Using as a starting point, the observer originally presented by Nicosia and Tomei in [27], the control input is designed based on the observed integrator backstepping technique [22]. Since the backstepping approach is used for the control synthesis, the observer/controller developed in this section can be extended to rigid mechanical systems that include the actuator dynamics (e.g., electrical dynamics and joint flexibilities[3]).

Since velocity measurements (i.e., $\dot{q}(t)$) are assumed to be unavailable, we will estimate this quantity with a model-based velocity observer. To quantify the difference between the actual and observed velocities, we define the velocity observation error $\dot{\tilde{q}}(t) \in \mathbb{R}^n$ as follows:

$$\dot{\tilde{q}} = \dot{q} - \dot{\hat{q}}, \tag{4.5}$$

where $\dot{\hat{q}}(t) \in \mathbb{R}^n$ denotes the observed velocity. In addition, to simplify the development of the control input, we define an *observed* filtered tracking error signal, denoted by $\hat{r}(t) \in \mathbb{R}^n$, as follows:

$$\hat{r} = \dot{q}_d - \dot{\hat{q}} + \alpha e, \tag{4.6}$$

where $\alpha \in \mathbb{R}$ is a positive control gain. The expression for $\hat{r}(t)$ is similar to the filtered tracking error $r(t)$ defined in (3.7), but with the actual velocity replaced by the observed velocity.

4.3.1 Velocity Observer Formulation

The model-based velocity observer has the following form [27]:

$$\dot{\hat{q}} = y + k_o \tilde{q}, \qquad y(0) = -k_o \tilde{q}(0) \tag{4.7}$$

and

$$\dot{y} = M^{-1}(q)\left(\tau - V_m(q, \dot{\hat{q}})\,\dot{\hat{q}} - G(q) - F_d\,\dot{\hat{q}}\right), \tag{4.8}$$

where $\tilde{q}(t) \in \mathbb{R}^n$ is the position observation error denoted as

$$\tilde{q} = q - \hat{q}, \tag{4.9}$$

[3] See [25] for such extensions.

$y(t) \in \mathbb{R}^n$ is an auxiliary variable, and k_o is a positive control gain. To facilitate the subsequent Lyapunov-type stability analysis, the control gain k_o is defined as follows:

$$k_o = \frac{1}{m_1}\left(\zeta_{c1}\zeta_{d2} + \zeta_{c1}k_1 + \zeta_{c1}\alpha k_1 + \alpha + 2k_n\right), \tag{4.10}$$

where α is the same control gain defined in (4.6), k_1 is an additional positive control gain, k_n is a positive nonlinear damping gain, and $m_1, \zeta_{c1}, \zeta_{d2}$ are the bounding constants defined in (3.2), (4.2), and (4.4), respectively. Note that (4.8) is basically the system model of (3.1) with $\dot{q}(t)$ replaced by $\dot{\hat{q}}(t)$. That is, the observer is designed with the intent of mimicking the system dynamics. The form of (4.7) and (4.8), in particular the auxiliary variable $y(t)$, allows the observer to be implemented without the need to measure velocity. The selection of the initial condition shown in (4.7) (i.e., $y(0) = -k_o\tilde{q}(0)$ or $\dot{\hat{q}}(0) = 0$) is motivated by the subsequent stability analysis.

To give some insight into the design of the above observer, we now develop the observation error dynamics. We begin by writing the velocity observer in a single equation format. This is done by differentiating (4.7) with respect to time, premultiplying both sides of the resulting expression by $M(q)$, and substituting $\dot{y}(t)$ of (4.8) to produce

$$M(q)\ddot{\hat{q}} + V_m(q,\dot{\hat{q}})\dot{\hat{q}} + G(q) + F_d\dot{\hat{q}} - k_oM(q)\dot{\tilde{q}} = \tau. \tag{4.11}$$

Subtracting (4.11) from (3.1) yields the following observer error system:

$$M(q)\ddot{\tilde{q}} + V_m(q,\dot{q})\dot{q} - V_m(q,\dot{\hat{q}})\dot{\hat{q}} + F_d\dot{\tilde{q}} + k_oM(q)\dot{\tilde{q}} = 0. \tag{4.12}$$

With the intent of preparing the error dynamics of (4.12) to enable the use of (3.3) during the stability analysis, we add and subtract the term $V_m(q,\dot{q})\dot{\hat{q}}$ to (4.12), and then apply Property 4.1 and (4.5) to obtain

$$M(q)\ddot{\tilde{q}} = -V_m(q,\dot{q})\dot{\tilde{q}} - V_m(q,\dot{\hat{q}})\dot{\tilde{q}} - F_d\dot{\tilde{q}} - k_oM(q)\dot{\tilde{q}}. \tag{4.13}$$

To analyze the stability of the above observer error dynamics, we use the following non-negative function:

$$V_o = \frac{1}{2}\dot{\tilde{q}}^T M(q)\dot{\tilde{q}}. \tag{4.14}$$

The time derivative of $V_o(t)$ along (4.13) is given by

$$\dot{V}_o = -\dot{\tilde{q}}^T\left(V_m(q,\dot{\hat{q}}) + F_d + k_oM(q)\right)\dot{\tilde{q}}, \tag{4.15}$$

where (3.3) has been utilized. From (4.15) and the fact that F_d is positive-definite, the following upper bound for $\dot{V}_o(t)$ can be obtained:

$$\dot{V}_o \leq \left(\zeta_{c1}\left\|\dot{\hat{q}}\right\| - k_o m_1\right)\left\|\dot{\tilde{q}}\right\|^2, \tag{4.16}$$

where (3.2) and Property 4.2 have been utilized. To facilitate the subsequent composite observer/controller stability analysis, we use (4.6) to substitute for $\dot{\hat{q}}(t)$ in (4.16) to form a new upper bound for $\dot{V}_o(t)$ as follows:

$$\dot{V}_o \leq \left(\zeta_{c1}\zeta_{d2} + \zeta_{c1}\left\|\hat{r}\right\| + \zeta_{c1}\alpha\left\|e\right\| - k_o m_1\right)\left\|\dot{\tilde{q}}\right\|^2, \tag{4.17}$$

where (4.4) has been used. From the form of (4.17), we are motivated to design a controller that ensures that the $\|e\|$ and $\|\hat{r}\|$ terms in (4.17) are both driven to zero.

4.3.2 Controller Formulation

The dynamics for the position tracking error system can be formed by differentiating (4.3) with respect to time to yield

$$\dot{e} = \dot{q}_d - \dot{q}. \tag{4.18}$$

From the form of above position tracking error dynamics, we can see that if $\dot{q}(t)$ were measurable, a backstepping procedure could be used to determine the required control input $\tau(t)$. However, since this is not the case, we add and subtract the measurable term $\dot{\hat{q}}(t)$ to the right-hand side of (4.3) to produce

$$\dot{e} = \dot{q}_d - \dot{\hat{q}} - \dot{\tilde{q}}, \tag{4.19}$$

where (4.5) has been utilized. Since there is no control input in the above position tracking error dynamics, we add and subtract a *fictitious* control input to the right-hand side of (4.19) to yield

$$\dot{e} = \dot{q}_d - \left[\dot{q}_d + \alpha e\right] + \left[\dot{q}_d + \alpha e\right] - \dot{\hat{q}} - \dot{\tilde{q}}, \tag{4.20}$$

where α is the same control gain defined in (4.6). By using (4.6), we can simplify the expression given in (4.20) as shown below:

$$\dot{e} = -\alpha e + \hat{r} - \dot{\tilde{q}}. \tag{4.21}$$

To analyze the stability of the position tracking error dynamics, we use the following non-negative function:

$$V_{c1} = \frac{1}{2}e^T e. \tag{4.22}$$

The time derivative of $V_{c1}(t)$ along (4.21) yields

$$\dot{V}_{c1} = e\left(-\alpha e + \hat{r} - \dot{\tilde{q}}\right), \tag{4.23}$$

from which we can obtain

$$\dot{V}_{c1} \leq -\alpha \left\|e\right\|^2 + \left\|e\right\| \left\|\hat{r}\right\| + \left\|e\right\| \left\|\dot{\tilde{q}}\right\|. \tag{4.24}$$

From the form of (4.24) and the fact that (4.17) indicates that $\dot{\tilde{q}}(t)$ can be driven to zero, we are motivated to design a control input that ensures that $\hat{r}(t)$ is driven to zero.

With the goal of determining the open-loop dynamics for $\hat{r}(t)$, we differentiate (4.6), premultiply both sides of the resulting expression by $M(q)$, and then use (4.11) to substitute for $M(q)\ddot{\hat{q}}$ to produce

$$M(q)\dot{\hat{r}} = M(q)\ddot{q}_d + \alpha M(q)(\dot{q}_d - \dot{q}) - k_o M(q)\dot{\tilde{q}} + V_m(q,\dot{\hat{q}})\dot{\hat{q}} + G(q) + F_d\dot{\hat{q}} - \tau. \tag{4.25}$$

After utilizing (4.6) and (4.5), the right-hand side of (4.25) can be rewritten as

$$M(q)\dot{\hat{r}} = w_e(q,\dot{\hat{q}},t) - (\alpha + k_o) M(q)\dot{\tilde{q}} - V_m(q,\dot{\hat{q}})\hat{r} - \tau, \tag{4.26}$$

where the auxiliary term $w_e(q,\dot{\hat{q}},t) \in \mathbb{R}^n$ is defined as

$$w_e = M(q)\left(\ddot{q}_d + \alpha(\dot{q}_d - \dot{\hat{q}})\right) + V_m(q,\dot{\hat{q}})(\dot{q}_d + \alpha e) + G(q) + F_d\dot{\hat{q}}. \tag{4.27}$$

Given the structure of the open-loop dynamics of (4.26), the control input $\tau(t)$ is designed as follows:[4]

$$\tau = w_e(q,\dot{\hat{q}},t) + (\alpha + k_c)\hat{r}, \tag{4.28}$$

where k_c is a positive control gain defined as

$$k_c = 2k_n + \zeta_{c1}k_1 + (\alpha + k_o)^2 m_2^2 k_n \tag{4.29}$$

to facilitate the subsequent stability analysis. Substituting the control input of (4.28) into (4.26) gives the following closed-loop dynamics for $\hat{r}(t)$:

$$M(q)\dot{\hat{r}} = -(\alpha + k_c)\hat{r} - (\alpha + k_o) M(q)\dot{\tilde{q}} - V_m(q,\dot{\hat{q}})\hat{r}. \tag{4.30}$$

---

[4]It is easy to see from the form of (4.27) that the control input of (4.28) is the same as the nonadaptive version of the controller presented in Section 3.2, except that $\dot{\hat{q}}(t)$ has been used in lieu of $\dot{q}(t)$.

To enable the use of (3.3) during the stability analysis, Property 4.1 and (4.5) can be applied to the term $V_m(q, \dot{\hat{q}})\hat{r}$ in (4.30) to produce

$$M(q)\dot{\hat{r}} = -V_m(q,\dot{q})\hat{r} - (\alpha + k_c)\hat{r} - (\alpha + k_o)M(q)\dot{\tilde{q}} + V_m(q,\dot{\tilde{q}})\hat{r}. \quad (4.31)$$

To analyze the stability of the filtered tracking error closed-loop dynamics, we use the following non-negative function:

$$V_{c2} = \frac{1}{2}\hat{r}^T M(q)\hat{r}. \quad (4.32)$$

The time derivative of $V_{c2}(t)$ along (4.31) is given by

$$\dot{V}_{c2} = -(\alpha + k_c)\hat{r}^T\hat{r} - (\alpha + k_o)\hat{r}^T M(q)\dot{\tilde{q}} + \hat{r}^T V_m(q,\dot{\tilde{q}})\hat{r}, \quad (4.33)$$

where (3.3) has been utilized. From (4.33), we can obtain the following upper bound for $\dot{V}_{c2}(t)$:

$$\dot{V}_{c2} \leq -(\alpha + k_c)\|\hat{r}\|^2 + (\alpha + k_o)m_2\|\hat{r}\|\left\|\dot{\tilde{q}}\right\| + \zeta_{c1}\|\hat{r}\|^2\left\|\dot{\tilde{q}}\right\|, \quad (4.34)$$

where (3.2) and Property 4.2 have been used.

4.3.3 Composite Stability Result

The combination of error systems given by (4.13), (4.21), and (4.31) yields the following stability result for the velocity observation error and position tracking error.

Theorem 4.1 *The velocity observer of (4.7) and (4.8), and the control law of (4.28) ensure that the closed-loop observer/controller system is semiglobally exponentially stable as illustrated by*

$$\|x(t)\| \leq \sqrt{\frac{\lambda_2}{\lambda_1}}\|x(0)\| \, exp\left(-\frac{\lambda_3}{\lambda_2}t\right) \quad (4.35)$$

provided the observer/controller gains satisfy the following sufficient conditions

$$\alpha > \frac{1}{k_n} \qquad k_1 > \sqrt{\frac{\lambda_2}{\lambda_1}}\|x(0)\|, \quad (4.36)$$

where

$$x = \begin{bmatrix} \dot{\tilde{q}}^T & e^T & \hat{r}^T \end{bmatrix}^T \in \mathbb{R}^{3n}, \quad (4.37)$$

$$\lambda_1 = \min\{m_1, 1\}, \quad \lambda_2 = \max\{m_2, 1\}, \quad \lambda_3 = \alpha - \frac{1}{k_n}. \quad (4.38)$$

Proof. To prove the above result, we use the composite non-negative function:

$$V = V_o + V_{c1} + V_{c2}, \tag{4.39}$$

where $V_o(t)$, $V_{c1}(t)$, and $V_{c2}(t)$ were defined in (4.14), (4.22), and (4.32), respectively. From the form of (4.39), it is not difficult to show that the following bounds hold:

$$\frac{1}{2}\lambda_1 \|x\|^2 \leq V \leq \frac{1}{2}\lambda_2 \|x\|^2, \tag{4.40}$$

where $x(t)$ and λ_1, λ_2 were defined in (4.37) and (4.38), respectively. After taking the time derivative of (4.39), we can utilize (4.17), (4.24), (4.34), (4.10), and (4.29) to form the following upper bound on $\dot{V}(t)$:

$$\begin{aligned}\dot{V} \leq\ & -\alpha\|e\|^2 - \alpha\|\hat{r}\|^2 - \alpha\left\|\dot{\tilde{q}}\right\|^2 \\ & + \left[\|e\|\|\hat{r}\| - 2k_n\|\hat{r}\|^2\right] + \left[\|e\|\left\|\dot{\tilde{q}}\right\| - 2k_n\left\|\dot{\tilde{q}}\right\|^2\right] \\ & + \left[(\alpha + k_o) m_2 \left\|\dot{\tilde{q}}\right\|\|\hat{r}\| - k_n(\alpha + k_o)^2 m_2^2\|\hat{r}\|^2\right] \\ & + \zeta_{c1}\zeta_{d2}\left\|\dot{\tilde{q}}\right\|^2 - \zeta_{c1}\zeta_{d2}\left\|\dot{\tilde{q}}\right\|^2 \\ & - (k_1 - \|x\|)\left(\zeta_{c1}\left\|\dot{\tilde{q}}\right\|^2 + \zeta_{c1}\alpha\left\|\tilde{q}\right\|^2 + \zeta_{c1}\|\hat{r}\|^2\right),\end{aligned} \tag{4.41}$$

where we have used the fact derived from (4.37) that $\|x\| \geq \|e\|, \|\hat{r}\|, \left\|\dot{\tilde{q}}\right\|$. Since the bracketed terms are nonlinear damping pairs, we can apply Lemma A.10 in Appendix A to further upper bound $\dot{V}(t)$ as follows:

$$\begin{aligned}\dot{V} \leq\ & -\left(\alpha - \frac{1}{k_n}\right)\|e\|^2 - \alpha\|\hat{r}\|^2 - \left(\alpha - \frac{1}{k_n}\right)\left\|\dot{\tilde{q}}\right\|^2 \\ & - (k_1 - \|x\|)\left(\zeta_{c1}\left\|\dot{\tilde{q}}\right\|^2 + \zeta_{c1}\alpha\left\|\tilde{q}\right\|^2 + \zeta_{c1}\|\hat{r}\|^2\right).\end{aligned} \tag{4.42}$$

From the form of (4.42), we can state that

$$\dot{V} \leq -\lambda_3\|x\|^2 \qquad \text{for} \qquad k_1 > \|x(t)\|, \tag{4.43}$$

where $x(t)$ and λ_3 were defined in (4.37) and (4.38), respectively. From (4.40), we can obtain the following sufficient condition for (4.43):

$$\dot{V} \leq -\frac{2\lambda_3}{\lambda_2} V \qquad \text{for} \qquad k_1 > \sqrt{\frac{2V(t)}{\lambda_1}}. \tag{4.44}$$

Since $\dot{V}(t) \leq 0 \; \forall t \in [0, \infty)$, $V(t)$ is decreasing or constant $\forall t \in [0, \infty)$ (i.e., $V(t) \leq V(0) \; \forall t \in (0, \infty)$). Hence, a sufficient condition for (4.44) is given by

$$\dot{V} \leq -\frac{2\lambda_3}{\lambda_2} V \qquad \text{for} \qquad k_1 > \sqrt{\frac{2V(0)}{\lambda_1}}. \tag{4.45}$$

Applying Lemma A.4 in Appendix A to (4.45) produces

$$V(t) \leq V(0) \exp\left(-\frac{2\lambda_3}{\lambda_2} t\right) \qquad \text{for} \qquad k_1 > \sqrt{\frac{2V(0)}{\lambda_1}}, \tag{4.46}$$

which yields the result delineated by (4.35) and (4.36) upon the application of (4.40). □

Remark 4.1 *From the above theorem, we know that* $x(t) \in \mathcal{L}_\infty$*; hence, we can use (4.3), (4.6), (4.5), and the fact that the desired motion trajectory is bounded to show that* $q(t)$, $\dot{\hat{q}}(t), \dot{q}(t) \in \mathcal{L}_\infty$. *Since* $\dot{\tilde{q}}(t)$ *is exponentially stable as dictated by (4.35), we can utilize Lemma A.11 in Appendix A to show that* $\tilde{q}(t) \in \mathcal{L}_\infty$. *As a consequence, we can use (4.9) and the fact that* $q(t) \in \mathcal{L}_\infty$ *to show that* $\hat{q}(t) \in \mathcal{L}_\infty$. *Using the above information, we can easily show from (3.1), (4.7), (4.28), (4.8), and the fact that* $M^{-1}(q)$ *exists and is bounded (see Property 3.1) that* $\ddot{q}(t), y(t), \tau(t), \dot{y}(t) \in \mathcal{L}_\infty$. *Thus, all system signals remain bounded during closed-loop operation.*

Remark 4.2 *From (4.36), we can see that the control gain* k_1 *can be increased to cover any set of initial conditions; therefore, the stability result is referred to as semiglobal. It is interesting to note the importance of the initial condition* $y(0) = -k_o\tilde{q}(0)$ *given in (4.7) on this semiglobal result. Specifically, if* $y(0) \neq -k_o\tilde{q}(0)$, *it is easy to see from (4.7), (4.5), and (4.37) that* $x(0)$ *would be a function of* k_1 *owing to (4.10). Hence, it would not be possible to satisfy (4.36) for all initial conditions. It is also important to note that the control gains* α *and* k_n *can be increased to speed up the transient response of the position tracking error, as illustrated by (4.35), (4.37), and (4.38).*

Remark 4.3 *After adding and subtracting $\dot{q}(t)$ to the right-hand side of (4.6) and rearranging terms, we can formulate the following inequality:*

$$\|\dot{e}\| = \|\dot{q}_d - \dot{q}\| \leq \|\hat{r}\| + \alpha\|e\| + \left\|\dot{\tilde{q}}\right\|. \tag{4.47}$$

Since each of the terms on the right-hand side of (4.47) is exponentially stable, as illustrated by Theorem 4.1, it is easy to see from (4.47) that the velocity tracking error $\dot{e}(t)$ is also exponentially stable despite the fact that the control law does not explicitly require velocity measurements.

Remark 4.4 *As illustrated by the form of (4.8) and (4.28), the observer-controller requires exact model knowledge of the mechanical dynamics; hence, one may question the robustness of proposed controller. It is important to note that the results given in [3, 30, 32] all confirmed the robustness of an OFB linear control loop used in conjunction with a linear observer (or a linear filter). Specifically, these results indicate that a semiglobal uniform ultimate boundedness stability result can be obtained even in the presence of parametric uncertainty and additive bounded disturbances.*

4.3.4 Experimental Results

We now discuss the real-time experimental implementation of the observer-controller on the two-link, direct-drive, rigid robot manipulator test-bed described in Section 3.5.3. A PC hosting a Texas Instruments TMS320C30 DSP board served as the computational engine while *WinMotor* provided the environment to write the control algorithm in the C programming language.

The OFB control law given by the velocity observer of (4.7) and (4.8) and the controller of (4.28) was applied to the two-link manipulator with the gains set to

$$k_o = 145, \qquad k_c = 350, \qquad \alpha = 125. \tag{4.48}$$

The desired position trajectory for the two links was selected to be

$$q_d(t) = \begin{bmatrix} 40.11\sin(t)(1-\exp\left(-0.3t^3\right)) \\ 68.75\sin(t)(1-\exp\left(-0.3t^3\right)) \end{bmatrix} \text{deg}. \tag{4.49}$$

All experimental runs were performed at a sampling period of 200 μsec. The results of the experiment are shown in Figure 4.1. One can see that the position tracking error for both joints is approximately 0.04 deg or smaller. To examine the robustness of the OFB control law under payload variations, a 5 lbs mass was added to the robot end-effector and the experiment was repeated under the same set of conditions. Figure 4.2 shows the results of the experiment with the 5 lbs payload. It is clear from the plots that there is no significant difference in the magnitude of the position tracking errors, which indicates that the OFB control law is indeed robust, to a certain extent, against payload variations.

To provide a means for comparing the proposed OFB control law with a typical method of implementing a robot controller, we implemented the controller of (4.28) (i.e., the same controller given in Section 3.4 with the system parameters being exactly known) under exactly the same conditions as stated above, but with $\hat{q}$ calculated by the following standard backwards difference algorithm:

$$\dot{\hat{q}} = \frac{q(pT) - q((p-1)T)}{T} \quad \text{for } p = 1, \cdots, n, \tag{4.50}$$

where p represents the sampling instant, T represents the sampling period, and nT represents the total experiment time. We found that the best position tracking performance was obtained with the following gain selection

$$k_c = 5, \qquad \alpha = \mathrm{diag}\{75, 45\}. \tag{4.51}$$

Although the control gain α was defined as a scalar in (4.6) to facilitate the stability analysis, α was specified as a matrix in (4.51) to provide more flexibility during controller tuning. We note that the control gains for the backwards difference/controller could not be adjusted to values as high as those given by (4.48) without saturating the power amplifiers. From Figure 4.3, we see that there is not a large difference in the magnitude of the position tracking errors between the model-based observer/controller and the backwards difference/controller. However, when the 5 lbs payload was added to the robot end-effector, the differences of the two approaches became apparent. Comparing Figures 4.2 and 4.4, it is clear that while the model-based observer/controller shows no change in the magnitude of the position tracking error with the additional load, the performance of the backwards difference/controller degraded significantly. These results seem to indicate that the proposed OFB control law is more robust than the backwards difference/controller.

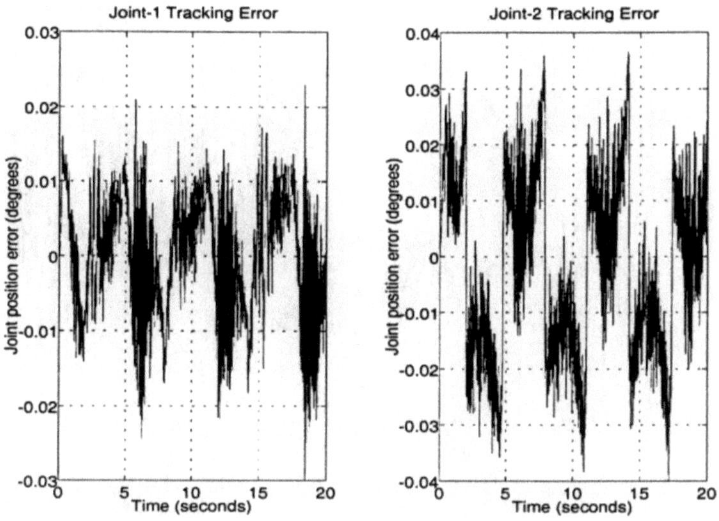

FIGURE 4.1. Position tracking errors of observer/controller without payload.

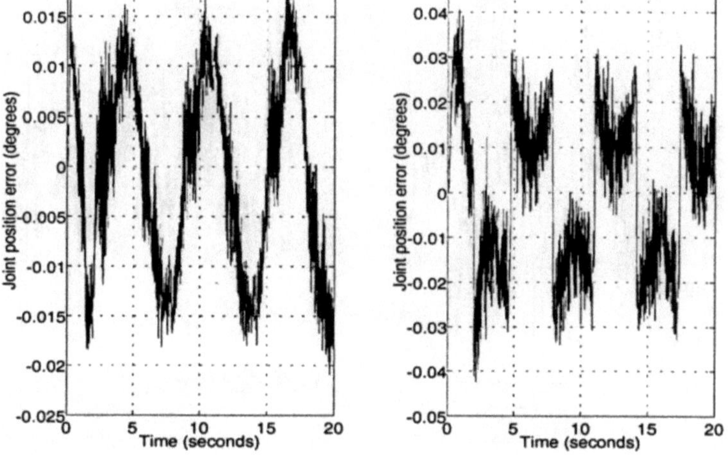

FIGURE 4.2. Position tracking errors of observer/controller with 5 [lbs] payload.

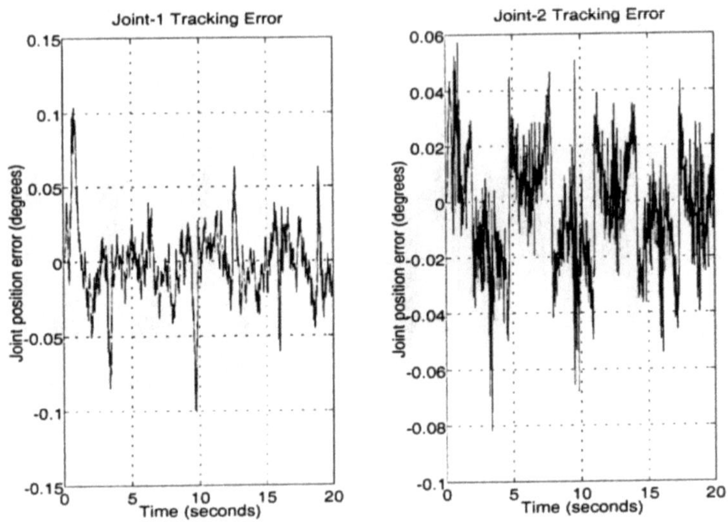

FIGURE 4.3. Position tracking errors of backwards difference/controller without payload.

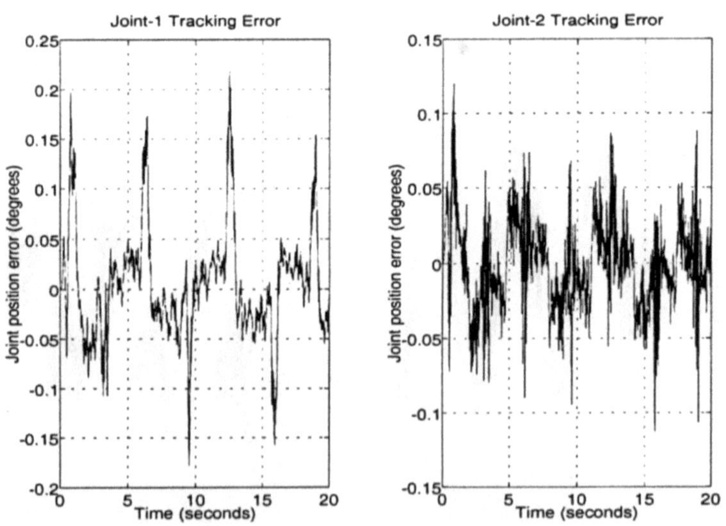

FIGURE 4.4. Position tracking errors of backwards difference/controller with 5 [lbs] payload.

4.4 Linear Filter-Based Adaptive Control

The model-based observer/controller of Section 4.3 exhibits the drawback of requiring exact model knowledge, i.e., it is unable to compensate for system uncertainty while guaranteeing the convergence of the tracking errors to zero. In this section, we will illustrate how an OFB control law can be designed given uncertain knowledge of the constant, mechanical system parameters. Specifically, the DCAL controller of Section 3.5 is combined with a non-model-based filter that generates a replacement signal for velocity using position measurements. A Lyapunov-type analysis along with a judicious definition of the tracking error terms are used to find the appropriate modifications to the DCAL controller.

As in Chapter 3, in order to compensate for the lack of exact knowledge of the system parameters, an adaptation law will be developed to generate on-line parameter estimates. The difference between the actual and estimated parameters is defined by

$$\tilde{\theta}(t) = \theta - \hat{\theta}(t), \tag{4.52}$$

where $\tilde{\theta}(t) \in \mathbb{R}^p$ denotes the parameter estimation error vector, and $\hat{\theta}(t) \in \mathbb{R}^p$ denotes the dynamic estimate of the unknown, constant, parameter vector θ defined in (3.4).

4.4.1 Filter Formulation

In order to compensate for the lack of direct measurements of the velocity signal, we will introduce a filter to generate a velocity tracking error related signal. This filter can be thought of as a high-pass filter with the position tracking error $e(t)$ as the input and a pseudo-velocity-tracking error signal as the output. This pseudo-velocity signal attempts to capture the behavior of $\dot{e}(t)$. The filter is defined by the following dynamic relationship:

$$e_f = -ke + p \tag{4.53}$$

and

$$\dot{p} = -(k+1)p + (k^2+1)e, \qquad p(0) = ke(0), \tag{4.54}$$

where $e_f(t) \in \mathbb{R}^n$ is the output of the filter and will be used as a surrogate for the actual velocity error, $p(t) \in \mathbb{R}^n$ is an auxiliary variable that allows the filter to be implemented without velocity measurements, and k is a positive control gain. The selection of the initial condition given in (4.54) (i.e., $p(0) = ke(0)$ or $e_f(0) = 0$) is motivated by the subsequent stability analysis.

To give some insight into the design of the above filter while also motivating the subsequent control input design, we now derive the dynamics of the filter output $e_f(t)$. To this end, we first differentiate (4.53) with respect to time and then use (4.54) to substitute for $\dot{p}(t)$ to produce

$$\dot{e}_f = -k\dot{e} - (k+1)\,p + \left(k^2+1\right)e. \tag{4.55}$$

After solving for $p(t)$ from (4.53), and substituting the resulting expression into (4.55), we obtain the dynamics for $e_f(t)$ as follows:

$$\dot{e}_f = -e_f - k\eta + e, \tag{4.56}$$

where $\eta(t) \in \mathbb{R}^n$ is a filtered tracking error-like variable defined as

$$\eta = \dot{e} + e_f + e. \tag{4.57}$$

The filtered tracking error-like signal $\eta(t)$ will be the starting point for the formulation of the control input. In addition, it should be noted that (4.57) can simply be rearranged as

$$\dot{e} = -e + \eta - e_f \tag{4.58}$$

to give the dynamics for $e(t)$.

4.4.2 Controller Formulation

We begin the control formulation by determining the open-loop dynamics for $\eta(t)$ defined in (4.57). To this end, we take the time derivative of (4.57) and premultiply the resulting expression by $M(q)$ to produce

$$\begin{aligned} M(q)\dot{\eta} &= M(q)\left(\ddot{q}_d - \ddot{q}\right) + M(q)\dot{e}_f + M(q)\dot{e} \\ &= M(q)\ddot{q}_d + V_m(q,\dot{q})\dot{q} + F_d\dot{q} + G(q) - \tau \\ &\quad + M(q)\left(-e_f - k\eta + e\right) + M(q)\left(-e + \eta - e_f\right), \end{aligned} \tag{4.59}$$

where (3.1), (4.56), and (4.58) were used to substitute for $M(q)\ddot{q}$, $\dot{e}_f(t)$, and $\dot{e}(t)$, respectively. At this point, the desired linear parameterization term $Y_d\theta$ defined in (3.22) is added and subtracted to the right-hand side of (4.59) and the relationship $\dot{q} = \dot{q}_d - \dot{e}$ is used, where needed, to yield

$$\begin{aligned} M(q)\dot{\eta} = {} & Y_d\theta - \tau + M(q)\left(-e_f - k\eta\right) + M(q)\left(\eta - e_f\right) \\ & - V_m(q,\dot{q})\dot{e} + M(q)\ddot{q}_d + V_m(q,\dot{q})\dot{q}_d + F_d\dot{q} + G(q) - Y_d\theta. \end{aligned} \tag{4.60}$$

After again utilizing (4.58) to substitute for $\dot{e}(t)$ in (4.60), we have

$$\begin{aligned} M(q)\dot{\eta} = \quad & -V_m(q,\dot{q})\eta - kM(q)\eta + Y_d\theta - \tau + \tilde{Y} \\ & +M(q)\eta - 2M(q)e_f + V_m(q,\dot{q})\left(e_f + e\right), \end{aligned} \tag{4.61}$$

where the auxiliary variable $\tilde{Y}(e, e_f, \eta, t) \in \mathbb{R}^n$ is defined as

$$\tilde{Y} = M(q)\ddot{q}_d + V_m(q,\dot{q})\dot{q}_d + F_d\dot{q} + G(q) - Y_d\theta. \tag{4.62}$$

To simplify the notation, an additional auxiliary variable $\chi(e, e_f, \eta, t) \in \mathbb{R}^n$ is defined as follows:

$$\chi = \tilde{Y} + M(q)\eta - 2M(q)e_f + V_m(q,\dot{q})\left(e_f + e\right), \tag{4.63}$$

and applied to (4.61) to rewrite the open-loop dynamics for $\eta(t)$ in the following final form:

$$M(q)\dot{\eta} = -V_m(q,\dot{q})\eta - kM(q)\eta + Y_d\theta - \tau + \chi. \tag{4.64}$$

Based on the structure of (4.64) and the subsequent stability analysis, the control input $\tau(t)$ is designed as

$$\tau = Y_d\hat{\theta} - ke_f + e, \tag{4.65}$$

where k is the same control gain used in (4.54) and (4.53), and the dynamic parameter estimate vector $\hat{\theta}(t)$ is defined as follows:

$$\hat{\theta}(t) = \Gamma\int_0^t Y_d^T(\sigma)\left(e(\sigma) + e_f(\sigma)\right)d\sigma - \Gamma\int_0^t \frac{d}{d\sigma}\left\{Y_d^T(\sigma)\right\}e(\sigma)d\sigma + \Gamma Y_d^T e, \tag{4.66}$$

with $\Gamma \in \mathbb{R}^{p\times p}$ being a diagonal, positive-definite, adaptation gain matrix. After substituting (4.65) into (4.64), we obtain the following closed-loop dynamics for $\eta(t)$:

$$M(q)\dot{\eta} = -V_m(q,\dot{q})\eta - kM(q)\eta + Y_d\tilde{\theta} + \chi + ke_f - e, \tag{4.67}$$

where the definition given in (4.52) has been used.

Remark 4.5 *It is not difficult to show that an upper bound can be placed on the norm of the auxiliary variable $\chi(\cdot)$ defined in (4.63) as follows (see Appendix B for proof):*

$$\|\chi\| \le \zeta_1\|x\| + \zeta_2\|x\|^2, \tag{4.68}$$

where $x(t) \in \mathbb{R}^{3n}$ is defined as

$$x = \left[\begin{array}{ccc} e_f^T & e^T & \eta^T \end{array} \right]^T, \tag{4.69}$$

and ζ_1, ζ_2 are some positive bounding constants that depend on the physical parameters of the mechanical system and on the bounding constants defined in (4.4). The bound of (4.68) will be utilized during the stability analysis.

Remark 4.6 *As will be seen in the stability analysis, the parameter estimate vector in (4.66) is based on the need to have $\hat{\theta}(t)$ updated according to*

$$\dot{\hat{\theta}} = \Gamma Y_d^T \eta. \tag{4.70}$$

The implementation of the above form would, however, require $\dot{e}(t)$ (see the definition of (4.57)), which would in turn contain the unmeasurable velocity. To circumvent this problem, the expression given by (4.70) can be integrated and segregated into measurable and unmeasurable terms as follows:

$$\hat{\theta} = \Gamma \int_0^t \left(Y_d^T(\sigma) \left(e(\sigma) + e_f(\sigma) \right) \right) d\sigma + \Gamma \int_0^t Y_d^T(\sigma) \dot{e}(\sigma) d\sigma. \tag{4.71}$$

The first integral of (4.71) can be directly implemented, while the second integral can be rewritten using integration by parts to yield the update law given in (4.66). Note that this last operation introduces the constraint that $Y_d(q_d, \dot{q}_d, \ddot{q}_d)$ must be differentiable with respect to time, which leads to the constraint that $\dddot{q}_d(t)$ be bounded (see (4.4)).

4.4.3 Composite Stability Result

The combination of the error dynamics of (4.56), (4.58), (4.67) and (4.70) gives the following stability result for the position tracking error.

Theorem 4.2 *The filter of (4.54) and (4.53) and the adaptive control law given by (4.65) and (4.70) ensure the position tracking error is semiglobally asymptotically stable in the sense that*

$$\lim_{t \to \infty} x(t) = 0, \tag{4.72}$$

where $x(t)$ was defined in (4.69). This result holds provided the control gain k is defined as follows:

$$k = \frac{1}{m_1} \left(1 + \zeta_1^2 k_n + \zeta_2^2 k_n \right), \tag{4.73}$$

with k_n being a nonlinear damping gain selected to satisfy the following sufficient condition:

$$k_n > \frac{\lambda_2}{\lambda_1} \|z(0)\|^2 + 1, \tag{4.74}$$

where m_1 was defined in (3.2), ζ_1, ζ_2 were defined in (4.68),

$$z = \begin{bmatrix} x^T & \tilde{\theta}^T \end{bmatrix}^T \in \mathbb{R}^{3n+p} \tag{4.75}$$

and

$$\lambda_1 = \frac{1}{2} \min\left\{1,\ m_1,\ \lambda_{\min}\left\{\Gamma^{-1}\right\}\right\} \qquad \lambda_2 = \frac{1}{2} \max\left\{1,\ m_2,\ \lambda_{\max}\left\{\Gamma^{-1}\right\}\right\}. \tag{4.76}$$

Note that $\lambda_{\min}\{\cdot\}$ and $\lambda_{\max}\{\cdot\}$ in (4.76) are used to denote the minimum and maximum eigenvalues of a given matrix, respectively.

Proof. We begin by introducing a non-negative function of the form

$$V = \frac{1}{2} e^T e + \frac{1}{2} e_f^T e_f + \frac{1}{2} \eta^T M(q) \eta + \frac{1}{2} \tilde{\theta}^T \Gamma^{-1} \tilde{\theta}. \tag{4.77}$$

From (4.77), $V(t)$ can be bounded as

$$\lambda_1 \|x\|^2 \leq \lambda_1 \|z\|^2 \leq V \leq \lambda_2 \|z\|^2, \tag{4.78}$$

where $x(t)$ and $z(t)$ were defined in (4.69) and (4.75), respectively, and the positive constants λ_1, λ_2 were defined in (4.76). Differentiating (4.77) with respect to time produces

$$\dot{V} = e^T \dot{e} + e_f^T \dot{e}_f + \frac{1}{2} \eta^T \dot{M}(q) \eta + \eta^T M(q) \dot{\eta} + \tilde{\theta}^T \Gamma^{-1} \dot{\tilde{\theta}}. \tag{4.79}$$

Substitution from the three error systems defined in (4.56), (4.58), and (4.67) yields

$$\dot{V} = -e^T e - e_f^T e_f - k \eta^T M(q) \eta + \eta^T \chi + \tilde{\theta}^T \left(Y_d^T \eta - \Gamma^{-1} \dot{\hat{\theta}} \right), \tag{4.80}$$

where (3.3) has been applied, and we have used the fact from (4.52) that $\dot{\tilde{\theta}} = -\dot{\hat{\theta}}$ (i.e., θ is a vector of constants). If the parameter estimate update law of (4.70) is now substituted into (4.80), we obtain

$$\dot{V} = -e^T e - e_f^T e_f - k \eta^T M(q) \eta + \eta^T \chi. \tag{4.81}$$

After applying (3.2) to (4.81), we can form an upper bound on $\dot{V}(t)$ as shown below:

$$\dot{V} \leq -\|e\|^2 - \|e_f\|^2 - k m_1 \|\eta\|^2 + \|\eta\| \|\chi\|. \tag{4.82}$$

After substituting (4.68) and the definition of the control gain k given in (4.73) into (4.82), we obtain the following new upper bound on $\dot{V}(t)$:

$$\begin{aligned} \dot{V} &\leq -\|e\|^2 - \|e_f\|^2 - \|\eta\|^2 + \left[\zeta_1 \|x\| \|\eta\| - k_n \zeta_1^2 \|\eta\|^2\right] \\ &\quad + \left[\zeta_2 \|x\|^2 \|\eta\| - k_n \zeta_2^2 \|\eta\|^2\right]. \end{aligned} \tag{4.83}$$

Since the bracketed terms form nonlinear damping pairs, we can apply Lemma A.10 in Appendix A to further upper bound $\dot{V}(t)$ as

$$\dot{V} \leq -\left[1 - \frac{1}{k_n}\left(1 + \|x\|^2\right)\right] \|x\|^2, \tag{4.84}$$

where (4.69) was applied to the first three terms on the right-hand side of (4.83). The sign of the upper bound on $\dot{V}(t)$ is determined by the bracketed term of (4.84), which must be positive in order to ensure the negative semidefiniteness of $\dot{V}(t)$. That is, we must have

$$1 - \frac{1}{k_n}\left(1 + \|x(t)\|^2\right) > 0 \tag{4.85}$$

for $\dot{V}(t)$ to be negative semidefinite. From (4.78), a sufficient condition for (4.85) can be derived as follows:

$$1 - \frac{1}{k_n}\left(1 + \frac{V(t)}{\lambda_1}\right) > 0; \tag{4.86}$$

hence, the analysis to this point can be summarized as

$$\dot{V} \leq -\beta \|x\|^2 \qquad \text{for} \qquad k_n > \frac{V(t)}{\lambda_1} + 1, \tag{4.87}$$

where β is some positive constant. From (4.87), we know that since $\dot{V}(t) \leq 0$ $\forall t \in [0, \infty)$, $V(t)$ is decreasing or constant $\forall t \in [0, \infty)$; therefore, a sufficient condition for (4.87) is given by

$$\dot{V} \leq -\beta \|x\|^2 \qquad \text{for} \qquad k_n > \frac{V(0)}{\lambda_1} + 1, \tag{4.88}$$

or by (4.78), we have a sufficient condition for (4.88) as

$$\dot{V} \leq -\beta \|x\|^2 \qquad \text{for} \qquad k_n > \frac{\lambda_2}{\lambda_1} \|z(0)\|^2 + 1. \tag{4.89}$$

The direct implication of (4.77) and (4.89) is that $z(t) \in \mathcal{L}_\infty$, (i.e., $e(t)$, $e_f(t)$, and $\eta(t) \in \mathcal{L}_\infty$ and $\tilde{\theta}(t) \in \mathcal{L}_\infty$). The bound on $e(t)$ along with the required bound on $q_d(t)$ implies that $q(t) \in \mathcal{L}_\infty$. From the boundedness of $e(t), e_f(t)$, and $\eta(t)$, we can use (4.56) and (4.58) to show that $\dot{e}_f(t), \dot{e}(t) \in \mathcal{L}_\infty$; hence, $\dot{q}(t) \in \mathcal{L}_\infty$ owing to the boundedness of $\dot{q}_d(t)$. Since $\tilde{\theta}(t)$ is bounded, the definition in (4.52) and the fact that θ is a constant vector can be used to show that $\hat{\theta}(t) \in \mathcal{L}_\infty$. From (4.53) and (4.54), it is clear that $p(t), \dot{p}(t) \in \mathcal{L}_\infty$. From the above boundedness statements and the fact that $Y_d(t)$ is only a function of the bounded desired trajectory, we can use (4.65) to state that the control input $\tau(t) \in \mathcal{L}_\infty$. Finally, we can utilize (3.1), (4.67), and the fact that $M^{-1}(q)$ exists and is bounded to illustrate that $\ddot{q}(t), \dot{\eta}(t) \in \mathcal{L}_\infty$.

Using the above information, we can state from the definition of (4.69) that $x(t), \dot{x}(t) \in \mathcal{L}_\infty$. In addition, it is not difficult to see from (4.89) that $x(t) \in \mathcal{L}_2$ (see (2.20) to (2.22)); hence, Lemma A.3 in Appendix A can now be invoked to obtain the result of (4.72). □

Remark 4.7 *Similar to Theorem 4.1, the semiglobal nature of Theorem 4.2 stems from the fact that the control gain k_n can be made arbitrarily large to encompass any set of initial conditions as illustrated by (4.74). This semiglobal result is facilitated by the filter initial condition $p(0) = ke(0)$ given in (4.54) (see Remark 4.2 for a similar motivation). Moreover, it is obvious from the form of (4.74) and (4.75) that we must have a priori information on bounds for the unknown parameters and on the initial velocity to facilitate the calculation of k_n.*

Remark 4.8 *An extension to Theorem 4.2 can be made to describe velocity tracking. The equality in (4.58) can be used to form the following inequality:*

$$\|\dot{e}\| \leq \|\eta\| + \|e_f\| + \|e\| . \tag{4.90}$$

From (4.72) and (4.69), we know that the terms on the right-hand side of the above inequality are driven to zero; hence, we can conclude that

$$\lim_{t\to\infty} \dot{e}(t) = 0; \tag{4.91}$$

i.e., the controller ensures semiglobal asymptotic velocity tracking.

Remark 4.9 *An interesting aside to the above developments lies in the similarity between the filter used to generate the velocity information and a continuous time approximation of the backwards difference algorithm. That is, the Laplace domain transfer function for producing a velocity tracking error estimate from the position tracking error by a backwards difference approach can be obtained using the bilinear transformation as $H_{bd}(s) = \frac{2s}{Ts+2}$ (where T is the sample time), while the Laplace domain* *transfer function for the velocity error generating filter in (4.54) and (4.53) is* $H_f(s) = \frac{e_f(s)}{e(s)} = -\frac{ks+(k-1)}{s+(k+1)}$. *The similarity in the magnitude frequency responses of these two transfer functions suggests that the velocity error generating filter acts much like the discrete differentiator over a range of frequencies. This similarity also seems to suggest that since the filter dynamics can be incorporated into the closed-loop stability analysis, the same tracking result might be extrapolated to the backwards difference approach (i.e., the proven use of this velocity error generating filter validates to some degree the use of a backwards difference algorithm).*

4.4.4 Experimental Results

The linear filter/controller scheme was implemented on the same experimental test-bed described in Section 4.3.4. Based on (3.4) and (3.58), the unknown parameter vector θ was constructed as

$$\theta = \begin{bmatrix} p_1 & p_2 & p_3 & f_{d1} & f_{d2} \end{bmatrix}^T . \tag{4.92}$$

The experiment was performed using the following desired position trajectories

$$q_d(t) = \begin{bmatrix} 45.84\sin(2t)\left(1-\exp\left(-0.3t^3\right)\right) \\ 45.84\sin(2t)\left(1-\exp\left(-0.3t^3\right)\right) \end{bmatrix} \text{deg.} \tag{4.93}$$

The experiment was run several times and it was found that the best tracking performance was achieved using the following set of gains:

$$k = \operatorname{diag}\{32.25, 28.0\}, \qquad \Gamma = \operatorname{diag}\{22.8, 0.8, 1.25, 100.6, 40.2\}.$$

where, for implementation purposes, the control gain k was specified as a matrix as opposed to the scalar definition of (4.73). In the experiment, all parameter estimates were initialized to zero (i.e., $\hat{\theta}(0) = 0$). The integrations required in the parameter estimator of (4.66) and the filter of (4.54) and (4.53) were performed using a standard trapezoidal algorithm with a sampling period of 0.5 msec. The results of the experiment are shown in Figure 4.5 and Figure 4.6. Figure 4.5 shows the two position tracking errors, while the parameter estimates are shown in Figure 4.6.

4.4.5 Nonadaptive Extensions

Similar to the nonadaptive extensions to the adaptive FSFB controller discussed in Section 3.5.4, it is possible to use fixed parameter estimates in the feedforward component of the OFB controller of (4.65) as an alternative to the dynamic, parameter adaptation law of (4.66). The choice of the fixed

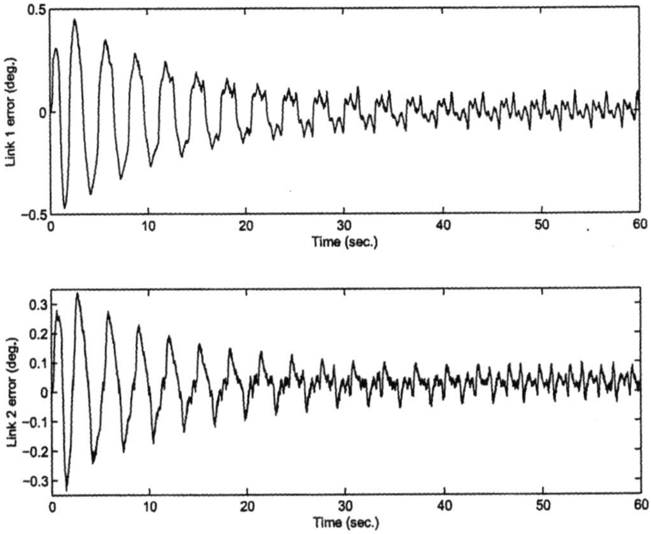

FIGURE 4.5. Position tracking errors of linear filter/controller.

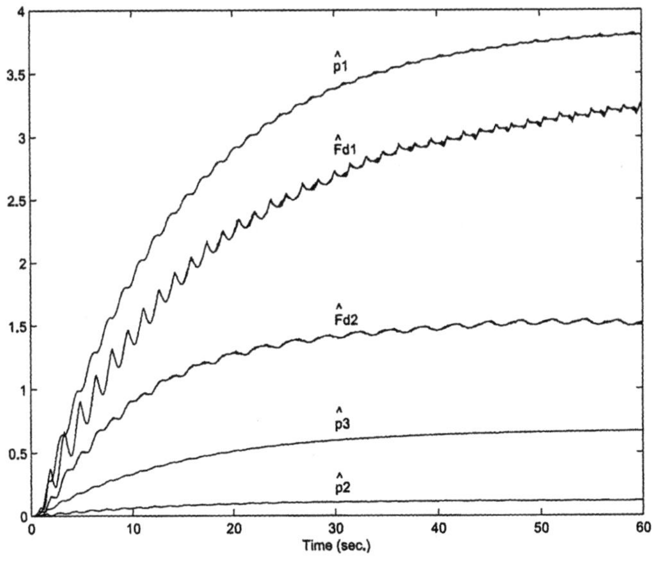

FIGURE 4.6. Parameter estimates of linear filter/controller.

parameter estimates will dictate the type of position tracking result that can be achieved as described by the following two corollaries.

Corollary 4.1 (Constant Best Guess Parameter Estimate) *Let* $\hat{\theta}$ *in (4.65) contain constant, best guess estimates of the system parameters and the control gain* k *of (4.73) be redefined as*

$$k = \frac{1}{m_1}\left(1 + \zeta_1^2 k_n + \zeta_2^2 k_n + \frac{\zeta_3^2}{\epsilon}\right), \tag{4.94}$$

where ϵ *is a positive nonlinear damping gain, and* ζ_3 *is a positive bounding constant defined as*

$$\left\| Y_d \tilde{\theta} \right\| \leq \zeta_3, \tag{4.95}$$

which depends only on the bounds on the desired trajectory and the magnitude of the difference between the constant, best guess parameter estimates and the actual parameters. If the nonlinear damping gain k_n *in (4.94) is selected such that*

$$k_n > \frac{\lambda_2}{\lambda_1} \left\| x(0) \right\|^2 + 1 + \frac{\lambda_2 \epsilon}{\lambda_1 \beta}, \tag{4.96}$$

then the position tracking error is semiglobally uniformly ultimately bounded (SGUUB) in the sense that

$$\left\| x(t) \right\| \leq \sqrt{\frac{\lambda_2}{\lambda_1} \left\| x(0) \right\|^2 \exp\left(-\frac{\beta}{\lambda_2} t\right) + \frac{\lambda_2 \epsilon}{\beta \lambda_1}\left(1 - \exp\left(-\frac{\beta}{\lambda_2} t\right)\right)}, \tag{4.97}$$

where $x(t)$ *was defined in (4.69),* β *is some positive constant less than one, and*

$$\lambda_1 = \frac{1}{2}\min\{1,\ m_1\} \qquad \lambda_2 = \frac{1}{2}\max\{1,\ m_2\}. \tag{4.98}$$

Proof. The proof of Corollary 4.1 follows directly from the combination of the proofs of Corollary 3.1 and Theorem 4.2. □

Remark 4.10 *The rate of convergence of* $\|x(t)\|$ *in (4.97) to the steady-state bound given in (3.75) is limited by the maximum value of* β*, which will be smaller than one. This rate of convergence can be increased if the filtered tracking error-like variable of (4.57) is redefined to include the weighted sum of* $e_f(t)$ *and* $e(t)$ *in the follow manner:*

$$\eta = \dot{e} + k_\alpha e + k_\beta e_f,$$

in which k_α*,* k_β *are positive control gains. This change in* $\eta(t)$ *will require minor changes to the velocity error generating filter in (4.54) and (4.53) and to the selection of the control gains* k *and* k_n *(see [31] for details).*

Remark 4.11 *It should be noted that the controller of (4.65) with $\hat{\theta}$ set to a constant best guess estimate vector is directly related to the approach given by Yuan and Stepanenko in [32]. Furthermore, Corollary 4.1 also includes the case where $\hat{\theta}= 0$. That is, even if there is no feedforward compensation in the control law, the SGUUB tracking result of (4.97) remains valid. This controller can be seen as a simple linear controller that is directly related to the controllers proposed by Berghuis and Nijmeijer [3] and Qu et al. [30]. While the linear controllers proposed in [3, 30] used second-order filters to generate velocity information, the approach presented here uses a first-order filter.*

Corollary 4.2 (Exact Parameter Knowledge) *Let $\hat{\theta}=\theta$ in (4.65) and the control gain k be selected according to (4.73). If the nonlinear damping gain k_n in (4.73) is selected such that*

$$k_n > \frac{\lambda_2}{\lambda_1} \|x(0)\|^2 + 1, \tag{4.99}$$

then the position tracking error is semiglobally exponentially stable in the sense that

$$\|x(t)\| \leq \sqrt{\frac{\lambda_2}{\lambda_1}} \|x(0)\| \exp\left(-\frac{\beta t}{2\lambda_2}\right), \tag{4.100}$$

where $x(t)$ was defined in (4.69), λ_1, λ_2 were defined in (4.98), and β is some positive constant less than one.

Proof. The proof follows directly from the proof of Corollary 4.1. □

Remark 4.12 *It follows from (4.100) that the rate of convergence of $\|x(t)\|$ to zero can be increased if the controller is modified in accordance with Remark 4.10.*

4.5 Nonlinear Filter-Based Adaptive Control

A theoretical limitation of the linear filter-based, OFB, tracking controller of Section 4.4 is the semiglobal nature of the stability result. In this section, we demonstrate how to construct a *nonlinear* filter-based controller such that a *global* stability result is obtained for the position tracking error. The control law structure is composed of (i) a DCAL feedforward term to compensate for parametric uncertainty, and (ii) a nonlinear feedback term coupled to a nonlinear dynamic filter to compensate for the lack of velocity measurements and the difference between the actual system dynamics and the DCAL feedforward term.

To aid the control design and analysis of this section, we define the vector function $Tanh(\cdot) \in \mathbb{R}^n$ and the matrix function $Cosh(\cdot) \in \mathbb{R}^{n \times n}$ as follows:

$$Tanh(\xi) = [\tanh(\xi_1), \ \ ..., \ \ \tanh(\xi_n)]^T \tag{4.101}$$

and

$$Cosh(\xi) = diag\{\cosh(\xi_1), \cdots, \cosh(\xi_n)\}, \tag{4.102}$$

where $\xi(t) = [\xi_1, \ ..., \ \xi_n]^T \in \mathbb{R}^n$. Based on the definition of (4.101), it can easily be shown that the following inequalities hold for $\forall \xi(t), \nu(t) \in \mathbb{R}^n$:

$$\begin{aligned} \frac{1}{2}\tanh^2(\|\xi\|) \leq \ln(\cosh(\|\xi\|)) \leq \sum_{i=1}^{n} \ln(\cosh(\xi_i)) \leq \|\xi\|^2 \\ \tanh^2(\|\xi\|) \leq \|Tanh(\xi)\|^2 = Tanh^T(\xi)\,Tanh(\xi) \end{aligned} \tag{4.103}$$

and

$$\begin{aligned} \left|\cos\left(\sum_{i=1}^{n}\xi_i\right) - \cos\left(\sum_{i=1}^{n}\nu_i\right)\right| \leq 8\sum_{i=1}^{n}|\tanh(\xi_i - \nu_i)| \\ \left|\sin\left(\sum_{i=1}^{n}\xi_i\right) - \sin\left(\sum_{i=1}^{n}\nu_i\right)\right| \leq 8\sum_{i=1}^{n}|\tanh(\xi_i - \nu_i)|, \end{aligned} \tag{4.104}$$

where $\xi_i(t), \nu_i(t)$ denote the i-th elements of the vectors $\xi(t), \nu(t)$. Based on the inequalities of (4.104), we will assume that the dynamic model of the mechanical system considered in this section satisfies the following property (in addition to the properties given in Sections 3.2 and 4.2).

Property 4.3: The following bounds for the inertia, centripetal-Coriolis, and gravity terms of (3.1) are assumed to exist:

$$\begin{aligned} \|M(\xi) - M(\nu)\|_{i\infty} \leq \zeta_m \|Tanh(\xi - \nu)\| \\ \|G(\xi) - G(\nu)\| \leq \zeta_g \|Tanh(\xi - \nu)\| \\ \|V_m(\xi, \dot{q}) - V_m(\nu, \dot{q})\|_{i\infty} \leq \zeta_{c2} \|\dot{q}\| \, \|Tanh(\xi - \nu)\|, \end{aligned} \tag{4.105}$$

where $\zeta_m, \zeta_g, \zeta_{c2}$ are positive bounding constants, and $\xi(t), \nu(t) \in \mathbb{R}^n$. An outline of the proof for the bounds given in (4.105) is presented in Appendix C for a specific mechanical system (the 6 degrees-of-freedom Puma robot [23]).

4.5.1 Filter/Controller Formulation

Let us redefine the filtered tracking error-like variable $\eta(t)$ of (4.57) as follows:

$$\eta = \dot{e} + Tanh\,(e) + Tanh\,(e_f)\,, \tag{4.106}$$

where $Tanh(\cdot)$ was defined in (4.101), and $e_f(t)$ is the filter output which is redefined to have the following dynamics

$$\dot{e}_f = -Tanh\,(e_f) + Tanh\,(e) - k\,Cosh^2\,(e_f)\,\eta, \qquad e_f(0) = 0, \tag{4.107}$$

with $k \in \mathbb{R}$ being a positive control gain, and $Cosh(\cdot)$ being defined in (4.102). Note that now the dynamics for $e\,(t)$ are given by rearranging (4.106) to yield

$$\dot{e} = -Tanh\,(e) + \eta - Tanh\,(e_f)\,. \tag{4.108}$$

Following similar derivations as those outlined in Section 4.4.2, the open-loop dynamics for $\eta\,(t)$ can be obtained, as shown below:

$$M(q)\dot{\eta} = -V_m(q,\dot{q})\eta - kM(q)\eta + Y_d\theta + \chi + \tilde{Y} - \tau, \tag{4.109}$$

where $\chi\,(e, e_f, \eta, t)\,, \tilde{Y}\,(e, e_f, \eta, t) \in \mathbb{R}^n$ are defined as

$$\begin{aligned}\chi = \;& M(q)\,Cosh^{-2}\,(e)\,(\eta - Tanh(e_f) - Tanh(e)) \\ & + M(q)\,Cosh^{-2}\,(e_f)\,(-Tanh(e_f) + Tanh(e)) \\ & + V_m(q, \dot{q}_d + Tanh(e_f) + Tanh(e))\,(Tanh(e_f) + Tanh(e)) \\ & + V_m(q, \dot{q}_d)\,(Tanh(e_f) + Tanh(e)) \\ & - V_m(q, \eta)\,(\dot{q}_d + Tanh(e_f) + Tanh(e))\end{aligned} \tag{4.110}$$

and

$$\tilde{Y} = M(q)\ddot{q}_d + V_m(q, \dot{q}_d)\dot{q}_d + G(q) + F_d\dot{q} - Y_d\theta. \tag{4.111}$$

Note that in the formulation of (4.110), Property 4.1 has been utilized. From the form of (4.109) and the subsequent stability analysis, the following control input is formulated:

$$\tau = Y_d\hat{\theta} - k\,Cosh^2\,(e_f)\;Tanh(e_f) + Tanh(e), \tag{4.112}$$

where k is the same control gain defined in (4.107), and the parameter estimate vector $\hat{\theta}(t) \in \mathbb{R}^p$ is generated according to the following adaptation algorithm:

$$\dot{\hat{\theta}} = \Gamma Y_d^T \eta, \tag{4.113}$$

with $\Gamma \in \mathbb{R}^{p\times p}$ being a diagonal, positive-definite, adaptation gain matrix. After substituting (4.112) into (4.109), we can form the closed-loop dynamics for $\eta(t)$ as given below:

$$\begin{aligned} M(q)\dot{\eta} &= -V_m(q,\dot{q})\eta - kM(q)\eta + Y_d\tilde{\theta} + \tilde{Y} + \chi \\ &\quad + k\,Cosh^2\left(e_f\right) Tanh(e_f) - Tanh(e). \end{aligned} \tag{4.114}$$

Remark 4.13 *By exploiting (3.3), (4.2), and the properties of the hyperbolic functions, we can place an upper bound on χ of (4.110) as follows:*

$$\|\chi\| \le \zeta_1 \|x\|, \tag{4.115}$$

where ζ_1 is some positive bounding constant that depends on the mechanical system parameters and the desired trajectory, and $x(t) \in \mathbb{R}^{3n}$ is defined as

$$x = \begin{bmatrix} Tanh^T(e) & Tanh^T(e_f) & \eta^T \end{bmatrix}^T. \tag{4.116}$$

Furthermore, by utilizing Property 4.3, it can be shown that $\tilde{Y}(\cdot)$ of (4.111) can be upper bounded as follows:

$$\left\|\tilde{Y}\right\| \le \zeta_2 \|x\|, \tag{4.117}$$

where ζ_2 is also some positive bounding constant that depends on the mechanical system parameters and the desired trajectory.

Remark 4.14 *Based on the definition of $\eta(t)$ given in (4.106), it seems that velocity measurements are required for control implementation in equations (4.107), (4.112), and (4.113). However, in a subsequent section, we will illustrate how the control scheme has an equivalent form that depends only on position measurements.*

4.5.2 Composite Stability Result

Theorem 4.3 *The filter of (4.107) and the adaptive control law of (4.112) and (4.113) ensure global asymptotic position tracking in the sense that*

$$\lim_{t\to\infty} e(t) = 0 \tag{4.118}$$

provided the control gain k of (4.112) is selected as follows:

$$k = \frac{1}{m_1}\left(1 + k_n\left(\zeta_1 + \zeta_2\right)^2\right), \tag{4.119}$$

where m_1, ζ_1, and ζ_2 were defined in (3.3), (4.115), and (4.117), respectively, and k_n is a nonlinear damping gain that must satisfy the following sufficient condition:

$$k_n > 1. \tag{4.120}$$

Proof. We start with the following non-negative function:[5]

$$V = \sum_{i=1}^{n} \ln(\cosh(e_i)) + \sum_{i=1}^{n} \ln(\cosh(e_{fi}) + \frac{1}{2}\eta^T M(q)\eta + \frac{1}{2}\tilde{\theta}^T \Gamma^{-1}\tilde{\theta}, \quad (4.121)$$

where $e_i(t)$, $e_{fi}(t)$ are the ith elements of the vectors $e(t)$ and $e_f(t)$. After taking the time derivative of (4.121), we obtain the following expression for $\dot{V}(t)$:

$$\dot{V} = Tanh^T(e)\dot{e} + Tanh^T(e_f)\dot{e}_f + \eta^T M(q)\dot{\eta} + \frac{1}{2}\eta^T \dot{M}(q)\eta - \tilde{\theta}^T \Gamma^{-1}\dot{\hat{\theta}}. \quad (4.122)$$

We can now utilize (4.106), (4.107), (4.114), and (4.113) in (4.122) to simplify the expression for $\dot{V}(t)$ as follows:

$$\dot{V} = -Tanh^T(e)Tanh(e) - Tanh^T(e_f)Tanh(e_f) + \eta^T\left(\tilde{Y} + \chi\right) - k\eta^T M(q)\eta, \quad (4.123)$$

where (3.3) has been employed. After applying (4.115), (4.117), (3.2), and (4.119) to (4.123), we obtain the following upper bound for $\dot{V}(t)$:

$$\begin{aligned} \dot{V} \leq\; & -\|Tanh(e)\|^2 - \|Tanh(e_f)\|^2 - \|\eta\|^2 \\ & + \left[(\zeta_1 + \zeta_2)\|x\|\|\eta\| - k_n(\zeta_1 + \zeta_2)^2\|\eta\|^2\right]. \end{aligned} \quad (4.124)$$

Application of the Lemma A.10 in Appendix A to the bracketed term in (4.124) yields the following new upper bound on $\dot{V}(t)$:

$$\dot{V} \leq -\|x\|^2 + \frac{1}{k_n}\|x\|^2, \quad (4.125)$$

where $x(t)$ was defined in (4.116). Finally, if k_n is selected according to (4.120), we can rewrite (4.125) as

$$\dot{V} \leq -\beta\|x\|^2, \quad (4.126)$$

where β is some positive constant.

We note that $\ln(\cosh(0)) = 0$ and that $\ln(\cosh(\cdot))$ is a radially unbounded, globally positive-definite function. Hence, owing to the structure of $V(t)$ given in (4.121), $V(t)$ is a radially unbounded, globally positive-definite function for all $e(t)$, $e_f(t)$, $\eta(t)$, $\tilde{\theta}(t)$, and t. Since $\dot{V}(t)$ is negative semidefinite as illustrated by (4.126), we now know that $V(t) \in \mathcal{L}_\infty$,

[5] It should be noted that the first two terms in $V(t)$ are motivated by the work given in [26].

which implies that $e(t), e_f(t), \eta(t), \tilde{\theta}(t) \in \mathcal{L}_\infty$; hence, $q(t), \dot{q}(t), \hat{\theta}(t)$ and $\dot{\hat{\theta}}(t) \in \mathcal{L}_\infty$. We can now utilize (4.106), (4.107), and (4.114) to state that $\dot{e}(t), \dot{e}_f(t), \dot{\eta}(t) \in \mathcal{L}_\infty$. The above boundedness statements together with the fact that the desired trajectory is bounded allow us to conclude that $\ddot{q}(t),\ \tau(t) \in \mathcal{L}_\infty$.

Utilizing the above information, (4.116), and (4.126), we can state that $x(t), \dot{x}(t) \in \mathcal{L}_\infty$ and $x(t) \in \mathcal{L}_2$; hence, we can now invoke Lemma A.3 in Appendix A to conclude that $\lim_{t\to\infty} x(t) = 0$. From the definition given in (4.116), we can now see that $\lim_{t\to\infty} Tanh(e(t)) = 0$, which leads to the result given by (4.118) owing to the properties of the hyperbolic tangent function. □

4.5.3 OFB Form of Filter/Controller

We now illustrate how the control scheme formulated in Section 4.5.1 has an equivalent form that only requires position measurements. Note that the control input given by (4.112) and (4.113) does not actually need the computation of $e_f(t)$; rather, only computations of $Tanh(e_f)$ and $Cosh^2(e_f)$ are required. Also note that if we define the following relationship:

$$y_i = [Tanh(e_f)]_i = \tanh(e_{fi}), \tag{4.127}$$

then according to standard hyperbolic identities, we have

$$\cosh^2(e_{fi}) = \frac{1}{1-\tanh^2(e_{fi})} = \frac{1}{1-y_i^2}, \tag{4.128}$$

where $y_i(t)$ is the i-th element of the vector $y(t) \in \mathbb{R}^n$.

We will now show that $y_i(t)$ (and hence, $\tanh(e_{fi})$ and $\cosh^2(e_{fi})$) can be calculated with only position measurements. First, note that the filter given by (4.107) can be written element-wise as follows:

$$\dot{e}_{fi} = -tanh(e_{fi}) + tanh(e_i) - k cosh^2(e_{fi})\,\eta_i, \qquad e_{fi}(0) = 0, \tag{4.129}$$

where $\eta_i(t)$ is the i-th element of the vector $\eta(t)$ defined in (4.106). After taking the time derivative of (4.127), we can substitute (4.129) and (4.128) into the resulting expression to obtain

$$\dot{y}_i = \cosh^{-2}(e_{fi})\dot{e}_{fi} = -\left(1-y_i^2\right)\left(y_i - \tanh(e_i)\right) - k\left(\dot{e}_i + \tanh(e_i) + y_i\right)$$
$$y_i(0) = 0. \tag{4.130}$$

It is now straightforward to utilize (4.130) to construct the following filter, which also computes $y_i(t)$:

$$\begin{cases} \dot{p}_i = -\left(1 - (p_i - ke_i)^2\right)(p_i - ke_i - \tanh(e_i)) - k(\tanh(e_i) + p_i - ke_i) \\ y_i = p_i - ke_i \end{cases} \tag{4.131}$$

where $p_i(0) = ke_i(0)$. In (4.131), $p_i(t)$ is an auxiliary variable that allows $y_i(t)$ (and hence, $\tanh(e_{fi})$ and $\cosh^2(e_{fi})$) to be calculated with only position measurements.

We now illustrate how the controller given by (4.112) and (4.113) can be computed with only position measurements. First, we substitute the definition of $\eta(t)$ given by (4.106) into (4.113) and then integrate the resulting expression by parts to obtain

$$\hat{\theta}(t) = \Gamma Y_d^T(t)e(t) + \Gamma \int_0^t \left(Y_d^T(\sigma)\left(Tanh(e(\sigma)) + y(\sigma)\right) - \dot{Y}_d^T(\sigma)e(\sigma)\right) d\sigma, \tag{4.132}$$

where we have utilized the fact from (4.127) that $y(t) = Tanh(e_f(t))$. By utilizing (4.131) to compute $y(t)$, it is now easy to see that $\hat{\theta}(t)$ can be computed with only position measurements. After substituting (4.127) and (4.128) into (4.112), the i-th component of the control input can be written as follows

$$\tau_i = \left(Y_d\hat{\theta}\right)_i - k\frac{y_i}{1 - y_i^2} + tanh(e_i), \tag{4.133}$$

where $\tau_i(t)$ and $\left(Y_d\hat{\theta}\right)_i$ are the i-th elements of the vectors $\tau(t)$ and $Y_d\hat{\theta}$, $\hat{\theta}(t)$ is computed using (4.132), and $y_i(t)$ is computed using (4.131).

Remark 4.15 *It is clear from (4.127) and (4.133) that $y_i(t)$ must be restricted such that $|y_i(t)| < 1$ for all time. To illustrate that this does indeed occur, first, note that since $e_f(0) = 0$ (see (4.107)), we know from (4.127) that $y(0) = 0$. Secondly, from the proof of Theorem 4.3, it follows that $e_{fi}(t) \in \mathcal{L}_\infty$; hence, we can use the definition given by (4.127) and the properties of the hyperbolic functions to show that $|y_i(t)| < 1$ for $t > 0$. The proof of Theorem 4.3 also illustrates that $\hat{\theta}(t) \in \mathcal{L}_\infty$ and $e(t) \in \mathcal{L}_\infty$; thus, it is now easy to see that all of the signals in the OFB form of the control given by (4.131), (4.133), and (4.132) remain bounded for all time (i.e., $p_i(t)$, $\dot{p}_i(t)$, $\tau_i(t) \in \mathcal{L}_\infty$).*

4.5.4 Simulation Results

The adaptive OFB controller given by (4.133), (4.131), and (4.132) was simulated utilizing the dynamic model of the IMI manipulator [15] given in (3.58). The desired position trajectory for links 1 and 2 in the simulations were selected as follows:

$$q_d(t) = \begin{bmatrix} 0.7\sin(t)\left(1-\exp\left(-0.3t^3\right)\right) \\ 1.2\sin(t)\left(1-\exp\left(-0.3t^3\right)\right) \end{bmatrix} \text{ rad}, \tag{4.134}$$

while the parameter estimates were initialized to zero (i.e., $\hat{\theta}(0)=0$). First, the controller was simulated with the actual positions and velocities initialized to zero, and the control and adaptation gains set to

$$k = 10, \qquad \Gamma = \text{diag}\left\{25, 1.0, 5.0, 30, 30\right\}.$$

The position tracking errors for links 1 and 2 are shown in Figure 4.7. The control inputs are depicted in Figure 4.8 along with the variables $y_1(t)$ and $y_2(t)$ defined in (4.127). Figure 4.9 illustrates the parameter estimates. Next, the controller was run with $q(0) = [1.0, -1.0]^T$ rad and $\dot{q}(0) = [0,0]^T$ rad/sec, while the control and adaptation gains were selected as

$$k = 0.1, \qquad \Gamma = \text{diag}\left\{50, 1.0, 1.0, 150, 150\right\}.$$

Note that a small value was utilized for the control gain k in order to illustrate the global properties of the control law. That is, even for large initial conditions, the use of a small control gain still provides asymptotic tracking. For scaling purposes, Figure 4.10 shows magnified plots of the position tracking errors for $t \in [0, 10]$ sec and for $t \in [10, 120]$ sec with the idea of illustrating the asymptotic tracking performance. The control inputs and the variables $y_1(t)$ and $y_2(t)$ are shown in Figure 4.11, while the parameter estimates are shown in Figure 4.12. It is important to observe from Figures 4.8 and 4.11 that $|y_1(t)| < 1$ and $|y_2(t)| < 1$, as predicted by the proof of Theorem 4.3.

Since the definition of (4.119) and the condition of (4.120) arise from a conservative stability analysis, it comes to no surprise that even if the value chosen for k does not satisfy these sufficient conditions, we still obtain asymptotic tracking. This phenomenon can be explained by noting that the function $\cosh^2(\cdot)$ becomes big as its argument increases; hence, the dynamic control gain matrix $k\,Cosh^2\left(e_f\right)$ in (4.112) automatically increases to compensate for large tracking errors even for small values of k.

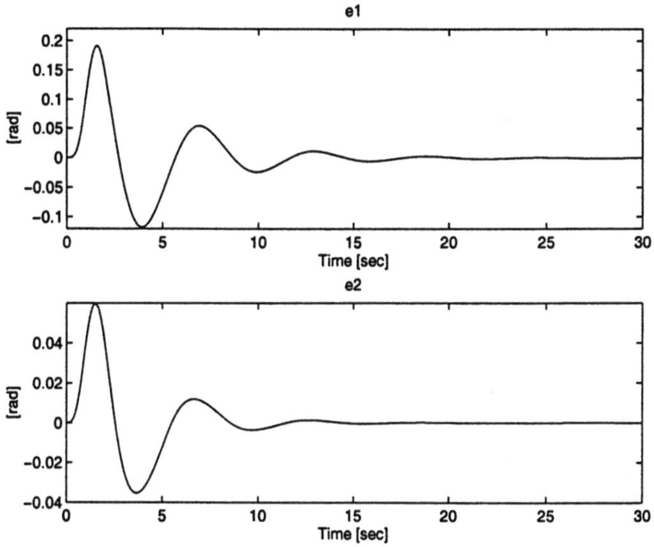

FIGURE 4.7. Position tracking errors of nonlinear filter/controller with zero initial conditions.

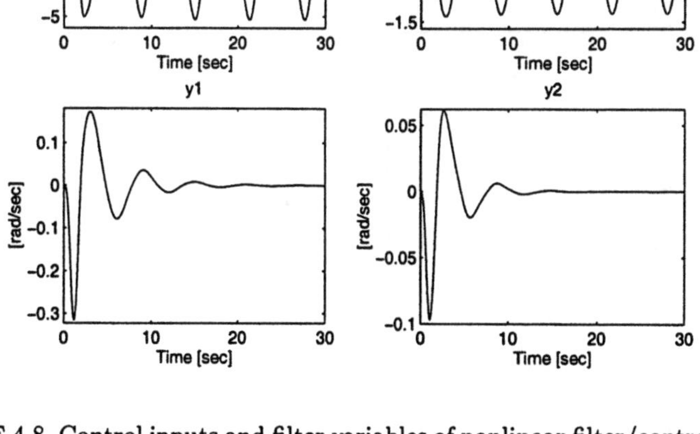

FIGURE 4.8. Control inputs and filter variables of nonlinear filter/controller with zero initial conditions.

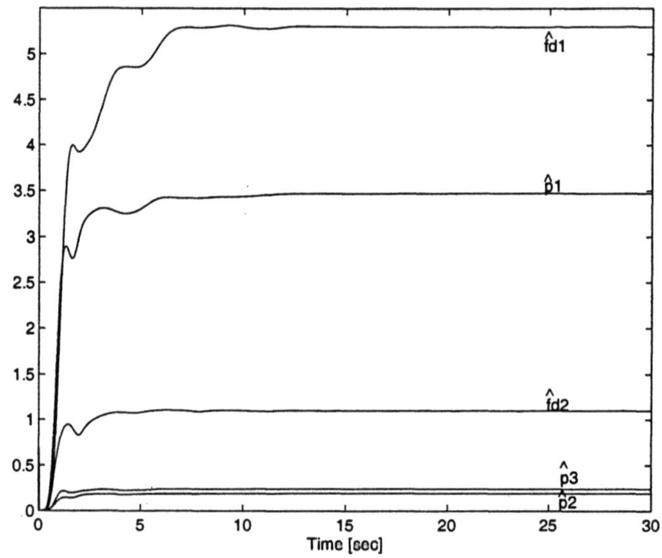

FIGURE 4.9. Parameter estimates of nonlinear filter/controller with zero initial conditions.

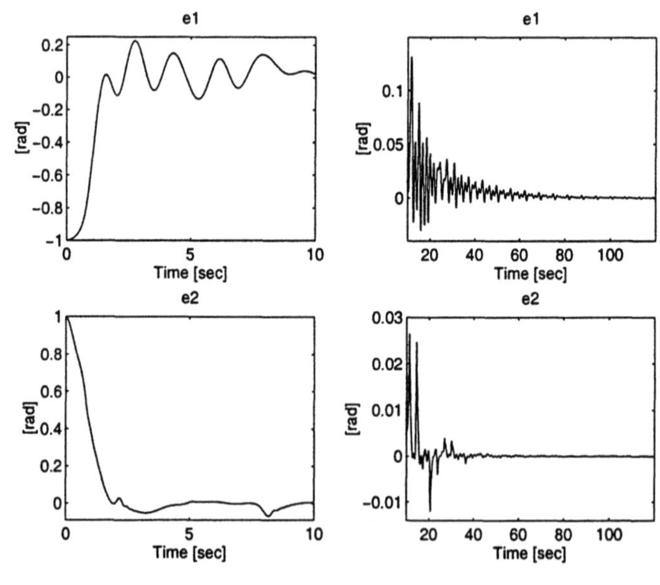

FIGURE 4.10. Position tracking errors of nonlinear filter/controller with non-zero initial conditions.

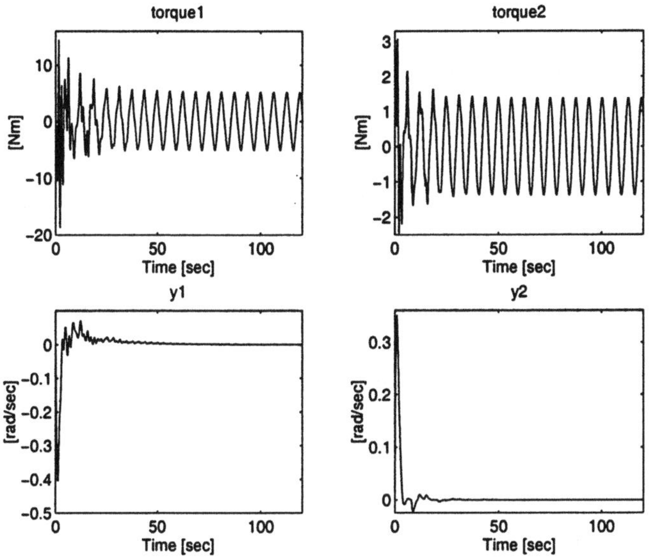

FIGURE 4.11. Control inputs and filter variables of nonlinear filter/controller with non-zero initial conditions.

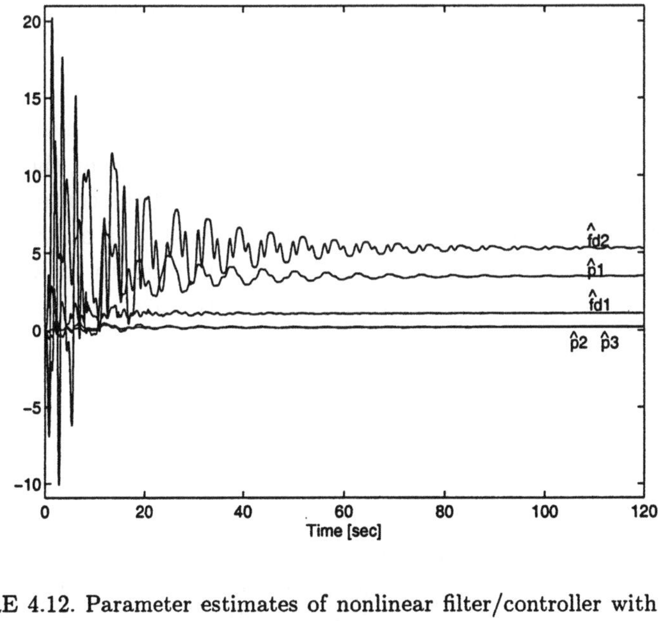

FIGURE 4.12. Parameter estimates of nonlinear filter/controller with non-zero initial conditions.

4.5.5 Extensions

Since (4.112) is a position *tracking* controller, a simplified version of the controller can be used for global position *setpoint* control (i.e., q_d =constant, $\dot{q}_d = \ddot{q}_d = \dddot{q}_d = 0$). Specifically, for the setpoint control problem, $Y_d\theta$ defined in (3.4) becomes

$$Y_d(q_d)\theta = G(q_d); \tag{4.135}$$

and hence, the parameter update law given by (4.132) simplifies to

$$\hat{\theta}(t) = \Gamma Y_d^T(t)e(t) + \Gamma \int_0^t Y_d^T(\sigma)\left(Tanh(e(\sigma)) + y(\sigma)\right) d\sigma. \tag{4.136}$$

The filter given by (4.131) and the control input given by (4.133) are utilized to complete the structure of the adaptive setpoint control law.

If we assume exact model knowledge, the control input of (4.112) can be redesigned as follows

$$\tau = M(q)\ddot{q}_d + V_m(q,\dot{q}_d)\dot{q}_d + G(q) + F_d\dot{q}_d - kCosh^2\left(e_f\right) Tanh(e_f) + Tanh(e); \tag{4.137}$$

hence, the closed-loop dynamics for $\eta(t)$ becomes

$$M(q)\dot{\eta} = -V_m(q,\dot{q})\eta - kM(q)\eta + \tilde{Y} + \chi + kCosh^2\left(e_f\right) Tanh(e_f) - Tanh(e), \tag{4.138}$$

where $\chi(\cdot)$ is still given by (4.110), but now $\tilde{Y}(t)$ is defined as

$$\tilde{Y} = -F_d\dot{e}. \tag{4.139}$$

Note that by using (4.108), $\tilde{Y}(\cdot)$ of (4.139) can now be bounded as in (4.117) but without the need for Property 4.3. By slightly modifying the stability analysis used to prove Theorem 4.3, we can show that the controller given by (4.137) yields global asymptotic position tracking. In a manner similar to that used for the adaptive controller, it is also easy to show that the controller given by (4.137) can be computed with only position measurements.

4.6 Notes

Global solutions to the OFB position setpoint control problem have been presented by several researchers. For example, model-based OFB controllers, composed of a dynamic linear feedback loop plus feedforward gravity compensation, were proposed by Berghuis and Nijmeijer [4], Burkov [7], and Kelly [21] to globally asymptotically stabilize a class of mechanical systems.

In [1], Arimoto et al. also presented a model-based, global regulating OFB controller; however, the gravity compensation term was dependent on the desired position setpoint as opposed to the actual position. With the intent of overcoming the requirement of exact model knowledge, Ortega et al. [29] designed an OFB, PID-like regulator that compensated for uncertain gravity effects; however, the stability result was semiglobal asymptotic. In [14], Colbaugh et al. proposed a global regulating OFB controller that compensated for uncertain gravity effects; however, the control strategy required the use of two different control laws (i.e., one control law was used to drive the setpoint error to a small value, then another control law was used to drive the setpoint error to zero).

With respect to the more general problem of OFB position tracking control, semiglobal results have dominated the scenario. For example, in Berghuis and Nijmeijer [4] and Lim et al. [24], model-based observers and controllers yielded semiglobal exponential position tracking, while in Nicosia and Tomei [27], a semiglobal asymptotic tracking result was achieved. Robust, filter-based control schemes were designed in Berghuis and Nijmeijer [3], Qu et al. [30], and Yuan and Stepanenko [32] to compensate for parametric uncertainty while producing semiglobal uniformly ultimately bounded position tracking. In Burg et al. [5, 6] and Kaneko and Horowitz [20], adaptive, filtered-based controllers were presented that yielded semiglobal asymptotic position tracking. Other semiglobal OFB tracking controllers can be found in Wit et al. [9]–[13], Erlic and Lu [16, 17], Hsu and Lizarralde [18], Nicosia et al. [28], and Zhu et al. [34].

To the best of our knowledge, the only previous work targeted at the global OFB tracking control problem is given in [26] and [8]. In [26], Loria developed a model-based controller that yielded global asymptotic position tracking; however, the control was designed for only SISO rigid mechanical systems. In [8], Burkov used singular perturbation analysis to show that a model-based controller, used in conjunction with a linear observer, can yield asymptotic tracking for any initial conditions; however, as pointed out in [26], no explicit bound on the singular perturbation parameter was given. Motivated by the structure of the stability analysis given in [26], Zhang et al. [33] designed a filter-based, adaptive controller for the MIMO case that compensated for parametric uncertainty and achieved global asymptotic position tracking.

In comparison to the above literature review, we first note that the observer presented in Section 4.3 is nearly identical to the observer presented in [27]; however, the controller given in (4.28) has some additional feedforward terms that are not included in the controller presented in [27]. Indeed, it is these additional terms along with the resulting stability analysis that allowed an exponential stability result to be obtained as opposed to the asymptotic stability result of [27]. Furthermore, we would like to point out that the observer/controller presented in Section 4.3 should be considered as secondary to the work of [2]–[4] since in [4] a general approach based on passivity techniques was used to generate a class of observer-controllers. However, the following differences in the work of [4] and Section 4.3 are worth mentioning. First, even though the same stability results are achieved, the design in Section 4.3 explicitly illustrates how the control gains can be increased to improve the transient response (see Remark 4.2). In addition, it is interesting to note that it is not necessary to drive $\tilde{q}(t)$ (defined in (4.9)) to zero to obtain the desired stability result. As for the linear filter-based, adaptive tracking controller presented in Section 4.4, we would like to mention that it yields, as a subresult, a linear regulator that semiglobally asymptotically stabilizes the mechanical system dynamics with uncertain gravity effects (similar to [29]).

As a final comparison note, we observe that while the structure of the nonlinear filter-based, adaptive controller of Section 4.5 resembles that of [26] in certain aspects, the following design/analysis characteristics were all instrumental in the extension of the model-based controller for SISO mechanical systems given in [26] to uncertain, MIMO mechanical systems: the use of a desired trajectory-based feedforward term, a different filter structure, the exploitation of an additional property of the mechanical systems dynamics, and a different error system development and analysis.

References

[1] S. Arimoto, V. Parra-Vega, and T. Naniwa, A Class of Linear Velocity Observers for Nonlinear Mechanical Systems, *Asian Control Conference*, pp. 633-636, Tokyo, Japan, 1994.

[2] H. Berghuis and H. Nijmeijer, Observer Design in the Tracking Control Problem for Robots, *Proceedings of the IFAC Symposium NOLCOS*, pp. 588-593, Bordeaux, France, June 1992.

[3] H. Berghuis and H. Nijmeijer, Robust Control of Robots via Linear Estimated State Feedback, *IEEE Transactions on Automatic Control*, Vol. 39, No. 10, pp. 740-754, Dec. 1993.

[4] H. Berghuis and H. Nijmeijer, A Passivity Approach to Controller-Observer Design for Robots, *IEEE Transactions on Robotics and Automation*, Vol. 9, No. 6, pp. 740-754, Dec. 1993.

[5] T. Burg, D. Dawson, J. Hu, and M. de Queiroz, An Adaptive Partial State Feedback Controller for RLED Robot Manipulators, *IEEE Transactions on Automatic Control*, Vol. 41, No. 7, pp. 1024-1031, July 1996.

[6] T. Burg, D. Dawson, and P. Vedagarbha, A Redesigned DCAL Controller without Velocity Measurements: Theory and Demonstration, *Robotica*, Vol. 15, pp. 337-346, 1997.

[7] I. Burkov, Asymptotic Stabilization of Nonlinear Lagrangian Systems without Measuring Velocities, *International Symposium on Active Control in Mechanical Engineering*, Lyon, France, 1993.

[8] I. V. Burkov, Mechanical System Stabilization via Differential Observer, *IFAC Conference on System Structure and Control*, pp. 532-535, Nantes, France, 1995.

[9] C. Canudas de Wit, K. Aström, and N. Fixot, Computed Torque Control via a Nonlinear Observer, *International Journal of Adaptive Control and Signal Processing*, Vol. 4, No. 6, pp. 443-452, 1990.

[10] C. Canudas de Wit and J. Slotine, Sliding Observers for Robot Manipulators, *Automatica,* Vol. 27, No. 5, pp. 859- 864, May 1991.

[11] C. Canudas de Wit and N. Fixot, Robot Control Via Robust Estimated State Feedback, *IEEE Transactions on Automatic Control*, Vol. 36, No. 12, pp. 1497-1501, Dec. 1991.

[12] C. Canudas de Wit, N. Fixot, and K. Aström, Trajectory Tracking in Robot Manipulators via Nonlinear Estimated State Feedback, *IEEE Transactions on Robotics and Automation*, Vol. 8, No. 1, pp. 138-144, Feb. 1992.

[13] C. Canudas de Wit and N. Fixot, Adaptive Control of Robot Manipulators via Velocity Estimated Feedback, *IEEE Transactions on Automatic Control*, Vol. 37, No. 8, pp. 1234-1237, Aug. 1992.

[14] R. Colbaugh, E. Barany, and K. Glass, Global Stabilization of Uncertain Manipulators Using Bounded Controls, *Proceedings of the American Control Conference*, pp. 86-91, Albuquerque, NM, June 1997.

[15] Integrated Motion, Inc., *Direct Drive Manipulator Research and Development Package Operations Manual*, Berkeley, CA, 1992.

[16] M. Erlic and W. Lu, Manipulator Control with an Exponentially Stable Velocity Observer, *Proceedings of the American Control Conference*, pp. 1241-1242, Chicago, IL, June 1992.

[17] M. Erlic and W. Lu, A Reduced-Order Adaptive Velocity Observer for Manipulator Control, *Proceedings of the IEEE Conference on Robotics and Automation*, pp. 328-333, Atlanta, GA, May 1993.

[18] L. Hsu and F. Lizarralde, Variable Structure Adaptive Tracking Control of Robot Manipulators without Joint Velocity Measurement, *IFAC World Congress*, pp. 145-148, Sydney, Australia, July 1993.

[19] T. Kailaith, *Linear Systems*, Englewood Cliffs, NJ: Prentice Hall, 1980.

[20] K. Kaneko and R. Horowitz, Repetitive and Adaptive Control of Robot Manipulators with Velocity Estimation, *IEEE Transactions on Robotics and Automation,* Vol. 13, No. 2, pp. 204-217, Apr. 1997.

[21] R. Kelly, A Simple Setpoint Controller by Using Only Position Measurements, *Preprint IFAC World Congress*, pp. 289-293, Sydney, Australia, July 1993.

[22] M. Krstić, I. Kanellakopoulos, and P. Kokotović, *Nonlinear and Adaptive Control Design*, New York: John Wiley & Sons, Inc., 1995.

[23] F. L. Lewis, C. T. Abdallah, and D. M. Dawson, *Control of Robot Manipulators*, New York, NY: Macmillan Publishing Co., 1993.

[24] S. Y. Lim, D. M. Dawson, and K. Anderson, Re-examining the Nicosia-Tomei Robot Observer-Controller from a Backstepping Perspective, *IEEE Transactions on Control Systems Technology*, Vol. 4, No. 3, pp. 304-310, May 1996.

[25] S. Y. Lim, *Partial State Feedback Link Position Tracking Controllers for Robotic Manipulator Systems*, Ph. D. Dissertation, Department of Electrical and Computer Engineering, Clemson University, Aug. 1994.

[26] A. Loria, Global Tracking Control of One Degree of Freedom Euler-Lagrange Systems Without Velocity Measurements, *European Journal of Control*, Vol. 2, No. 2, pp.144-151, June 1996.

[27] S. Nicosia and P. Tomei, Robot Control by Using Only Joint Position Measurements, *IEEE Transactions on Automatic Control*, Vol. 35, No. 9, pp. 1058-1061, Sept. 1990.

[28] S. Nicosia, A. Tornambe, and P. Valigi, Experimental Results in State Estimation of Industrial Robots, *Proceedings of the IEEE Conference on Decision and Control*, pp. 360-365, Honolulu, HI, Dec. 1990.

[29] R. Ortega, A. Loria, and R. Kelly, A Semiglobally Stable Output Feedback PI^2D Regulator for Robot Manipulators, *IEEE Transactions on Automatic Control*, Vol. 40, No. 8, pp. 1432-1436, Aug. 1995.

[30] Z. Qu, D. Dawson, J. Dorsey, and J. Duffie, Robust Estimation and Control of Robotic Manipulators , *Robotica*, Vol. 13, pp. 223-231, 1995.

[31] P. Vedagarbha, T. Burg, J. Hu, and D. Dawson,Development and Demonstration of a New Class of Adaptive Partial State Feedback Controllers for Electric Machines, *Mechatronics - An International Journal*, Vol. 6, No. 6, pp. 691-727, 1996.

[32] J. Yuan and Y. Stepanenko, Robust Control of Robotic Manipulators without Velocity Measurements, *International Journal of Robust and Nonlinear Control*, Vol. 1, pp. 203-213, 1991.

[33] F. Zhang, D. M. Dawson, M. S. de Queiroz, and W. Dixon, Global Adaptive Output Feedback Tracking Control of Robot Manipulators, *Proceedings of the IEEE Conference on Decision and Control*, pp. 3634-3639, San Diego, CA, Dec. 1997.

[34] W. Zhu, H. Chen, and Z. Zhang, A Variable Structure Robot Control Algorithm with an Observer, *IEEE Transactions on Robotics and Automation*, Vol. 8, No. 4, pp. 486-492, Aug. 1992.

5

Strings and Cables

5.1 Introduction

In the previous chapters, controllers were designed for mechanical systems that are modeled by nonlinear ODEs. For the remainder of this book, we will focus our attention on the development of control algorithms for mechanical systems that are assumed to be modeled by PDEs.

The vibrating string phenomenon is one of the most interesting and simple examples of a distributed parameter system because its dynamics can be described by a linear, second-order PDE. In the first part of this chapter, we will use Lyapunov-type design and analysis arguments to develop boundary control strategies for the most commonly used string model — the wave equation. We then discuss how the controller can be modified to compensate for relatively large string deflections and time-varying tension effects by using a nonlinear, distributed parameter model for control design purposes. The boundary control laws are divided into two classes: model-based controller and adaptive controller. The model-based controller provides a stepping stone towards the synthesis of the adaptive controller, which is designed to cope with constant, parametric uncertainty.

While practical application of controllers for the wave equation is somewhat limited, a thorough understanding of the control design methodology for the wave equation serves as a platform for developing controllers for more complex distributed parameter systems. As such, in the latter part of this chapter, we illustrate how the boundary control technique developed for the wave equation can be applied to mechanical systems that suffer from cable-induced vibrations. In contrast to the limited practical importance of the wave equation, cables are used in many engineering applications owing to their inherent low weight, flexibility, strength, and storability. The transverse stiffness of a cable depends on its tension and length; however, long, sagged cables can vibrate excessively in response to relatively small disturbances. This vibration degrades the performance of the cable system and ultimately leads to failure. This problem has motivated, in the past, the development of passive vibration control methods. One such method for reducing cable vibration is to increase the cable tension; however, this remedy induces high stress and eventually reduces the life of the cable. Another approach is the use of passive cable dampers, which can reduce the resonant forced response but may have little effect on the transient response. Because of its inherent cost and/or feasibility advantages, active boundary control provides an alternative, practical approach for reducing cable vibration. To this end, we will apply Lyapunov-type design and analysis arguments to develop model-based and adaptive boundary control strategies for a distributed parameter model of a cable system.

5.2 Actuator-String System

In this section, we present controllers for a boundary-actuated string model. Specifically, we first design a linear, model-based boundary controller that exponentially regulates the vibration of a small-displacement string. Next, we redesign the model-based boundary controller as an adaptive boundary controller that compensates for parametric uncertainty and asymptotically regulates the string displacement. Finally, we discuss a large-displacement extension of the linear, model-based controller and adaptive controller.

5.2.1 System Model

The out-of-plane dynamic model for the actuator-string system in Figure 5.1 is assumed to be described by a PDE of the form

$$\rho u_{tt}(x,t) - T_0 u_{xx}(x,t) = 0, \tag{5.1}$$

with the following boundary conditions:[1]

$$u(0,t) = 0 \tag{5.2}$$

[1] Given the pinned boundary condition of (5.2), we also know that $u_t(0,t) = 0$.

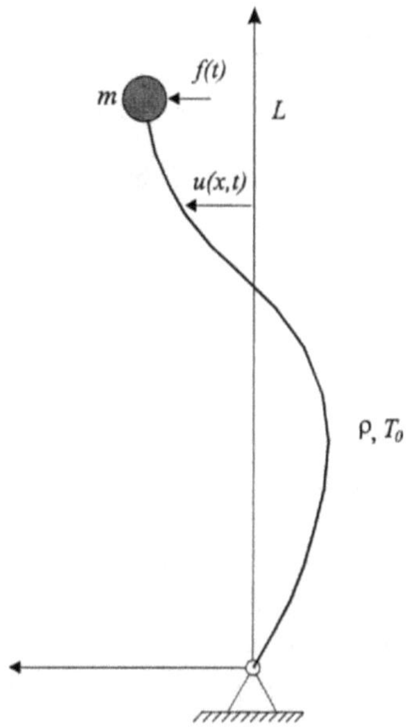

FIGURE 5.1. Schematic representation of the string-mass system.

$$mu_{tt}\left(L,t\right)+T_0u_x\left(L,t\right)=f\left(t\right), \tag{5.3}$$

where ρ denotes the mass per unit length of the string; m denotes the mass of the actuator at the free boundary; T_0 denotes the tension in the string; L denotes length of the string; $x \in [0, L]$ is the independent position variable; t is the independent time variable; $u(x,t)$ denotes the transverse displacement at position x for time t; the subscripts x, t denote the partial derivatives with respect to x, t, respectively; and $f\left(t\right)$ is the boundary control force. The string model described above is based on the following assumptions: (i) the amplitude of the string displacement $u(x,t)$ is assumed to be small, and (ii) the string is assumed to be under a constant tension T_0.

Since the control strategies will consist of relatively simple functions, we will assume the existence of a unique solution for the dynamics given by (5.1) through (5.3) under the control. In addition, we will assume that the distributed variable $u\left(x,t\right)$ and its time derivative $u_t\left(x,t\right)$ belong to a space of functions that possess the following properties.

Property 5.1: If the potential energy of the system given by (5.1) through (5.3), defined by

$$\Pi_p=\frac{1}{2}T_0\int_0^L u_\sigma^2\left(\sigma,t\right)d\sigma, \tag{5.4}$$

is bounded $\forall t \in [0, \infty)$ and $\forall x \in [0, L]$, then $\dfrac{\partial^n}{\partial x^n} u(x,t)$ is bounded for $n = 1, 2$, $\forall t \in [0, \infty)$, and $\forall x \in [0, L]$.

Property 5.2: If the kinetic energy of the system given by (5.1) through (5.3), defined by

$$\Pi_k = \frac{1}{2}\rho \int_0^L u_t^2(\sigma, t) d\sigma + \frac{1}{2} m u_t^2(L, t), \tag{5.5}$$

is bounded $\forall t \in [0, \infty)$, then $\dfrac{\partial^n}{\partial x^n} u_t(x,t)$ is bounded for $n = 0, 1$, $\forall t \in [0, \infty)$, and $\forall x \in [0, L]$.

Remark 5.1 *From a strict mathematical point of view, one might question the above boundedness properties. However, from an engineering point of view, it seems reasonable to assume for a real physical system that if the energy of the system is bounded, then all the signals that make up the governing dynamic equations will also remain bounded. We also note that the experimental results given in this chapter seem to lend credence to the fact that these engineer-based properties are indeed true. Finally, a review of previous boundary control work seems to indicate that there is not a well-established method for checking the boundedness of signals for linear distributed parameter systems much less for nonlinear distributed parameter systems.*

5.2.2 Problem Statement

The primary control objective is to design the boundary control force $f(t)$ such that the string displacement $u(x,t)$ is driven to zero $\forall x \in [0, L]$ as $t \to \infty$. To facilitate the control design and the subsequent Lyapunov-type stability analysis, we first define an auxiliary signal, denoted by $\eta(t)$, as follows:

$$\eta(t) = u_t(L, t) + u_x(L, t). \tag{5.6}$$

In order to rewrite the dynamic boundary condition of (5.3) in terms of $\eta(t)$, we first differentiate (5.6) with respect to time, multiply the resulting expression by m, and then utilize (5.3) to substitute for $mu_{tt}(L,t)$ to produce

$$m\dot{\eta}(t) = m u_{xt}(L, t) - T_0 u_x(L, t) + f(t). \tag{5.7}$$

The above equation denotes the open-loop dynamics for the auxiliary signal $\eta(t)$, and will form the basis for the design of the model-based and adaptive control laws.

5.2.3 Model-Based Control Law

Given exact knowledge of the system model, we will now develop an exponentially stabilizing controller for the system given by (5.1) through (5.3). Specifically, given the structure of the open-loop dynamics of (5.7), the boundary control force is designed as follows:

$$f(t) = -mu_{xt}(L,t) + T_0 u_x(L,t) - k_s \eta(t), \tag{5.8}$$

where k_s is a positive control gain. After substituting (5.8) into (5.7), we obtain the following closed-loop dynamics for $\eta(t)$:

$$m\dot{\eta}(t) = -k_s \eta(t). \tag{5.9}$$

Theorem 5.1 *The model-based boundary control law given by (5.8) ensures that the string displacement is exponentially regulated in the following sense:*

$$|u(x,t)| \leq \sqrt{\frac{2\lambda_2 L}{\lambda_1 T_0}\kappa_o \exp\left(-\frac{\lambda_3}{\lambda_2}t\right)} \qquad \forall x \in [0,L], \tag{5.10}$$

provided the control gain k_s is selected to satisfy the following inequality:

$$k_s > \frac{T_0}{2}, \tag{5.11}$$

where λ_1, λ_2, and λ_3 are some positive bounding constants, and the positive constant κ_o is given by

$$\kappa_o = \frac{1}{2}\rho \int_0^L u_t^2(\sigma,0)\,d\sigma + \frac{1}{2}T_0 \int_0^L u_\sigma^2(\sigma,0)\,d\sigma + \eta^2(0). \tag{5.12}$$

Proof. To prove the above result, we define the following function:

$$V(t) = E_s(t) + \frac{1}{2}m\eta^2(t) + E_c(t), \tag{5.13}$$

where

$$E_s(t) = \frac{1}{2}\rho \int_0^L u_t^2(\sigma,t)d\sigma + \frac{1}{2}T_0 \int_0^L u_\sigma^2(\sigma,t)\,d\sigma, \tag{5.14}$$

and

$$E_c(t) = 2\beta\rho \int_0^L \sigma u_t(\sigma,t)\, u_\sigma(\sigma,t)\,d\sigma, \tag{5.15}$$

with β being a positive weighting constant. While the form of the first two terms on the right-hand side of (5.13) is directly related to the total energy of the string system (see (5.4) and (5.5)), the cross term[2] $E_c(t)$, although apparently lacking a physical interpretation, is crucial in the proof of (5.10).

We now illustrate that for sufficiently small β, the function $V(t)$ in (5.13) will be a non-negative function. To this end, first note that the inequality (A.29) of Lemma A.13 in Appendix A can be used to upper bound $E_c(t)$ of (5.15) as follows:[3]

$$E_c = 2\beta\rho \int_0^L \sigma u_t u_\sigma d\sigma \leq 2\beta\rho L \int_0^L \left(u_t^2 + u_\sigma^2\right) d\sigma. \tag{5.16}$$

Also note that a lower bound can be placed on $E_s(t)$ of (5.14) as follows:

$$E_s \geq \frac{1}{2}\min\{\rho, T_0\} \int_0^L \left(u_t^2 + u_\sigma^2\right) d\sigma. \tag{5.17}$$

We can now use (5.16) and (5.17) to establish the following inequalities:

$$-\frac{4\beta\rho L}{\min\{\rho, T_0\}} E_s \leq E_c \leq \frac{4\beta\rho L}{\min\{\rho, T_0\}} E_s. \tag{5.18}$$

If β is selected according to

$$\beta < \frac{\min\{\rho, T_0\}}{4\rho L}, \tag{5.19}$$

we can utilize (5.18) to state that

$$0 \leq \xi_1 E_s \leq E_s + E_c \leq \xi_2 E_s, \tag{5.20}$$

where ξ_1, ξ_2 are some positive constants. From (5.20) and the definition of $V(t)$ given in (5.13), we can formulate the bounds on $V(t)$ as follows:

$$\lambda_1\left(E_s(t) + \eta^2(t)\right) \leq V(t) \leq \lambda_2\left(E_s(t) + \eta^2(t)\right), \tag{5.21}$$

where

$$\lambda_1 = \min\left\{1 - \frac{4\beta\rho L}{\min\{\rho, T_0\}}, \frac{m}{2}\right\} \qquad \lambda_2 = \max\left\{1 + \frac{4\beta\rho L}{\min\{\rho, T_0\}}, \frac{m}{2}\right\}, \tag{5.22}$$

---

[2]It is not clear who originally proposed the structure for the $E_c(t)$ term in (5.15). We first observed the use of this type of "cross" term in [10]; however, we have since noted that other PDE-based control work often makes use of a similar term during the stability analysis (e.g., [8, 15]).

[3]To reduce the notational complexity in most of the following derivations, the arguments x, t will be left out of all spatial/time-dependent variables (e.g., $u(x,t)$ will be denoted simply as u), while the argument t will be left out of the time-dependent variables (e.g., $u_x(L,t)$ will be denoted as $u_x(L)$).

with β satisfying (5.19). Note that as a result of (5.19), λ_1 and λ_2 will always be positive.

We will now illustrate that for sufficiently small β, the time derivative of $V(t)$ will be a nonpositive function. To this end, we differentiate (5.13) with respect to time to obtain

$$\dot{V}(t) = \dot{E}_s(t) + \dot{E}_c(t) - k_s\eta^2(t), \tag{5.23}$$

where (5.9) has been utilized. To determine $\dot{E}_s(t)$ in (5.23), we differentiate (5.14) with respect to time to obtain

$$\dot{E}_s = T_0 \int_0^L u_{\sigma\sigma} u_t d\sigma + T_0 \int_0^L u_\sigma u_{\sigma t} d\sigma, \tag{5.24}$$

where (5.1) has been utilized. Integrating by parts, the first term on the right-hand side of (5.24), we get

$$\dot{E}_s = T_0 u_x(L) u_t(L), \tag{5.25}$$

where the boundary condition given in (5.2) has been applied. Based on the definition of (5.6), we can rewrite (5.25) as follows:

$$\dot{E}_s = -\frac{T_0}{2}\left(u_t^2(L) + u_x^2(L)\right) + \frac{T_0}{2}\eta^2. \tag{5.26}$$

To obtain the expression for $\dot{E}_c(t)$, we differentiate (5.15) with respect to time, and then use (5.1) to produce

$$\dot{E}_c = A_1 + A_2, \tag{5.27}$$

where

$$A_1 = 2\beta\rho \int_0^L \sigma u_t u_{\sigma t} d\sigma \qquad A_2 = 2\beta T_0 \int_0^L \sigma u_\sigma u_{\sigma\sigma} d\sigma. \tag{5.28}$$

After integrating by parts, the right-hand side of the expression for A_1, we obtain

$$A_1 = 2\beta\rho\left(Lu_t^2(L) - \int_0^L u_t^2 d\sigma\right) - 2\beta\rho \int_0^L \sigma u_t u_{\sigma t} d\sigma, \tag{5.29}$$

where (5.2) has been utilized. After noting that the last term on the right-hand side of (5.29) is equal to A_1, we can rearrange (5.29) to obtain

$$A_1 = \beta\left(\rho L u_t^2(L) - \rho \int_0^L u_t^2 d\sigma\right). \tag{5.30}$$

We now follow the same procedure for A_2. That is, first integrating by parts, the right-hand side of the expression for A_2 yields

$$A_2 = 2\beta T_0 \left(L u_x^2 (L) - \int_0^L u_\sigma^2 d\sigma \right) - 2\beta T_0 \int_0^L \sigma u_\sigma u_{\sigma\sigma} d\sigma, \tag{5.31}$$

where (5.2) has been used. After noting that the last term on the right-hand side of (5.31) is equal to A_2, we can rearrange (5.31) to obtain

$$A_2 = \beta \left(T_0 L u_x^2 (L) - T_0 \int_0^L u_\sigma^2 d\sigma \right). \tag{5.32}$$

We can now substitute (5.30) and (5.32) into (5.27), and then substitute the resulting expression along with (5.26) into (5.23) to produce

$$\begin{aligned} \dot{V} = & -\left(\frac{T_0}{2} - \beta\rho L\right) u_t^2 (L) - \left(\frac{T_0}{2} - \beta T_0 L\right) u_x^2 (L) \\ & -\beta\rho \int_0^L u_t^2 d\sigma - \beta T_0 \int_0^L u_\sigma^2 d\sigma - \left(k_s - \frac{T_0}{2}\right)\eta^2. \end{aligned} \tag{5.33}$$

From (5.33), it is obvious that if k_s is selected to satisfy (5.11) and β is selected according to

$$\beta < \min\left\{\frac{T_0}{2\rho L}, \frac{1}{2L}\right\}, \tag{5.34}$$

then $\dot{V}(t)$ can be upper bounded by a nonpositive function as shown below:

$$\dot{V}(t) \leq -\lambda_3 \left(E_s(t) + \eta^2(t)\right), \tag{5.35}$$

where (5.17) has been used and λ_3 is a positive bounding constant defined as[4]

$$\lambda_3 = \min\left\{k_s - \frac{T_0}{2}, 2\beta\right\}. \tag{5.36}$$

From (5.21) and (5.35), we can obtain the following new upper bound for the time derivative of $V(t)$:

$$\dot{V}(t) \leq -\frac{\lambda_3}{\lambda_2} V(t). \tag{5.37}$$

After applying Lemma A.4 in Appendix A to (5.37), we have

$$V(t) \leq V(0) \exp\left(-\frac{\lambda_3}{\lambda_2} t\right) \leq \lambda_2 \left(E_s(0) + \eta^2(0)\right) \exp\left(-\frac{\lambda_3}{\lambda_2} t\right), \tag{5.38}$$

[4]Note that λ_3 is positive as a result of the condition given in (5.11).

where (5.21) has been used. After utilizing (5.14), (5.21), and (A.25) of Lemma A.12 in Appendix A, we have

$$\frac{1}{2L}T_0u^2(x,t) \leq \frac{1}{2}T_0\int_0^L u_\sigma^2(\sigma,t)\,d\sigma \leq E_s(t) \leq \frac{1}{\lambda_1}V(t) \qquad \forall x\in[0,L]. \tag{5.39}$$

The result given by (5.10) and (5.12) now directly follows by combining (5.38), (5.14), and (5.39). □

Remark 5.2 *Note that the proof of Theorem 5.1 relies on the fact that the weighting constant β has been selected sufficiently small such that both conditions of (5.19) and (5.34) are satisfied. Thus, the overall condition on β can be expressed as follows:*

$$\beta < \min\left\{\frac{\min\{\rho,T_0\}}{4\rho L},\frac{T_0}{2\rho L},\frac{1}{2L}\right\}. \tag{5.40}$$

Remark 5.3 *From (5.21) and (5.35), we can state that $E_s(t)$ and $\eta(t)$ are bounded $\forall t\in[0,\infty)$. Since $E_s(t)$ is bounded $\forall t\in[0,\infty)$, we can use (5.14) and (A.25) of Lemma A.12 in Appendix A to show that $u(x,t)$ is bounded $\forall t\in[0,\infty)$ and $\forall x\in[0,L]$. Since $E_s(t)$ of (5.14) is bounded $\forall t\in[0,\infty)$, the potential energy given by (5.4) is bounded $\forall t\in[0,\infty)$; hence, we can use Property 5.1 to show that $\frac{\partial^n}{\partial x^n}u(x,t)$ is bounded for $n=1,\,2$, $\forall t\in[0,\infty)$ and $\forall x\in[0,L]$. Since $\eta(t)$ and $u_x(L,t)$ are bounded $\forall t\in[0,\infty)$, we can use (5.6) to state that $u_t(L,t)$ is bounded $\forall t\in[0,\infty)$. From the boundedness of $E_s(t)$ and $u_t(L,t)$, we can see that the kinetic energy of the system defined in (5.5) is bounded $\forall t\in[0,\infty)$. Since the kinetic energy is bounded $\forall t\in[0,\infty)$, we can use Property 5.2 to conclude that $\frac{\partial^n}{\partial x^n}u_t(x,t)$ is bounded for $n=0,\,1$, $\forall t\in[0,\infty)$ and $\forall x\in[0,L]$. From the above information, we can now state that all of the signals in the control law of (5.8) are bounded $\forall t\in[0,\infty)$. Finally, we can use (5.1), (5.3), and the above boundedness statements to show that $u_{tt}(x,t)$ is bounded $\forall t\in[0,\infty)$ and $\forall x\in[0,L]$.*

5.2.4 Adaptive Control Law

The control law given in (5.8) requires exact knowledge of some of the system parameters. We will now illustrate how the controller of (5.8) can be redesigned to compensate for constant parametric uncertainty while asymptotically stabilizing the string displacement. After taking note of the fact that the open-loop dynamics for $\eta(t)$ given by (5.7) can be linearly parameterized, we rewrite (5.7) in the following advantageous form:

$$m\dot{\eta}(t) = Y\left(u_{xt}(L,t),u_x(L,t)\right)\theta + f(t), \tag{5.41}$$

where $Y(\cdot) \in \mathbb{R}^{1\times 2}$ is a known regression matrix, and $\theta \in \mathbb{R}^2$ is an unknown, constant parameter vector defined as follows:

$$Y(\cdot) = \begin{bmatrix} u_{xt}(L,t) & -u_x(L,t) \end{bmatrix} \qquad \theta = \begin{bmatrix} m & T_0 \end{bmatrix}^T. \tag{5.42}$$

Based on the form of (5.41), the boundary control force can now be designed as

$$f(t) = -Y(\cdot)\widehat{\theta}(t) - k_s\eta(t), \tag{5.43}$$

where $\eta(t)$ was defined in (5.6), and $\widehat{\theta}(t) \in \mathbb{R}^2$ is a dynamic, parameter estimate vector defined as follows:

$$\widehat{\theta}(t) = \begin{bmatrix} \widehat{m}(t) & \widehat{T_0}(t) \end{bmatrix}^T. \tag{5.44}$$

Based on the subsequent Lyapunov-type stability analysis, the parameter estimate vector is updated according to the following gradient update law:

$$\dot{\widehat{\theta}}(t) = \Gamma Y^T(\cdot)\eta(t), \tag{5.45}$$

where $\Gamma \in \mathbb{R}^{2\times 2}$ is a diagonal, positive-definite, adaptation gain matrix. If we define $\widetilde{\theta}(t) = \theta - \widehat{\theta}(t) \in \mathbb{R}^2$ as the parameter estimation error vector and then substitute (5.43) into (5.41), we obtain the following closed-loop dynamics:

$$m\dot{\eta}(t) = -k_s\eta(t) + Y(\cdot)\widetilde{\theta}(t) \qquad \dot{\widetilde{\theta}}(t) = -\Gamma Y^T(\cdot)\eta(t), \tag{5.46}$$

where (5.45) was used to obtain the parameter estimation error dynamics.

Theorem 5.2 *The adaptive boundary control law given by (5.43) and (5.45) ensures that the string displacement is asymptotically regulated in the following sense:*

$$\lim_{t\to\infty} |u(x,t)| = 0 \quad \forall x \in [0,L], \tag{5.47}$$

where the control gain k_s defined in (5.43) must be selected to satisfy (5.11).

Proof. The following proof is based on arguments similar to those used in the proof of Theorem 5.1; hence, some of the details will not be repeated. First, we define the following function:

$$V_a(t) = V(t) + \frac{1}{2}\widetilde{\theta}^T(t)\Gamma\widetilde{\theta}(t), \tag{5.48}$$

where $V(t)$ was defined in (5.13). If the constant β of (5.15) is selected sufficiently small according to the condition given in (5.19), we can formulate the following bounds on $V_a(t)$:

$$\lambda_{1a}\left(E_s(t) + \eta^2(t) + \left\|\widetilde{\theta}(t)\right\|^2\right) \le V_a(t) \le \lambda_{2a}\left(E_s(t) + \eta^2(t) + \left\|\widetilde{\theta}(t)\right\|^2\right) \tag{5.49}$$

for some positive constants λ_{1a} and λ_{2a}. After differentiating (5.48) with respect to time and substituting (5.1), (5.2), and (5.46), we can follow the derivations used in Theorem 5.1 to obtain the following upper bound for the time derivative of $V_a(t)$:

$$\dot{V}_a(t) \leq -\lambda_3 \left(E_s(t) + \eta^2(t)\right) + \tilde{\theta}^T(t) \left(Y^T(\cdot)\eta(t) + \Gamma^{-1}\dot{\tilde{\theta}}(t)\right), \quad (5.50)$$

where λ_3 was defined in (5.36) (the control gain k_s defined in (5.43) must be selected according to (5.11) to ensure that λ_3 is positive). After substituting the second equation of (5.46) into (5.50), we can simplify the upper bound on $\dot{V}_a(t)$ to

$$\dot{V}_a(t) \leq -\lambda_3 \left(E_s(t) + \eta^2(t)\right) \triangleq -g_a(t). \quad (5.51)$$

We can now use (5.49), (5.51), and arguments similar to those outlined in Remark 5.3 to state that all signals in the control law of (5.43) and (5.45) are bounded $\forall t \in [0, \infty)$, and that all signals in the system given by (5.1) through (5.3) remain bounded $\forall t \in [0, \infty)$ and $\forall x \in [0, L]$.

After differentiating $g_a(t)$ defined in (5.51) with respect to time, we have

$$\dot{g}_a(t) = \lambda_3 \left(\dot{E}_s(t) + 2\eta(t)\dot{\eta}(t)\right). \quad (5.52)$$

Since we already know that all the system signals remain bounded $\forall t \in [0, \infty)$, we can use (5.25) and (5.46) to show that $\dot{E}_s(t)$ and $\dot{\eta}(t)$ are bounded $\forall t \in [0, \infty)$. From (5.52), it is easy to see that $\dot{g}_a(t)$ is bounded $\forall t \in [0, \infty)$; hence, we can now apply Lemma A.6 in Appendix A to (5.51) to show that

$$\lim_{t \to \infty} E_s(t), \eta(t) = 0. \quad (5.53)$$

Finally, the result given by (5.53) and the inequalities developed in (5.39) can be combined to state the result given by (5.47). □

5.2.5 Extensions

The actuator-string system described by the linear field equation of (5.1) and linear boundary conditions of (5.2) and (5.3) have been formulated for small displacements $u(x,t)$. We now discuss how the basic control laws given in previous sections can be redesigned to account for relatively large displacements of the string.

A nonlinear PDE and nonlinear boundary conditions for the actuator-string system depicted in Figure 5.1 can be formulated as follows [12]:

$$\rho u_{tt}(x,t) = T(y(t)) \left[\frac{u_x(x,t)}{\sqrt{1 + u_x^2(x,t)}}\right]_x \qquad \forall x \in [0, L] \quad (5.54)$$

$$mu_{tt}(L,t) + T(y(t))\frac{u_x(L,t)}{\sqrt{1+u_x^2(x,t)}} + W(u_t(L,t))\phi = f(t), \tag{5.55}$$

where the term $W(u_t(L,t))\phi$ is a linear parameterization representing additional actuator dynamics (e.g., friction) with $W(\cdot) \in \mathbb{R}^{1\times p}$ being a regression matrix and $\phi \in \mathbb{R}^p$ being a constant parameter vector. The tension function $T(y(t))$ introduced in (5.54) and (5.55) is assumed to have the following properties: (i) $T(y(t))$ is a strictly positive function that satisfies

$$T(y(t)) \geq T_0 > 0, \tag{5.56}$$

where $y(t)$ is the string stretch represented by the following non-negative function:

$$y(t) = \int_0^L \left(-1 + \sqrt{1 + (u_\sigma(\sigma,t))^2}\right) d\sigma, \tag{5.57}$$

(ii) if $y(t)$ is bounded, then $T(y(t))$ is also bounded, and (iii) the potential energy stored in the string can be bounded as follows:

$$\alpha_l y(t) \leq \Pi_p = \int_0^y T(y(t))\,dy \leq \kappa_u(y(t)), \tag{5.58}$$

where α_l is a positive constant, and $\kappa_u(\cdot)$ is some continuous class K function.[5]

Based on a more involved stability analysis than that used in Section 5.2.3 (see [18] for details), we can redefine the model-based controller originally given in (5.8) as follows:

$$\begin{aligned} f(t) = & -\frac{mu_{xt}(L,t)}{\left(1+(u_x(L,t))^2\right)^{3/2}} + T(y(t))\frac{u_x(L,t)}{\sqrt{1+(u_x(L,t))^2}} \\ & + W(u_t(L,t))\phi - \left(k + \frac{k_r}{2}T(y(t))\right)\eta(t), \end{aligned} \tag{5.59}$$

where k, k_r are positive control gains, and the variable $\eta(t)$, originally defined in (5.6), is now defined as follows:

$$\eta(t) = u_t(L,t) + \frac{u_x(L,t)}{\sqrt{1+(u_x(L,t))^2}}. \tag{5.60}$$

As illustrated in [18], the controller given by (5.59) will asymptotically regulate the total energy of the system in contrast to the exponential result given in Theorem 5.1.

[5] A continuous function $\kappa(p)$ is said to be class K if: (i) $\kappa(0) = 0$, (ii) $\kappa(p) > 0\ \forall p > 0$, and (iii) $\kappa(p)$ is nondecreasing.

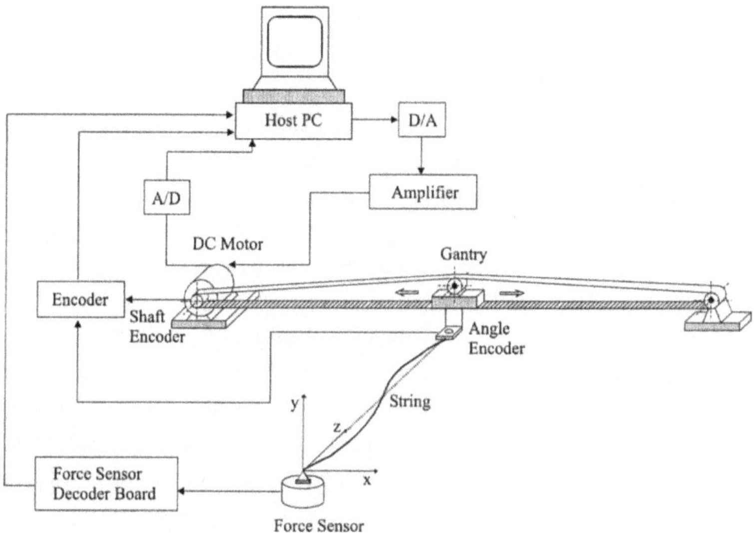

FIGURE 5.2. Schematic diagram of the experimental setup.

By following the adaptive control derivations of Section 5.2.4, we can redesign the control law given by (5.59) as an adaptive controller that compensates for parametric uncertainty and asymptotically stabilizes the total energy of the system. For this new adaptive controller, the regression matrix $Y(u_{xt}(L,t), u_t(L,t)) \in \mathbb{R}^{1\times(1+p)}$ and the unknown parameter vector $\theta \in \mathbb{R}^{1+p}$, originally defined in (5.42), are now defined as follows:

$$Y(\cdot) = \begin{bmatrix} \dfrac{u_{xt}(L,t)}{\left(1+(u_x(L,t))^2\right)^{3/2}} & -W(u_t(L,t)) \end{bmatrix} \qquad \theta = \begin{bmatrix} m & \phi^T \end{bmatrix}^T. \tag{5.61}$$

Based on (5.59), the adaptive control force and parameter update law are designed as shown below:

$$f(t) = -Y(\cdot)\hat{\theta} + T(y(t))\frac{u_x(L,t)}{\sqrt{1+(u_x(L,t))^2}} - \left(k + \frac{k_r}{2}T(y(t))\right)\eta(t) \tag{5.62}$$

$$\dot{\hat{\theta}}(t) = \Gamma_e Y^T(\cdot)\eta(t), \tag{5.63}$$

where $\hat{\theta}(t) \in \mathbb{R}^{1+p}$ is a dynamic estimate of the parameter vector θ and $\Gamma_e \in \mathbb{R}^{(1+p)\times(1+p)}$ is a diagonal, positive-definite, adaptation gain matrix.

5.2.6 Experimental Evaluation

Experimental Setup

Figure 5.2 shows a schematic diagram of the experimental setup used to implement the controllers. The setup consists of a string pinned at one end and attached to a linearly translating gantry at the other end. A brushed DC motor (Baldor model 3300) drives the gantry via a belt-pulley transmission. The gantry rides on two parallel 1-inch-diameter steel rods with linear bearings. The displacement of the gantry $u\,(L,t)$ is obtained from a 1,000-count rotary encoder attached to the motor shaft. The static and dynamic measurement of the string tension $T\,(t)$ is obtained from a force sensor attached to the pinned-end of the string. A hollow-shaft 1,000-count rotatory encoder is mounted on the gantry to measure the string deflection angle $u_x\,(L,t)$ at the free-end. A mounting bracket attached to the gantry ensured that the string was held tangent to the rotary encoder.

A Pentium 166 MHz PC running QNX hosted the control algorithm while *Qmotor* provided the environment to write the control algorithm in the C programming language. The Quanser *MultiQ* I/O board [14] provided for data transfer between the computer subsystem and the electrical interface. All the controllers were implemented using a sampling period of 0.5 msec. The velocity of the gantry and the time derivative of the string deflection angle were obtained by using a backwards difference algorithm applied to the gantry position and the string deflection angle, respectively. To eliminate quantization noise, the velocity signals were filtered using a second-order digital filter. The parameter values for the mechanical system were determined via standard test procedures to be

$$m = 3.5 \text{ kg}, \qquad \rho = 0.03 \text{ kg/m}, \qquad L = 1.3 \text{ m}.$$

Experimental Results

Five experiments were conducted to assess the performance of the aforementioned controllers. The transient response to a consistent initial displacement was studied with all control gains tuned to provide the best response. Figure 5.3 shows the string response to the initial displacement without control (i.e., $f(t) = 0$). The response decays under natural damping in approximately 10 sec. In Figure 5.4, a simple *damper* control law given by

$$f\,(t) = -k_d u_t\,(L,t) \tag{5.64}$$

achieved the best regulation results with $k_d = 4.25$, and decreased the settling time to approximately 7 sec.

The linear, small-displacement, model-based controller of (5.8) was implemented with $T_0 = 29.75$ N and $k_s = 47.23$, and its performance is shown in Figure 5.5. Note that the inclusion of string slope feedback in the control law reduced the settling time to 6 sec. The fourth experiment used the nonlinear, large-displacement, model-based control of (5.59) with

$$W(\cdot) = \left[\begin{array}{cc} sgn(u_t(L,t)) & u_t(L,t) \end{array} \right] \qquad \phi = \left[\begin{array}{cc} F_s & F_d \end{array} \right]^T, \tag{5.65}$$

where $sgn(\cdot)$ is the standard signum function, and F_s and F_d are the static and dynamic friction coefficients, respectively. The friction coefficients were experimentally determined to be $F_s = 1.05$ N and $F_d = 15.23$ N/m-sec. The best regulation results were achieved with $k = 3.0$ and $k_r = 3.12$. Figure 5.6 shows the angular deflection $u_x(L,t)$, the gantry displacement $u(L,t)$, the time-varying tension signal $T(t)$, and the control voltage resulting from the initial displacement. The nonlinear control response is slightly improved compared to the linear control. Finally, the adaptive controller defined by (5.62) and (5.63) was implemented (see Figure 5.7). The parameter estimates were initialized to 25% of their nominal values and the best regulation was achieved with $k = 3.0$, $k_r = 2.97$, and $\Gamma = diag\{2.55, 50, 0.69\}$. Again, the response is similar to that of the previous two controllers.

The following table compares the experimental performance of the controllers by computing the root mean square (RMS) of the string deflection angle over a 10 sec time interval. The trends in the RMS data mirror the settling time performance, showing that the adaptive controller provides a 12% improvement in performance with similar control effort as compared to the damper controller.

Controller Type	RMS $u_x(L,t)$ deg	RMS control voltage V
Open-loop system	2.75	–
Damper controller	1.98	4.91
Linear controller	1.94	6.25
Nonlinear controller	1.70	7.30
Nonlinear adaptive controller	1.77	5.32

Remark 5.4 *The control force at the actuator was actually applied by rescaling this force as a desired motor torque. The desired motor torque was then achieved by using a high-gain current feedback loop (for the brushed DC motor $\tau(t) \propto I(t)$ where $\tau(t)$ is the motor torque and $I(t)$ is the motor current). The current feedback loop and the torque-force conversion were incorporated into the control software and power was supplied to the motor by the single-channel, linear power amplifier is capable of outputting*

up to 1000 W at 100 V with a power bandwidth of 0 to 40 kHz. It should be noted that the backstepping paradigm [9] allows the electrical dynamics to be incorporated in the overall control solution for this class of hybrid partial/ordinary differential equation problems as illustrated in Baicu et al. [2].

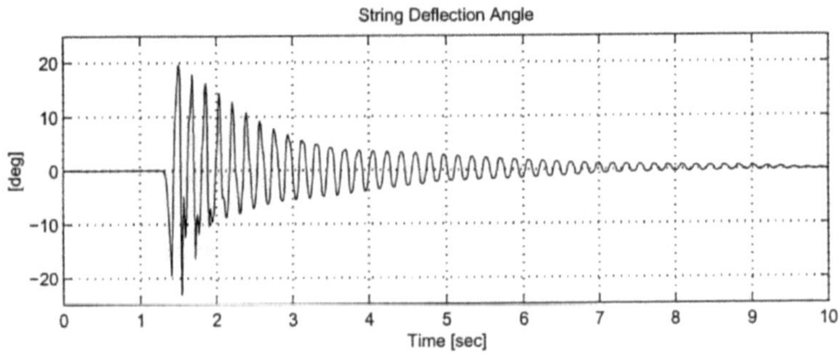

FIGURE 5.3. Open-loop response: $u_x(L,t)$.

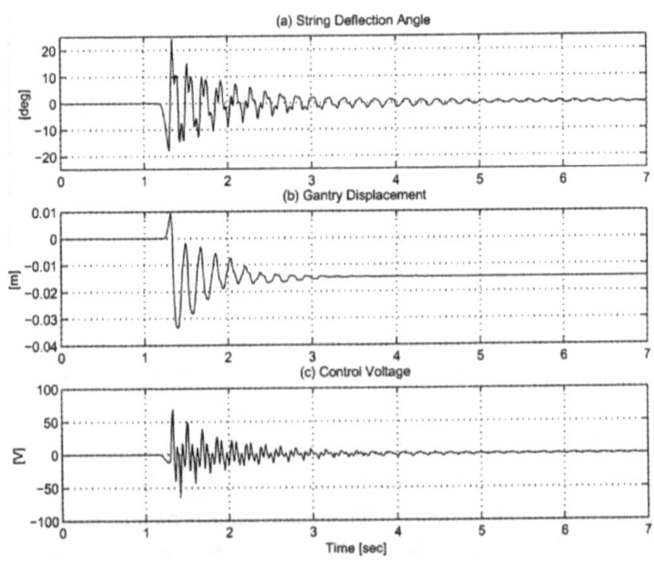

FIGURE 5.4. Damper controller: (a) $u_x(L,t)$, (b) $u(L,t)$, and (c) control voltage.

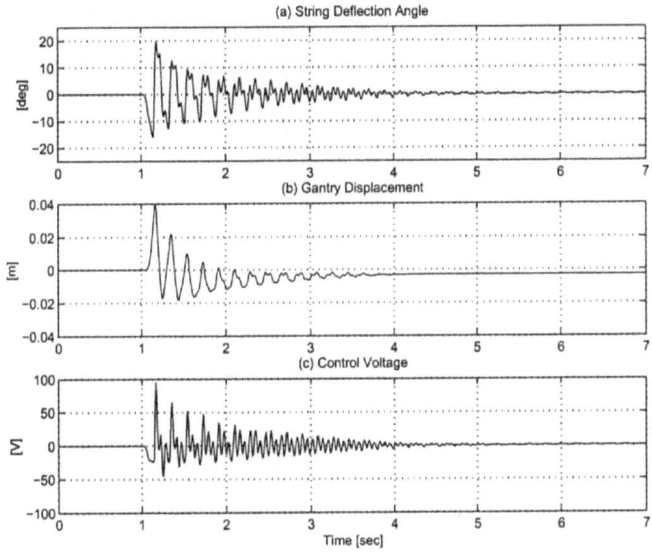

FIGURE 5.5. Linear, model-based controller: (a) $u_x(L,t)$, (b) $u(L,t)$, and (c) control voltage.

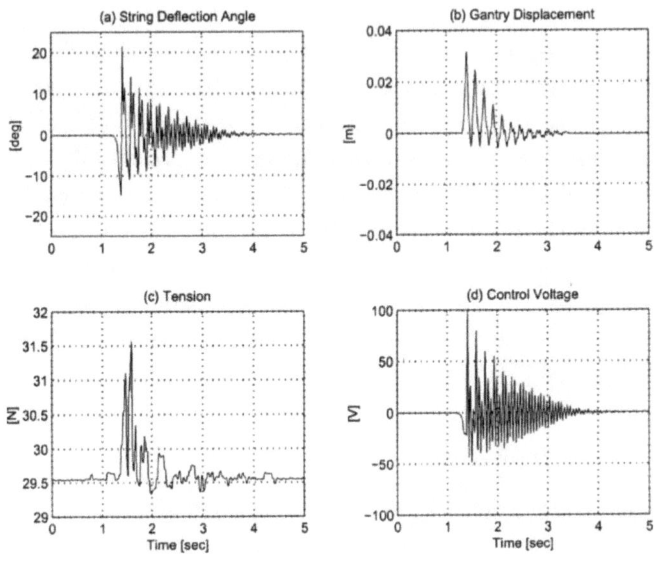

FIGURE 5.6. Nonlinear, model-based controller: (a) $u_x(L,t)$, (b) $u(L,t)$, (c) $T(t)$, and (d) control voltage.

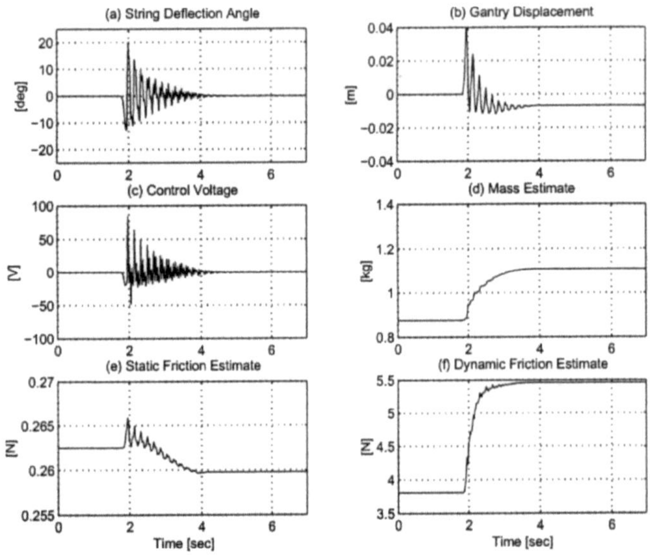

FIGURE 5.7. Nonlinear, adaptive controller: (a) $u_x(L,t)$, (b) $u(L,t)$, (c) control voltage, (d) $\hat{m}(t)$, (e) $\hat{F}_s(t)$, and (f) $\hat{F}_d(t)$.

5.3 Cable System

In this section, we develop control strategies for a distributed cable model. The control design and analysis described in the following will tread on the steps of those developed for the string system of Section 5.2. Specifically, we develop a model-based controller that exponentially stabilizes the displacement of the cable given exact knowledge of the mechanical system parameters and measurements of the slope, slope-rate, and velocity at the cable's actuated boundary. We then illustrate how the model-based controller can be redesigned as an adaptive controller, which asymptotically stabilizes the cable displacement, while compensating for parametric uncertainty.

5.3.1 System Model

The profile view of Figure 5.8 shows a cable sagging under gravity loading. The cable vibrates perpendicular to the plane formed by the equilibrium cable configuration. The top view in Figure 5.8 shows the out-of-plane displacement $U(S,T)$, where S is the arc length coordinate and T is time. The left boundary is pinned and the right boundary is free to translate in

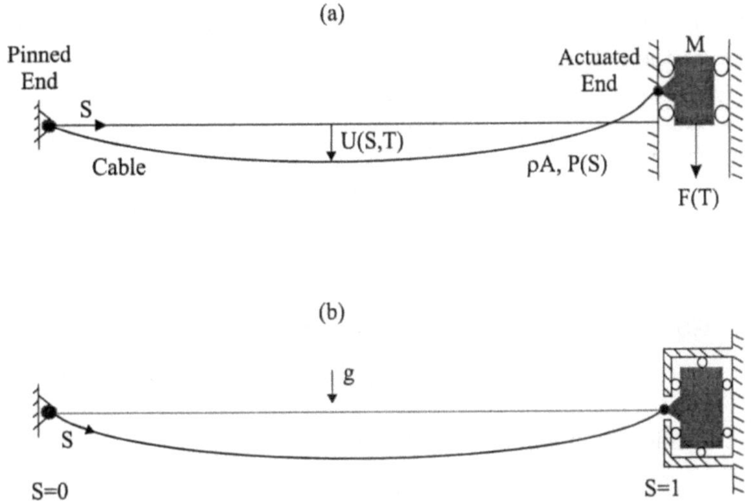

FIGURE 5.8. Schematic diagram of the cable system: (a) top view and (b) side view.

the out-of-plane direction. A control force $F(T)$ is applied in the out-of-plane direction at the left boundary. The linearized out-of-plane equation of motion for this cable system is given by [1]

$$\rho A U_{TT}(S,T) - [P(S)U_S(S,T)]_S = 0, \tag{5.66}$$

where ρA is the cable mass/length, and the subscripts S and T indicate partial differentiations with respect to the independent spatial and time variables S and T, respectively. The equilibrium tension $P(S)$ is defined as follows:

$$P(S) = \sqrt{P_0^2 + \left[\rho A g\left(S - \frac{L}{2}\right)\right]^2}, \tag{5.67}$$

where P_0 is the midspan ($S = L/2$) or horizontal component of the cable tension, g is the acceleration owing to gravity, and L is the cable length. The pinned boundary condition at $S = 0$ is given by

$$U(0,T) = 0, \tag{5.68}$$

and the dynamic boundary condition at $S = L$ is given by

$$MU_{TT}(S,T) + P(S)U_S(S,T) = F(T), \tag{5.69}$$

where M is the actuator mass. To simplify the control development, we substitute the following nondimensional variables:

$$x = \frac{S}{L}, \qquad t = T\sqrt{\frac{g}{L}}, \qquad u = \frac{U}{L}, \qquad m = \frac{M}{\rho AL}, \qquad f = \frac{F}{\rho AgL},$$

$$p(x) = \frac{P(S)}{\rho AgL} = \sqrt{p_0^2 + \left(x - \frac{1}{2}\right)^2}, \qquad p_0 = \frac{P_0}{\rho AgL} \tag{5.70}$$

into (5.66)–(5.69) to yield the following field equation:

$$u_{tt}(x,t) - [p(x)u_x(x,t)]_x = 0 \qquad \forall x \in [0,1], \tag{5.71}$$

with the following boundary conditions:[6]

$$u(0,t) = 0 \tag{5.72}$$

$$mu_{tt}(1,t) + p(1)u_x(1,t) + Y(u_t(1,t))\phi = f(t), \tag{5.73}$$

where the term $Y(u_t(1,t))\phi$ has been added to account for friction and/or damping in the actuator. We will assume that the actuator friction/damping can be linearly parameterized by a known regression matrix $Y(u_t(1,t)) \in \mathbb{R}^{1\times q}$ multiplied by a constant parameter vector $\phi \in \mathbb{R}^q$. For example, an actuator with viscous and Coulomb friction can be parameterized as follows:

$$Y(u_t(1,t)) = \begin{bmatrix} u_t(1,t) & sgn(u_t(1,t)) \end{bmatrix} \qquad \phi = \begin{bmatrix} b & \mu \end{bmatrix}^T, \tag{5.74}$$

where b and μ denote the viscous and Coulomb friction coefficients, respectively. For a general parametrization, we will assume that $Y(u_t(1,t))$ remains bounded $\forall t \in [0,\infty)$ if $u_t(1,t)$ is bounded $\forall t \in [0,\infty)$.

Since the control strategies will consist of relatively simple functions, we will assume the existence of a unique solution for the dynamics given by (5.71) through (5.73) under the control. In addition, based on the arguments stated in Remark 5.1, we will assume that the distributed variable $u(x,t)$ and its time derivative $u_t(x,t)$ belong to a space of functions which possess the following properties.

Property 5.3: If the potential energy of the system given by (5.71) through (5.73), defined by

$$\Pi_p = \frac{1}{2}\int_0^1 p(\sigma)\, u_\sigma^2(\sigma,t)\, d\sigma, \tag{5.75}$$

[6] The pinned boundary condition of (5.72) implies that $u_t(0,t) = 0$.

is bounded $\forall t \in [0,\infty)$, then $\frac{\partial^n}{\partial x^n}u(x,t)$ is bounded for $n = 1, 2$, $\forall t \in [0,\infty)$, and $\forall x \in [0,1]$.

Property 5.4: If the kinetic energy of the system given by (5.71) through (5.73), defined by

$$\Pi_k = \frac{1}{2}\int_0^1 u_t^2(\sigma,t)d\sigma + \frac{1}{2}mu_t^2(1,t), \tag{5.76}$$

is bounded $\forall t \in [0,\infty)$, then $\frac{\partial^n}{\partial x^n}u_t(x,t)$ is bounded for $n = 0, 1$, $\forall t \in [0,\infty)$, and $\forall x \in [0,1]$.

5.3.2 Problem Statement

As in Section 5.2, the primary control objective is to design the boundary control force $f(t)$ such that the cable displacement $u(x,t)$ is driven to zero $\forall x \in [0,1]$ as $t \to \infty$. To facilitate the control design and the subsequent stability analysis, we define an auxiliary signal, denoted by $\eta(t)$, as follows:

$$\eta(t) = u_t(1,t) + u_x(1,t). \tag{5.77}$$

In order to rewrite the dynamic boundary condition of (5.73) in terms of $\eta(t)$, we first differentiate (5.77) with respect to time, multiply the resulting expression by m, and then utilize (5.73) to substitute for $mu_{tt}(1,t)$ to produce

$$m\dot{\eta}(t) = mu_{xt}(1,t) - p(1)u_x(1,t) - Y(u_t(1,t))\phi + f(t). \tag{5.78}$$

The above equation denotes the open-loop dynamics of the auxiliary signal $\eta(t)$ and will form the basis for the design of the model-based and adaptive control laws.

5.3.3 Model-Based Control Law

Given exact knowledge of the system model given by (5.71) through (5.73) and the structure of the open-loop dynamics given by (5.78), the model-based control law is defined as follows:

$$f(t) = -mu_{xt}(1,t) + p(1)u_x(1,t) - k_s(u_t(1,t) + u_x(1,t)) + Y(u_t(1,t))\phi, \tag{5.79}$$

where k_s is a positive control gain. After substituting (5.79) into (5.78), we obtain the following closed-loop dynamics on the boundary:

$$m\dot{\eta}(t) = -k_s\eta(t). \tag{5.80}$$

Theorem 5.3 *The model-based boundary controller given by (5.79) ensures that cable displacement is exponentially regulated in the following sense:*

$$|u(x,t)| \leq \sqrt{\frac{2\lambda_2}{\lambda_1 p_0}\kappa_o \exp\left(-\frac{\lambda_3}{\lambda_2}t\right)} \qquad \forall x \in [0,1], \tag{5.81}$$

where λ_1, λ_2, and λ_3 are some positive bounding constants, the positive constant κ_o is given by

$$\kappa_o = \frac{1}{2}\int_0^1 u_t^2(\sigma,0)\,d\sigma + \frac{1}{2}\int_0^1 p(\sigma)\,u_\sigma^2(\sigma,0)\,d\sigma + \eta^2(0), \tag{5.82}$$

and the control gain k_s defined in (5.79) must be selected to satisfy the following inequality:

$$k_s > \frac{p(1)}{2}. \tag{5.83}$$

Proof. To prove the result of (5.81), we begin with the following function:

$$V(t) = E_s(t) + E_c(t) + \frac{1}{2}m\eta^2(t), \tag{5.84}$$

where $E_s(t)$ and $E_c(t)$ are defined as

$$E_s(t) = \frac{1}{2}\int_0^1 u_t^2(\sigma,t)d\sigma + \frac{1}{2}\int_0^1 p(\sigma)u_\sigma^2(\sigma,t)d\sigma \tag{5.85}$$

$$E_c(t) = 2\beta\int_0^1 g(\sigma)u_t(\sigma,t)u_\sigma(\sigma,t)d\sigma, \tag{5.86}$$

with β being a positive weighting constant and the weighting function $g(x)$ being defined as follows:

$$\begin{aligned} g(x) = \; & \frac{h(x)}{2p_0}\left[\tan^{-1}\left(\frac{2x-1}{2p_0}\right) + p_0\ln\left(2x-1+h(x)\right)\right. \\ & \left. + \tan^{-1}\left(\frac{1}{2p_0}\right) - p_0\ln\left(\sqrt{4p_0^2+1}-1\right)\right], \end{aligned} \tag{5.87}$$

where the auxiliary function $h(x)$ is given by

$$h(x) = \sqrt{4p_0^2 + 4x^2 - 4x + 1}.$$

First, note that the inequality (A.29) of Lemma A.13 in Appendix A can be used to upper bound $E_c(t)$ of (5.86) as follows:

$$
\begin{aligned}
E_c = 2\beta \int_0^1 g(\sigma) u_t u_\sigma d\sigma &\leq 2\beta g(1) \int_0^1 \left(u_t^2 + u_\sigma^2\right) d\sigma \\
&\leq \frac{4\beta g(1)}{\min(1, p_0)} \left[\frac{1}{2} \int_0^1 \left(u_t^2 + p(\sigma) u_\sigma^2\right) d\sigma\right],
\end{aligned}
\tag{5.88}
$$

where we have used the facts that $p(x)$ attains its minimum value at $x = 1/2$ for $x \in [0,1]$ and $g(x)$ of (5.87) attains its maximum value at $x = 1$ for $x \in [0,1]$. We can now use (5.85) and (5.88) to establish the following bounds on E_c:

$$
-\frac{4\beta g(1)}{\min(1, p_0)} E_s \leq E_c \leq \frac{4\beta g(1)}{\min(1, p_0)} E_s; \tag{5.89}
$$

hence, if β is selected according to the following sufficient condition:

$$
\beta < \frac{\min\{1, p_0\}}{4g(1)}, \tag{5.90}
$$

we have

$$
0 \leq \xi_1 E_s \leq E_s + E_c \leq \xi_2 E_s, \tag{5.91}
$$

where ξ_1 and ξ_2 are some positive bounding constants. Given the structure of $V(t)$ defined in (5.84) and the inequality given by (5.91), if the design constant β is selected according to (5.90), then we can formulate the following bounds on $V(t)$:

$$
\lambda_1 \left(E_s(t) + \eta^2(t)\right) \leq V(t) \leq \lambda_2 \left(E_s(t) + \eta^2(t)\right), \tag{5.92}
$$

where the positive constants λ_1 and λ_2 are given by

$$
\lambda_1 = \min\left\{1 - \frac{4\beta g(1)}{\min\{1, p_0\}}, \frac{m}{2}\right\} \qquad \lambda_2 = \max\left\{1 + \frac{4\beta g(1)}{\min\{1, p_0\}}, \frac{m}{2}\right\}. \tag{5.93}
$$

After differentiating (5.84) with respect to time, we have

$$
\dot{V}(t) = \dot{E}_s(t) + \dot{E}_c(t) - k_s \eta^2(t), \tag{5.94}
$$

where (5.80) has been used. To determine $\dot{E}_s(t)$ in (5.94), we differentiate (5.85) with respect to time to obtain

$$
\dot{E}_s = \int_0^1 u_t \left[p(\sigma) u_\sigma\right]_\sigma d\sigma + \int_0^1 p(\sigma) u_\sigma u_{\sigma t} d\sigma, \tag{5.95}
$$

where (5.71) has been used. If we integrate by parts, the first term on the right-hand side of (5.95), we obtain

$$\dot{E}_s = p(1)u_t(1,t)u_x(1,t) - p(0)u_t(0,t)u_x(0,t). \tag{5.96}$$

After applying the boundary condition given in (5.72) to (5.96), we have

$$\dot{E}_s = p(1)u_t(1,t)u_x(1,t), \tag{5.97}$$

which can be rewritten as

$$\dot{E}_s = -\frac{p(1)}{2}(u_t^2(1,t) + u_x^2(1,t)) + \frac{p(1)}{2}\eta^2 \tag{5.98}$$

upon application of (5.77).

To determine $\dot{E}_c(t)$ in (5.94), we differentiate (5.86) with respect to time to obtain

$$\dot{E}_c = A_1 + A_2, \tag{5.99}$$

where

$$A_1 = 2\beta \int_0^1 g(\sigma) u_t u_{\sigma t} d\sigma \qquad A_2 = 2\beta \int_0^1 g(\sigma) u_\sigma \left[p(\sigma)u_\sigma\right]_\sigma d\sigma, \tag{5.100}$$

and (5.71) has been used in the expression for A_2. After integrating by parts, the expression for A_1, we obtain

$$A_1 = 2\beta\left(g(1)u_t^2(1,t) - \int_0^1 g_\sigma(\sigma)u_t^2 d\sigma\right) - 2\beta \int_0^1 g(\sigma)u_{\sigma t}u_t d\sigma, \tag{5.101}$$

where we have used the fact from (5.87) that $g(0) = 0$. After noting that the last term in (5.101) is equal to A_1, we can write

$$A_1 = \beta\left(g(1)u_t^2(1,t) - \int_0^1 g_\sigma(\sigma)u_t^2 d\sigma\right), \tag{5.102}$$

where $g_x(x)$ can be calculated from (5.87) as follows:

$$\begin{aligned} g_x(x) = {} & \frac{8x-4}{4p_0 h(x)}\left[\tan^{-1}\left(\frac{2x-1}{2p_0}\right) + p_0 \ln\left(2x - 1 + h(x)\right)\right. \\ & \left. + \tan^{-1}\left(\frac{1}{2p_0}\right) - p_0 \ln\left(\sqrt{4p_0^2+1} - 1\right)\right] \\ & + \frac{h(x)}{2p_0}\left(\frac{1}{p_0\left(1 + \frac{(2x-1)^2}{4p_0^2}\right)} + \frac{p_0\left(2 + \frac{8x-4}{2h(x)}\right)}{2x - 1 + h(x)}\right). \end{aligned} \tag{5.103}$$

It is not difficult to show from (5.103) that $g_x(x) \geq 1$ for $x \in [0,1]$; hence, we can utilize (5.102) to construct the following upper bound for A_1:

$$A_1 \leq \beta \left(g(1)u_t^2(1,t) - \int_0^1 u_t^2 d\sigma \right). \tag{5.104}$$

After integrating by parts, the expression for A_2 given in (5.100), we obtain

$$A_2 = 2\beta \left(g(1)p(1)u_s^2(1,t) - \int_0^1 g_\sigma(\sigma)p(\sigma)u_\sigma^2 d\sigma - \int_0^1 g(\sigma)p(\sigma)u_\sigma u_{\sigma\sigma} d\sigma \right), \tag{5.105}$$

where we have used the fact that $g(0) = 0$. After noting that the expression for A_2 given in (5.100) can be expanded into the following form:

$$A_2 = 2\beta \left(\int_0^1 g(\sigma)u_\sigma^2 p_\sigma(\sigma) d\sigma + \int_0^1 g(\sigma)p(\sigma)u_\sigma u_{\sigma\sigma} d\sigma \right), \tag{5.106}$$

we can combine (5.105) and (5.106) to eliminate the last term in (5.105) as follows:

$$A_2 = \beta \left(g(1)p(1)u_x^2(1,t) - \int_0^1 \left[g_\sigma(\sigma)p(\sigma) - g(\sigma)p_\sigma(\sigma) \right] u_\sigma^2 d\sigma \right). \tag{5.107}$$

From the structure of $g(x)$ and $p(x)$, it is straightforward to show that

$$g_x(x)p(x) - g(x)p_x(x) = 1 + p(x); \tag{5.108}$$

thus, we can use (5.107) and (5.108) to construct the following upper bound for A_2:

$$A_2 \leq \beta \left(g(1)p(1)u_x^2(1,t) - \int_0^1 p(\sigma)u_\sigma^2 d\sigma \right). \tag{5.109}$$

After substituting (5.104) and (5.109) into (5.99), and then substituting the resulting expression along with (5.98) into (5.94), we have

$$\begin{aligned} \dot{V} \leq & -\left(\frac{p(1)}{2} - \beta g(1) \right) u_t^2(1,t) - \left(\frac{p(1)}{2} - \beta g(1)p(1) \right) u_x^2(1,t) \\ & -\left(k_s - \frac{p(1)}{2} \right) \eta^2 - 2\beta E_s, \end{aligned} \tag{5.110}$$

where the definition of (5.85) has been utilized. From (5.110), it is clear that if the controller gain k_s is selected according to (5.83) and the weighting constant β is selected according to

$$\beta < \min \left\{ \frac{p(1)}{2g(1)}, \frac{1}{2g(1)} \right\}, \tag{5.111}$$

then $\dot{V}(t)$ can be upper bounded by the following nonpositive function:

$$\dot{V}(t) \leq -\lambda_3 \left(E_s(t) + \eta^2(t)\right), \tag{5.112}$$

where the positive constant[7] λ_3 is defined by

$$\lambda_3 = \min\left\{2\beta, k_s - \frac{p(1)}{2}\right\}. \tag{5.113}$$

From (5.92) and (5.112), we obtain the following upper bound for the time derivative of $V(t)$:

$$\dot{V}(t) \leq -\frac{\lambda_3}{\lambda_2} V(t), \tag{5.114}$$

whose solution, according to Lemma A.4 in Appendix A, yields

$$V(t) \leq V(0) \exp\left(-\frac{\lambda_3}{\lambda_2} t\right) \leq \lambda_2 \kappa_o \exp\left(-\frac{\lambda_3}{\lambda_2} t\right). \tag{5.115}$$

Note that (5.85), (5.92), and (5.82) have been utilized to formulate the inequality on the right-hand side of (5.115). In addition, we can use (A.25) of Lemma A.12 in Appendix A, (5.85), and (5.92) to formulate the following inequality:

$$\frac{p_0}{2} u^2(x,t) \leq \frac{1}{2}\int_0^1 p(\sigma) u_\sigma^2(\sigma,t) d\sigma \leq E_s(t) \leq \frac{1}{\lambda_1} V(t) \qquad \forall x \in [0,1]. \tag{5.116}$$

The inequality given in (5.81) now directly follows by combining (5.115) and (5.116). □

Remark 5.5 *The proof of Theorem 5.3 relies on the fact that the weighting constant β has been selected sufficiently small such that the inequalities of (5.90) and (5.111) are satisfied. Thus, an overall condition on β can be obtained by combining both inequalities, similarly to what was done in Remark 5.2. We also note that the structure of the weighting function $g(x)$ defined in (5.87) has been crafted to facilitate the construction of the bounds given in (5.92) and (5.112).*

Remark 5.6 *Utilizing similar arguments as in Remark (5.3) and Properties 5.3 and 5.4, we can easily illustrate the boundedness of all systems signals during closed-loop operation.*

Remark 5.7 *It should be noted that some extensions to the basic control law given in (5.79) can be developed to improve the closed-loop performance.*

[7] The control gain k_s must satisfy (5.83) to ensure that λ_3 is positive.

For example, an additional control gain can be included in the definition of the tracking variable $\eta(t)$ defined in (5.77) as follows:

$$\eta(t) = u_t(1,t) + \alpha u_x(1,t), \tag{5.117}$$

where α is a positive control gain. As a result, we can redefine the control force as

$$f(t) = -m\alpha u_{xt}(1,t) + p(1)u_x(1,t) - k_s\eta(t) + Y(u_t(1,t))\phi. \tag{5.118}$$

5.3.4 Adaptive Control Law

In this section, we redesign the model-based controller of (5.79) to compensate for constant parametric uncertainty while asymptotically stabilizing the cable displacement.

First, note that the open-loop dynamics for $\eta(t)$ given in (5.78) can be rewritten as

$$m\dot{\eta}(t) = W(u_{xt}(1,t), u_x(1,t), u_t(1,t))\,\theta + f(t), \tag{5.119}$$

where $W(\cdot) \in \mathbb{R}^{1\times(q+2)}$ is a known regression matrix and $\theta \in R^{(q+2)\times 1}$ is the unknown, constant parameter vector defined as follows:

$$\begin{aligned} W(\cdot) &= \begin{bmatrix} u_{xt}(1,t) & -u_x(1,t) & -Y(u_t(1,t)) \end{bmatrix} \\ \theta &= \begin{bmatrix} m & p(1) & \phi^T \end{bmatrix}^T. \end{aligned} \tag{5.120}$$

From the structure of (5.78), the adaptive control law is designed as follows:

$$f(t) = -W(\cdot)\hat{\theta}(t) - k_s\eta(t), \tag{5.121}$$

where $\hat{\theta}(t) \in \mathbb{R}^{(q+2)\times 1}$ is a dynamic parameter estimate vector defined as

$$\hat{\theta}(t) = \begin{bmatrix} \hat{m}(t) & \widehat{p(1,t)} & \hat{\phi}^T(t) \end{bmatrix}^T. \tag{5.122}$$

The parameter estimates are updated according to

$$\dot{\hat{\theta}}(t) = \Gamma W^T(\cdot)\eta(t), \tag{5.123}$$

where $\Gamma \in \mathbb{R}^{(q+2)\times(q+2)}$ is a diagonal, positive-definite, adaptation gain matrix. After substituting (5.121) into (5.119), we obtain the following closed-loop dynamics on the boundary:

$$m\dot{\eta}(t) = -k_s\eta(t) + W(\cdot)\tilde{\theta}(t) \qquad \dot{\tilde{\theta}}(t) = -\Gamma W^T(\cdot)\eta(t), \tag{5.124}$$

where $\tilde{\theta}(t) = \theta - \hat{\theta}(t) \in \mathbb{R}^{(q+2)\times 1}$ denotes the parameter estimation error, and (5.123) has been used to obtain the parameter estimation error dynamics.

Theorem 5.4 *The adaptive boundary controller given by (5.121) and (5.123) ensures that the cable displacement is asymptotically regulated in the following sense:*

$$\lim_{t\to\infty} |u(x,t)| = 0 \qquad \forall x \in [0,1], \tag{5.125}$$

where the controller gain k_s defined in (5.121) must be selected to satisfy (5.83).

Proof. The following proof is based on arguments similar to those used in the proofs of Theorems 5.2 and 5.3; hence, most of the details will not be repeated. First, we define the following function:

$$V_a(t) = V(t) + \frac{1}{2}\tilde{\theta}^T(t)\Gamma^{-1}\tilde{\theta}(t), \tag{5.126}$$

where $V(t)$ was defined in (5.84). If the design constant β of (5.86) is selected to be sufficiently small according to the condition given in (5.90), we can formulate the following bounds on $V_a(t)$:

$$\lambda_{1a}\left[E_s(t) + \eta^2(t) + \left\|\tilde{\theta}(t)\right\|^2\right] \le V_a(t) \le \lambda_{2a}\left[E_s(t) + \eta^2(t) + \left\|\tilde{\theta}(t)\right\|^2\right], \tag{5.127}$$

for some positive constants λ_{1a} and λ_{2a}.

After differentiating (5.126) with respect to time, and substituting from (5.71), (5.72), and (5.124), we can proceed as in the proof of Theorem 5.3 to obtain the following upper bound for the time derivative of $V_a(t)$:

$$\dot{V}_a \le -\lambda_3\left(E_s(t) + \eta^2(t)\right) + \tilde{\theta}^T(t)\left(W^T(\cdot)\eta(t) + \Gamma^{-1}\dot{\tilde{\theta}}(t)\right), \tag{5.128}$$

where λ_3 was defined in (5.113) (the control gain k_s defined in (5.121) has to be selected according to (5.83) to ensure λ_3 is positive). After substituting the second equation of (5.124) into (5.128), we can simplify the upper bound on $\dot{V}_a(t)$ to

$$\dot{V}_a(t) \le -\lambda_3\left(E_s(t) + \eta^2(t)\right) \triangleq g_a(t). \tag{5.129}$$

The proof of Theorem 5.2 can now be directly followed to show that

$$\lim_{t\to\infty} E_s(t), \eta(t) = 0. \tag{5.130}$$

We can now use the result given by (5.130) and the inequality-type bound developed in (5.116) to state the result given by (5.125). □

5.3.5 Experimental Evaluation

Experimental Setup

The boundary controllers were implemented on an experimental setup similar to the one described in Section 5.2.6 with a braided polyester rope replacing the string. A 486 PC hosting a Texas Instruments TMS320C30 DSP board served as the computational engine. *WinMotor* provided the environment to write the control algorithm in the C programming language. An encoder interface card (Integrated Motions Inc. Model DS-2) allowed for quadrature extrapolation of the encoder signals. The values for the mechanical system parameter defined in (5.70) were determined via standard test procedures to be

$$M = 3.229\text{ kg},\quad L = 2.69\text{ m},\quad \rho A = 0.085\text{ kg/m},\quad P_0 = 0.127\text{ N},$$

$$m = 14.1,\quad p_0 = 0.056,\quad p(1) = 0.503.$$

To test the performance of the controllers, the cable was perturbed using a disturbance input applied to the cable's pinned-end. Since dimensional quantities were used in the experiment, the applied boundary control force was scaled according to (5.70) (i.e., $F = f\rho AgL$). This implies that the weight of the cable must be known in order to implement the controllers.

Experimental Results

First, the model-based controller given bý (5.118) was implemented using the following settings:

$$k_s = 5.0,\qquad \alpha = 4.0,\qquad Y(u_t(1,t))\phi = 0,$$

where the gains were tuned for the best performance. The results of these experiments appear in Figure 5.9. As is clear from the figure, the cable system exhibits excellent transient response under the controller. The uncontrolled swing of the cable to the disturbance input is shown for comparison purposes. The gantry position and the motor voltage signal are also shown. It should also be noted that the $Y(u_t(1,t))\phi$ term could have been used to model static friction on the rail of the actuator; however, the feedback portion of the controller seemed to adequately compensation for friction in this experiment.

The adaptive controller of (5.121) and (5.123) was then implemented with the parameter estimates initialized to 80% of their nominal values. The best performance was achieved using the following settings:

$$k_s = 5.0,\qquad \alpha = 4.0,\qquad Y(u_t(1,t))\phi = 0,\qquad \Gamma = \text{diag}\{5,2\}.$$

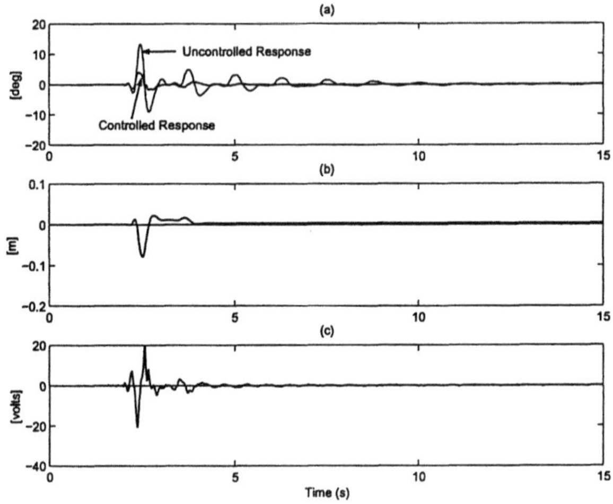

FIGURE 5.9. Model-based controller: (a) $u(1,t)$, (b) gantry position, and (c) motor voltage.

The results of this experiment appears in Figure 5.10. In the first subplot of Figure 5.10, the transient performance of the adaptive controller to the disturbance is compared to the uncontrolled swing of the cable. Subsequent subplots show the gantry position, the motor voltage and the parameter estimates.

5.4 Notes

To understand the vibration of cable-like structures, many researchers have developed sophisticated modeling and analysis techniques (see Irvine [7] for a review). Small sag approximation techniques (Soler [16]), lumped parameter models (West et al. [17]), finite element techniques (Fried [4]), and Galerkin methods (Perkins and Mote [13]) represent several approaches used to predict the behavior of cables in a variety of applications.

Active vibration control has only recently been applied to string and cable systems. For example, Fujino et al. [5] developed axial boundary force control laws for a single mode cable model and experimentally demonstrated a large reduction in peak resonant response. This approach was later extended to multiple cable modes in [6]. Baicu et al. [1] developed a linear boundary controller for the out-of-plane vibration of a distributed cable model that required boundary position, velocity, and slope feedback. An adaptive boundary control scheme for out-of-plane cable vibrations was

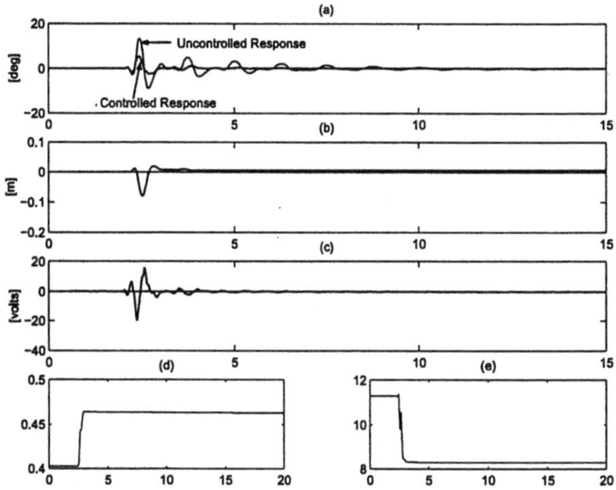

FIGURE 5.10. Adaptive controller: (a) $u(1,t)$, (b) gantry position, and (c) motor voltage, (d) $\hat{p}(1)$, and (e) $\hat{m}$.

proposed in Canbolat et al. [3] that compensated for unknown, constant boundary mass, tension, and friction effects. Boundary control work for string-like dynamic models can be found in Morgül [11], Shahruz et al. [15], and Zhang et al. [18]. Interestingly, the model-based control law presented in Section 5.3 reduces to a form that is similar to the one given in Morgül [11] if the cable tension is assumed to be constant.

References

[1] C. F. Baicu, C. D. Rahn, and B. D. Nibali, Active Boundary Control of Elastic Cables: Theory and Experiment, *Journal of Sound and Vibration*, Vol. 198, No. 1, pp. 17-26, 1996.

[2] C. F. Baicu, C. D. Rahn, and D. M. Dawson, Backstepping Boundary Control of a Flexible-Link Electrically Driven Gantry Robots, *IEEE Transactions on Mechatronics*, Vol. 3, No. 1, pp. 60-66, Mar. 1998.

[3] H. Canbolat, D. Dawson, C. Rahn, and S. Nagarkatti, Adaptive Control of Out-of-Plane Cable Vibration, *ASME Journal of Applied Mechanics*, Vol. 65, pp. 963-969, Dec. 1998.

[4] I. Fried, Large Deformation Static and Dynamic Finite Element Analysis of Extensible Cables, *Computers and Structures,* Vol. 15, pp. 315-319, 1982.

[5] Y. Fujino, P. Warnitchai, and B. Pacheco, Active Stiffness Control of Cable Vibration, *Journal of Applied Mechanics*, Vol. 60, pp. 948-953, Dec. 1993.

[6] Y. Fujino and T. Susumpow, Active Control of Multimodal Cable Vibrations by Axial Support Motion, *Smart Materials and Structures*, Vol. 4, pp. A41-51, 1995.

[7] H. Irvine, *Cable Structure*, Cambridge, MA: MIT Press, 1981.

[8] J. U. Kim and Y. Renardy, Boundary Control of the Timoshenko Beam, *SIAM Journal of Control and Optimization*, Vol. 25, No. 6, Nov. 1987.

[9] M. Krstić, I. Kanellakopoulos, and P. Kokotović, *Nonlinear and Adaptive Control Design*, New York, NY: Wiley Interscience, 1995.

[10] O. Morgül, Control and Stabilization of a Flexible Beam Attached to a Rigid Body, *International Journal of Control,* Vol. 51, No. 1, pp. 11-31, Jan. 1990.

[11] O. Morgül, B. Rao, and F. Conrad, On the Stabilization of a Cable with a Tip Mass, *IEEE Transactions on Automatic Control,* Vol. 39, No. 10, pp. 2140-2145, Oct. 1994.

[12] D. Oplinger, Frequency Response of a Nonlinear Stretched String, *Journal of the Acoustical Society of America*, Vol. 32, No. 12, pp. 1529-1538, Dec. 1960.

[13] N. Perkins and C. Mote, Three-dimensional Vibration of Travelling Elastic Cables, *Journal of Sound and Vibration*, Vol. 114, No. 2, pp. 325-340, 1987.

[14] http://www.quanser.com.

[15] S. M. Shahruz and L. G. Krishna, Boundary Control of a Nonlinear String, *Proceedings of the ASME Dynamics Systems and Control Division,* DSC-Vol. 58, pp.831-835, 1996.

[16] A. Soler, Dynamic Response of Single Cable with Initial Sag, *Journal of the Franklin Institute*, Vol. 190, pp. 377-387, 1970.

[17] H. West, L. Geschwinder, and J. Suchoski, Natural Vibrations of Suspension Cables, *ASCE Journal of the Structural Division*, Vol. 101, pp. 2277-2291, 1975.

[18] F. Zhang, D. M. Dawson, S. P. Nagarkatti, and D. V. Haste, Boundary Control for a General Class of Nonlinear Actuator-String Systems, *Proceedings of the IEEE Conference on Decision and Control*, pp. 3484-3489, Tampa, FL, Dec. 1998.

6
Cantilevered Beams

6.1 Introduction

In many distributed parameter mechanical systems, the flexible element can be modeled as a beam-type structure (e.g., space structures, flexible link robots, helicopter rotor/blades, turbine blades, etc.). The most commonly used beam model is based on the classical Euler-Bernoulli theory, which provides a good description of the beam's dynamic behavior when the beam's cross-sectional dimensions are small in comparison to its length (i.e., this model neglects the rotary inertia of the beam). A more accurate beam model is provided by the Timoshenko theory, which takes into account not only the rotary inertial energy but also the beam's deformation owing to shear. As discussed in [15], the Timoshenko beam model has been shown to have a broader applicability than the Euler-Bernoulli model. In [1], Aldraihem et al. compared the accuracy and validity of these two beam models by simulating a cantilevered beam under distributed piezoelectric sensoring/actuation. The results provided in [1] indicated that the Timoshenko model is superior to the Euler-Bernoulli model in predicting the beam's response. While the Timoshenko model may be more accurate at predicting the beam's response in comparison to the Euler-Bernoulli model, the Timoshenko model is more difficult to utilize for control design purposes owing to its higher order. For this reason, the design of boundary controllers for flexible-beam-type structures has been based mainly on the Euler-Bernoulli model.

In this chapter, we develop boundary control strategies for the Euler-Bernoulli and Timoshenko models of a cantilevered flexible beam. To provide a more accurate description of the system model, the PDEs for the beam dynamics are augmented to include the ODEs for the mass/inertial dynamics of a rigid payload and/or boundary actuator located at the beam's free end-point. The control strategies, which explicitly exploit the structure of the *hybrid* PDE-ODE dynamic model, are designed with the objective of regulating the beam vibration. As in Chapter 5, both model-based and adaptive control laws are formulated. The model-based controller provides a stepping stone towards the synthesis of the adaptive controller, which is designed to cope with constant, parametric uncertainty.

6.2 Euler-Bernoulli Beam

In this section, we consider the problem of stabilizing the distributed displacement of a cantilevered Euler-Bernoulli beam with actuator dynamics at the beam's free end-point. The control strategy is composed of a boundary control force applied to the beam's free end-point and requires measurements of the beam's end-point shear, shear-rate, and velocity. A model-based controller is first developed that exponentially regulates the displacement of the beam. We then illustrate how the model-based controller can be redesigned as an adaptive controller that asymptotically regulates the displacement of the beam while compensating for parametric uncertainty.

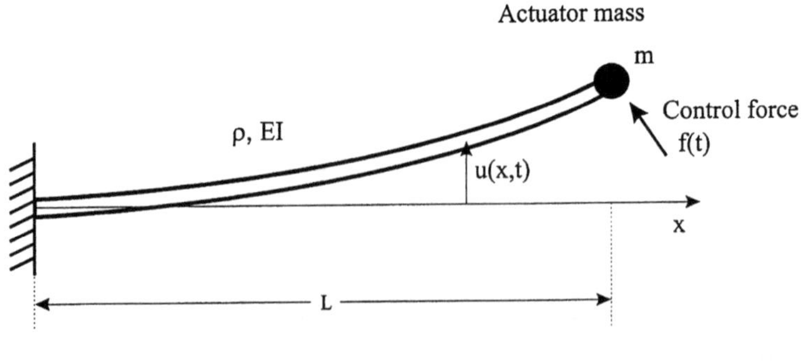

FIGURE 6.1. Schematic representation of a cantilevered Euler-Bernoulli beam with end-point payload/actuator mass.

6.2.1 System Model

The model for the cantilevered Euler-Bernoulli beam system shown in Figure 6.1 is described by a PDE of the form [12]

$$\rho u_{tt}(x,t) + EIu_{xxxx}(x,t) = 0, \tag{6.1}$$

with the following boundary conditions:[1]

$$u(0,t) = u_x(0,t) = u_{xx}(L,t) = 0 \tag{6.2}$$

and

$$mu_{tt}(L,t) - EIu_{xxx}(L,t) = f(t), \tag{6.3}$$

where x and t represent the independent spatial and time variables, respectively; $u(x,t)$ denotes the beam displacement at the position x for time t; ρ is the mass/length of the beam; EI is the bending stiffness of the beam; L is the length of the beam; m represents the payload/actuator mass attached to the free end-point of the beam; the subscripts x, t denote the partial derivatives with respect to x, t, respectively; and $f(t)$ denotes the boundary control input force.

Since the control strategies will consist of relatively simple functions, we will assume the existence of a unique solution for the dynamics given by (6.1) through (6.3) under the control. In addition, based on the arguments outlined in Remark 5.1, we will assume that the distributed variable $u(x,t)$ and its time derivative $u_t(x,t)$ belong to a space of functions that have the following properties.

Property 6.1 If the potential energy of the system given by (6.1) through (6.3), which is given by

$$\Pi_p = \frac{1}{2} EA \int_0^L u_{\sigma\sigma}^2 (\sigma,t)\, d\sigma, \tag{6.4}$$

is bounded $\forall t \in [0,\infty)$ and $\forall x \in [0,L]$, then $\dfrac{\partial^n}{\partial x^n} u(x,t)$ is bounded for $n = 2$, 3, 4, $\forall t \in [0,\infty)$, and $\forall x \in [0,L]$.

Property 6.2 If the kinetic energy of the system of (6.1) through (6.3), which is given by

$$\Pi_k = \frac{1}{2}\rho \int_0^L u_t^2(\sigma,t) d\sigma + \frac{1}{2} m u_t^2(L,t), \tag{6.5}$$

[1] Given the clamped-end boundary conditions of (6.2), we also know that $u_t(0,t) = u_{xt}(0,t) = 0$.

is bounded $\forall t \in [0, \infty)$, then $\frac{\partial^n}{\partial x^n} u_t(x,t)$ is bounded for $n = 0, 1, 2, 3$, $\forall t \in [0, \infty)$, and $\forall x \in [0, L]$.

6.2.2 Problem Statement

The primary control objective is to design the boundary control force $f(t)$ such that the beam displacement $u(x,t)$ is driven to zero $\forall x \in [0, L]$ as $t \to \infty$. To facilitate the control design and the subsequent Lyapunov-type closed-loop stability analysis, we first define an auxiliary signal, denoted by $\eta(t)$, as follows:

$$\eta(t) = u_t(L,t) - u_{xxx}(L,t). \tag{6.6}$$

We now rewrite the dynamic boundary condition of (6.3) in terms of $\eta(t)$. To this end, we differentiate (6.6) with respect to time, multiply the resulting expression by m, and then utilize (6.3) to substitute for $mu_{tt}(L,t)$ to produce

$$m\dot{\eta}(t) = -mu_{xxxt}(L,t) + EIu_{xxx}(L,t) + f(t). \tag{6.7}$$

The above open-loop equation will form the basis for the design of the model-based and adaptive control laws.

6.2.3 Model-Based Control Law

Given exact model knowledge, we will now design an exponentially stabilizing controller for the system given by (6.1) through (6.3). Specifically, given the structure of the open-loop dynamics of (6.7), the control force is designed as follows:

$$f(t) = mu_{xxxt}(L,t) - EIu_{xxx}(L,t) - k_s\eta(t), \tag{6.8}$$

where k_s is a positive control gain. After substituting (6.8) into (6.7), we obtain the following closed-loop dynamics for $\eta(t)$:

$$m\dot{\eta}(t) = -k_s\eta(t). \tag{6.9}$$

Theorem 6.1 *The model-based boundary control law given by (6.8) ensures that the beam displacement is exponentially regulated in the following sense:*

$$|u(x,t)| \le \sqrt{\frac{2\lambda_2 L^3}{\lambda_1 EI}\kappa_o \exp\left(-\frac{\lambda_3}{\lambda_2}t\right)} \qquad \forall x \in [0, L], \tag{6.10}$$

provided the control gain k_s is selected to satisfy the following inequality:

$$k_s > \frac{EI}{2}, \tag{6.11}$$

where λ_1, λ_2, and λ_3 are some positive bounding constants, and the positive constant κ_o is given by

$$\kappa_o = \frac{1}{2}\rho\int_0^L u_t^2(\sigma,0)d\sigma + \frac{1}{2}EI\int_0^L u_{\sigma\sigma}^2(\sigma,0)d\sigma + (u_t(L,0) - u_{xxx}(L,0))^2. \tag{6.12}$$

Proof. To prove the result given by (6.10), we begin by defining the following function:

$$V(t) = E_b(t) + \frac{1}{2}m\eta^2(t) + E_c(t), \tag{6.13}$$

where the beam's energy-related term $E_b(t)$ and the "cross" term $E_c(t)$ are defined as

$$E_b(t) = \frac{1}{2}\rho\int_0^L u_t^2(\sigma,t)d\sigma + \frac{1}{2}EI\int_0^L u_{\sigma\sigma}^2(\sigma,t)d\sigma \tag{6.14}$$

and

$$E_c(t) = 2\beta\rho\int_0^L \sigma u_t(\sigma,t)u_\sigma(\sigma,t)d\sigma, \tag{6.15}$$

with β being a positive weighting constant, which can be made sufficiently small to ensure $V(t)$ is always non-negative. To see this, first note that the inequality (A.29) of Lemma A.13 in Appendix A can be used to bound $E_c(t)$ of (6.15) as follows:[2]

$$E_c = 2\beta\rho\int_0^L \sigma u_t u_\sigma d\sigma \le 2\beta\rho L\int_0^L \left(u_t^2 + u_\sigma^2\right)d\sigma. \tag{6.16}$$

After applying the inequality (A.27) of Lemma A.12 in Appendix A to the u_σ^2 term on the right-hand side of (6.16), we obtain

$$\begin{aligned} E_c &\le 2\beta\rho L\max\left\{1,L^2\right\}\int_0^L \left(u_t^2 + u_{\sigma\sigma}^2\right)d\sigma \\ &\le 4\beta\rho L\frac{\max\left\{1,L^2\right\}}{\min\left\{\rho,EI\right\}}\frac{1}{2}\int_0^L \left(\rho u_t^2 + EIu_{\sigma\sigma}^2\right). \end{aligned} \tag{6.17}$$

We can now use (6.14) and (6.17) to establish the following inequalities:

$$-4\beta\rho L\frac{\max\left\{1,L^2\right\}}{\min\left\{\rho,EI\right\}}E_b \le E_c \le 4\beta\rho L\frac{\max\left\{1,L^2\right\}}{\min\left\{\rho,EI\right\}}E_b. \tag{6.18}$$

---

[2] To reduce the notational complexity in most of the following derivations, the arguments x,t will be left out of all spatial/time-dependent variables (e.g., $u(x,t)$ will be denoted simply as u), while the argument t will be left out of the time-dependent variables (e.g., $u_x(L,t)$ will be denoted as $u_x(L)$).

If β is selected according to

$$\beta < \frac{\min\{\rho, EI\}}{4\rho L \max\{1, L^2\}}, \tag{6.19}$$

we can utilize (6.18) to state that

$$0 \le \xi_1 E_b \le E_b + E_c \le \xi_2 E_b \tag{6.20}$$

for some positive constants ξ_1 and ξ_2. Given the definition of (6.13) and the inequality given by (6.20), we can formulate the following bounds on $V(t)$:

$$\lambda_1 \left(E_b(t) + \eta^2(t)\right) \le V(t) \le \lambda_2 \left(E_b(t) + \eta^2(t)\right), \tag{6.21}$$

where λ_1 and λ_2 are defined as follows:

$$\begin{aligned} \lambda_1 &= \min\left\{1 - 4\beta\rho L \frac{\max\{1, L^2\}}{\min\{\rho, EI\}}, \tfrac{1}{2}m\right\} > 0 \\ \lambda_2 &= \max\left\{1 + 4\beta\rho L \frac{\max\{1, L^2\}}{\min\{\rho, EI\}}, \tfrac{1}{2}m\right\} > 0, \end{aligned} \tag{6.22}$$

with β satisfying (6.19).

Next, we illustrate that for sufficiently large k_s and sufficiently small β, the time derivative of $V(t)$ will be a nonpositive function. If we differentiate (6.13) with respect to time, we have

$$\dot{V}(t) = \dot{E}_b(t) + \dot{E}_c(t) - k_s \eta^2(t), \tag{6.23}$$

where (6.9) has been utilized. To determine $\dot{E}_b(t)$ in (6.23), we differentiate (6.14) with respect to time to obtain

$$\dot{E}_b = -EI \int_0^L u_t u_{\sigma\sigma\sigma\sigma} d\sigma + EI \int_0^L u_{\sigma\sigma} u_{\sigma\sigma t} d\sigma, \tag{6.24}$$

where (6.1) has been utilized. If we integrate by, parts twice, the first term on the right-hand side of (6.24), we obtain

$$\dot{E}_b = -EI u_t(L) u_{xxx}(L), \tag{6.25}$$

where the boundary conditions given in (6.2) have been applied. Finally, upon the application of (6.6), (6.25) can be rewritten as

$$\dot{E}_b = -\frac{EI}{2}\left(u_t^2(L) + u_{xxx}^2(L)\right) + \frac{EI}{2}\eta^2. \tag{6.26}$$

To determine $\dot{E}_c(t)$ in (6.23), we differentiate (6.15) with respect to time and then apply (6.1) to produce

$$\dot{E}_c = A_1 + A_2, \tag{6.27}$$

where

$$A_1 = 2\beta\rho \int_0^L \sigma u_t u_{\sigma t} d\sigma \qquad A_2 = -2\beta EI \int_0^L \sigma u_\sigma u_{\sigma\sigma\sigma\sigma} d\sigma. \tag{6.28}$$

After integrating, by parts, the expression for A_1 given in (6.28), we obtain

$$A_1 = 2\beta\rho \left(L u_t^2(L) - \int_0^L u_t^2 d\sigma \right) - 2\beta\rho \int_0^L \sigma u_t u_{\sigma t} d\sigma, \tag{6.29}$$

where (6.2) has been used. After noting that the last term on the right-hand side of (6.29) is equal to A_1, we can rearrange (6.29) to yield

$$A_1 = \beta \left(L\rho u_t^2(L) - \rho \int_0^L u_t^2 d\sigma \right). \tag{6.30}$$

After integrating, by parts, the expression for A_2 given in (6.28), we obtain

$$A_2 = -2\beta EI \left(L u_x(L) u_{xxx}(L) - \int_0^L u_\sigma u_{\sigma\sigma\sigma} d\sigma - \int_0^L \sigma u_{\sigma\sigma} u_{\sigma\sigma\sigma} d\sigma \right) \tag{6.31}$$

upon application of (6.2). After integrating, by parts, the first integral on the right-hand side of (6.31), we obtain

$$A_2 = -2\beta EI \left(L u_x(L) u_{xxx}(L) + \int_0^L u_{\sigma\sigma}^2 d\sigma - \int_0^L \sigma u_{\sigma\sigma} u_{\sigma\sigma\sigma} d\sigma \right) \tag{6.32}$$

upon application of (6.2). After integrating, by parts, the last integral on the right-hand side of (6.32), we obtain

$$A_2 = -2\beta EI \left(L u_x(L) u_{xxx}(L) + 2 \int_0^L u_{\sigma\sigma}^2 d\sigma + \int_0^L \sigma u_{\sigma\sigma} u_{\sigma\sigma\sigma} d\sigma \right), \tag{6.33}$$

where (6.2) has been again used. After adding the expressions given by (6.32) and (6.33), we obtain

$$A_2 = -\beta EI \left(2 L u_x(L) u_{xxx}(L) + 3 \int_0^L u_{\sigma\sigma}^2 d\sigma \right). \tag{6.34}$$

We can now substitute (6.30) and (6.34) into (6.27), and then substitute the resulting expression along with (6.26) into (6.23) to produce

$$\begin{aligned}\dot{V} = & -\left(\frac{EI}{2}-\beta L\rho\right)u_t^2(L)-\frac{EI}{2}u_{xxx}^2(L)-\left(k_s-\frac{EI}{2}\right)\eta^2-2\beta E_b \\ & -2\beta EI\int_0^L u_{\sigma\sigma}^2 d\sigma-2\beta EILu_x(L)u_{xxx}(L),\end{aligned} \tag{6.35}$$

where (6.14) has been utilized. After applying (A.24) of Lemma A.12 in Appendix A and (A.30) of Lemma A.13 in Appendix A to the second line of (6.35), we can obtain the following upper bound for $\dot{V}(t)$:

$$\begin{aligned}\dot{V} \leq & -\left(\frac{EI}{2}-\beta L\rho\right)u_t^2(L)-\frac{EI}{2}u_{xxx}^2(L)-\left(k_s-\frac{EI}{2}\right)\eta^2-2\beta E_b \\ & -2\beta EI\frac{1}{L}u_x^2(L)+2EIL\beta\delta u_x^2(L)+\frac{2EIL\beta}{\delta}u_{xxx}^2(L),\end{aligned} \tag{6.36}$$

where δ is an arbitrary positive constant. We now rearrange (6.36) into the following advantageous form:

$$\begin{aligned}\dot{V} \leq & -\left(\frac{EI}{2}-\beta L\rho\right)u_t^2(L)-\left(k_s-\frac{EI}{2}\right)\eta^2-2\beta E_b \\ & -2EIL\beta\left(\frac{1}{L^2}-\delta\right)u_x^2(L)-EI\left(\frac{1}{2}-\frac{2L\beta}{\delta}\right)u_{xxx}^2(L).\end{aligned} \tag{6.37}$$

From (6.37), it is not difficult to see that if the control gain k_s and the constants δ, β are selected to satisfy the following conditions:

$$k_s > \frac{EI}{2}, \qquad \delta < \frac{1}{L^2}, \qquad \beta < \min\left\{\frac{EI}{2\rho L}, \frac{\delta}{4L}\right\}, \tag{6.38}$$

then $\dot{V}(t)$ can be upper bounded by a nonpositive scalar function as shown below:

$$\dot{V}(t) \leq -\lambda_3\left(E_b(t)+\eta^2(t)\right), \tag{6.39}$$

where λ_3 is defined as follows:[3]

$$\lambda_3 = \min\left\{k_s-\frac{EI}{2}, 2\beta\right\} > 0. \tag{6.40}$$

[3]Note that λ_3 is positive as a result of the condition given in (6.38).

From (6.21) and (6.39), we can obtain the following upper bound for the time derivative of $V(t)$:

$$\dot{V}(t) \leq -\frac{\lambda_3}{\lambda_2}V(t). \tag{6.41}$$

Upon application of Lemma A.4 in Appendix A to (6.41), we have

$$V(t) \leq V(0)\exp\left(-\frac{\lambda_3}{\lambda_2}t\right). \tag{6.42}$$

From (6.21), (A.28) of Lemma A.12 in Appendix A, and (6.14), we have that

$$V(0) \leq \lambda_2\left(E_b(0) + \eta^2(0)\right) \tag{6.43}$$

and

$$\frac{1}{2L^3}EIu^2(x,t) \leq \frac{1}{2}EI\int_0^L u_{\sigma\sigma}^2(\sigma,t)d\sigma \leq E_b(t) \leq \frac{1}{\lambda_1}V(t) \qquad \forall x \in [0,L]. \tag{6.44}$$

The result given by (6.10) and (6.12) now directly follows by combining (6.42), (6.43), and (6.44), and then using the definitions given by (6.6) and (6.14). □

Remark 6.1 *Note that the proof of Theorem 6.1 relies on the weighting constant β being chosen sufficiently small such that both conditions (6.19) and (6.38) are satisfied. Thus, the overall condition on β can be written as*

$$\beta < \min\left\{\frac{\min\{\rho, EI\}}{4\rho L\max\{1, L^2\}}, \frac{EI}{2\rho L}, \frac{\delta}{4L}\right\}, \tag{6.45}$$

where δ must be selected to satisfy the second inequality given in (6.38).

Remark 6.2 *From (6.13) and (6.42), we can state that $E_b(t)$ and $\eta(t)$ are bounded $\forall t \in [0,\infty)$. Since $E_b(t)$ is bounded $\forall t \in [0,\infty)$, we can use (6.14) and (A.28) in Lemma A.12 in Appendix A to show that $u(x,t)$ is bounded $\forall t \in [0,\infty)$ and $\forall x \in [0,L]$. Since $E_b(t)$ defined in (6.14) is bounded $\forall t \in [0,\infty)$, the potential energy given by (6.4) is bounded $\forall t \in [0,\infty)$; hence, we can use Property 6.1 to show that $\frac{\partial^n}{\partial x^n}u(x,t)$ is bounded for $n = 2, 3, 4$, $\forall t \in [0,\infty)$, and $\forall x \in [0,L]$. Since $\eta(t)$ and $u_{xxx}(L,t)$ are bounded $\forall t \in [0,\infty)$, we can use (6.6) to state that $u_t(L,t)$ is bounded $\forall t \in [0,\infty)$. From the boundedness of $E_b(t)$ and $u_t(L,t)$, we can see that the kinetic energy of the system defined in (6.5) is bounded $\forall t \in [0,\infty)$. Since the kinetic energy is bounded $\forall t \in [0,\infty)$, we can use Property 6.2 to conclude that $\frac{\partial^n}{\partial x^n}u_t(x,t)$ is bounded for $n = 0, 1, 2, 3$, $\forall t \in [0,\infty)$, and $\forall x \in [0,L]$. From the above information, we can now state that all of* *the signals in the control law of (6.8) are bounded* $\forall t \in [0, \infty)$. *Finally, we can use (6.1), (6.3), and the above boundedness statements to show that* $u_{tt}(x,t)$ *is bounded* $\forall t \in [0, \infty)$ *and* $\forall x \in [0, L]$.

6.2.4 Adaptive Control Law

The control law given by (6.8) requires exact knowledge of the payload mass and the beam bending stiffness. We will now illustrate how the controller of (6.8) can be redesigned to compensate for constant, parametric uncertainty while asymptotically stabilizing the beam displacement. First, note that (6.7) can be rewritten into the following advantageous form:

$$m\dot{\eta}(t) = Y\left(\left(u_{xxxt}\left(L,t\right), u_{xxx}\left(L,t\right)\right)\right)\theta + f(t), \tag{6.46}$$

where $Y(\cdot) \in \mathbb{R}^{1\times 2}$ is a known regression matrix, and $\theta \in \mathbb{R}^2$ is an unknown, constant, parameter vector defined as follows:

$$Y(\cdot) = \begin{bmatrix} -u_{xxxt}(L,t) & u_{xxx}(L,t) \end{bmatrix} \qquad \theta = \begin{bmatrix} m & EI \end{bmatrix}^T. \tag{6.47}$$

Based on the form of (6.46), the control force is now designed as

$$f(t) = -Y(\cdot)\hat{\theta}(t) - k_s\eta(t), \tag{6.48}$$

where $\eta(t)$ was defined in (6.6), and $\hat{\theta}(t) \in \mathbb{R}^2$ is a dynamic, parameter estimate vector defined as follows:

$$\hat{\theta}(t) = \begin{bmatrix} \hat{m}(t) & \widehat{EI}(t) \end{bmatrix}^T. \tag{6.49}$$

Motivated by the subsequent Lyapunov-type stability analysis, the parameter estimate vector is updated according to

$$\dot{\hat{\theta}}(t) = \Gamma Y^T(\cdot)\eta(t), \tag{6.50}$$

where $\Gamma \in \mathbb{R}^{2\times 2}$ is a diagonal, positive-definite, adaptation gain matrix. If we define $\tilde{\theta}(t) = \theta - \hat{\theta}(t) \in \mathbb{R}^2$ as the parameter estimation error vector and then substitute (6.48) into (6.46), we obtain the closed-loop system dynamics as shown below:

$$m\dot{\eta}(t) = -k_s\eta(t) + Y(\cdot)\tilde{\theta}(t) \qquad \dot{\tilde{\theta}}(t) = -\Gamma Y^T(\cdot)\eta(t), \tag{6.51}$$

where (6.50) was used to obtain the parameter estimation error dynamics.

Theorem 6.2 *The adaptive boundary control law given by (6.48) and (6.50) ensures that the beam displacement is asymptotically regulated in the following sense:*

$$\lim_{t\to\infty} |u(x,t)| = 0 \qquad \forall x \in [0, L], \tag{6.52}$$

where the control gain k_s defined in (6.48) must be selected to satisfy (6.11).

Proof. The following proof is based on similar arguments as those used in the proof of Theorem 6.1; hence, some of the details will not be repeated. First, we define the following function:

$$V_a(t) = V(t) + \frac{1}{2}\tilde{\theta}^T(t)\Gamma^{-1}\tilde{\theta}(t), \tag{6.53}$$

where $V(t)$ was given in (6.13). If the constant β of (6.15) is selected sufficiently small according to the condition of (6.19), we can formulate the following bounds on $V_a(t)$:

$$\lambda_{1a}\left(E_b(t) + \eta^2(t) + \left\|\tilde{\theta}(t)\right\|^2\right) \le V_a(t) \le \lambda_{2a}\left(E_b(t) + \eta^2(t) + \left\|\tilde{\theta}(t)\right\|^2\right) \tag{6.54}$$

for some positive constants λ_{1a} and λ_{2a}. After differentiating (6.53) with respect to time and substituting (6.1), (6.2), and the first equation of (6.51), we can follow the derivations used in the proof of Theorem 6.1 to obtain the following upper bound for the time derivative of $V_a(t)$:

$$\dot{V}_a(t) \le -\lambda_3\left(E_b(t) + \eta^2(t)\right) + \tilde{\theta}^T(t)\left(Y^T(\cdot)\eta(t) + \Gamma^{-1}\dot{\tilde{\theta}}(t)\right), \tag{6.55}$$

where λ_3 was defined in (6.40). Substituting the second equation of (6.51) into (6.55) simplifies the upper bound on $\dot{V}_a(t)$ to

$$\dot{V}_a(t) \le -\lambda_3\left(E_b(t) + \eta^2(t)\right) \triangleq -g_a(t), \tag{6.56}$$

where $g_a(t)$ is a non-negative function. We can now use (6.54), (6.56), and the arguments outlined in Remark 6.2 to state that all signals in the control law of (6.48) and (6.50) are bounded $\forall t \in [0,\infty)$, and that all signals in the system given by (6.1) and (6.3) remain bounded $\forall x \in [0, L]$ and $\forall t \in [0,\infty)$.

After differentiating $g_a(t)$ defined in (6.56) with respect to time, we have

$$\dot{g}_a(t) = \lambda_{3a}\left(\dot{E}_b(t) + 2\eta(t)\dot{\eta}(t)\right). \tag{6.57}$$

Since we know that all system signals remain bounded $\forall t \in [0,\infty)$, we can use (6.25) and (6.51) to show that $\dot{E}_b(t)$ and $\dot{\eta}(t)$ are bounded $\forall t \in [0,\infty)$.

Hence, we can see from (6.57) that $\dot{g}_a(t)$ is bounded $\forall t \in [0,\infty)$. We can now apply Lemma A.6 in Appendix A to (6.56) to show that

$$\lim_{t\to\infty} E_b(t), \eta(t) = 0. \tag{6.58}$$

Finally, the result given by (6.58) and the inequalities developed in (6.44) can be combined to state the result given by (6.52). □

6.2.5 Extensions

We now illustrate how some extensions to the basic control laws given in the previous section can be developed.

Integral Feedback

Using the basic control structure given by (6.8), we can obtain a new control law with an additional integral feedback loop and an additional control gain in the definition of the tracking variable $\eta(t)$ defined in (6.6). That is, we redefine the model-based control force originally given in (6.8) as follows:[4]

$$f(t) = \alpha m u_{xxxt}(L,t) - EIu_{xxx}(L,t) - k_p\eta(t) - k_i \int_0^t \eta(\tau)d\tau. \tag{6.59}$$

where α, k_p, and k_i are positive control gains, and the variable $\eta(t)$ originally defined in (6.6) is now defined as follows:

$$\eta(t) = u_t(L,t) - \alpha u_{xxx}(L,t). \tag{6.60}$$

After deriving the open-loop dynamics for $\eta(t)$ in (6.60) in a similar fashion as was done in (6.7), we can substitute (6.59) to obtain the following closed-loop dynamics for $\eta(t)$:

$$m\dot{\eta}(t) = -k_p\eta(t) - k_i \int_0^t \eta(\tau)d\tau. \tag{6.61}$$

To prove the exponential stability result of Theorem 6.1 for the new controller of (6.59), we modify the function given in (6.13) as follows:

$$V_e(t) = E_b(t) + \frac{1}{2}z^T(t)\begin{bmatrix} \frac{1}{m}k_i + \frac{1}{2m}k_p & \frac{1}{2} \\ \frac{1}{2} & 1 \end{bmatrix} z(t) + E_c(t), \tag{6.62}$$

[4]The form of this new control law was motivated, in part, by experimental implementations of the controllers. These experimental results indicated that improved system performance could be potentially achieved by modifying (6.8) to include the modifications described by (6.59) and (6.60).

where $E_b(t)$ and $E_c(t)$ were defined in (6.14) and (6.15), respectively, and $z(t) \in \mathbb{R}^2$ is defined as

$$z(t) = \begin{bmatrix} \int_0^t \eta(\tau)d\tau & \eta(t) \end{bmatrix}^T. \tag{6.63}$$

If the constant β defined in (6.15) is selected to be sufficiently small, and the control gain k_p is selected to satisfy the following condition:

$$k_p > m, \tag{6.64}$$

we can follow similar steps as those used in the proof of Theorem 6.1 to formulate the following bounds on $V_e(t)$ of (6.62):

$$\lambda_{1e}\left(E_b(t) + \|z(t)\|^2\right) \leq V_e(t) \leq \lambda_{2e}\left(E_b(t) + \|z(t)\|^2\right) \tag{6.65}$$

for some positive constants λ_{1e} and λ_{2e}. After differentiating (6.62) with respect to time and substituting (6.1), (6.2), and (6.61), we can obtain the following upper bound[5] for the time derivative of $V_e(t)$ of (6.62):

$$\begin{aligned} \dot{V}_e(t) \leq & -\left(\frac{EI}{2\alpha} - \beta L\rho\right) u_t^2(L,t) + \frac{EI}{2\alpha}\eta^2(t) - z^T(t)Qz(t) \\ & -\beta E_b(t) - 2EIL\beta\left(\frac{1}{L^2} - \delta\right) u_x^2(L,t) \\ & -EI\left(\frac{\alpha}{2} - \frac{2L\beta}{\delta}\right) u_{xxx}^2(L,t), \end{aligned} \tag{6.66}$$

where $Q \in \mathbb{R}^{2\times 2}$ is given by

$$Q = \begin{bmatrix} \dfrac{k_i}{2m} & 0 \\ 0 & \dfrac{k_p}{m} - \dfrac{1}{2} \end{bmatrix}. \tag{6.67}$$

Since $k_i > 0$ and k_p has to satisfy (6.64), it is easy to see that the matrix Q is positive-definite. Hence, we can use (6.66) to obtain the following new

[5] To arrive at (6.66), we have used (6.60) to manipulate the expression for $\dot{E}_b(t)$ originally given in (6.26) into the following form $\dot{E}_b = -\frac{EI}{2\alpha}\left(u_t^2(L,t) + \alpha^2 u_{xxx}^2(L,t)\right) + \frac{EI}{2\alpha}\eta^2(t)$.

upper bound for $\dot{V}_e(t)$:

$$\begin{aligned}\dot{V}_e(t) \;\le\; & -\left(\frac{EI}{2\alpha}-\beta L\rho\right)u_t^2(L,t)-EI\left(\frac{\alpha}{2}-\frac{2L\beta}{\delta}\right)u_{xxx}^2(L,t)\\ & -\left(\min\left\{\frac{k_i}{2m},\left(\frac{k_p}{m}-\frac{1}{2}\right)\right\}-\frac{EI}{2\alpha}\right)\|z(t)\|^2\\ & -\beta E_b(t)-2EIL\beta\left(\frac{1}{L^2}-\delta\right)u_x^2(L,t).\end{aligned} \tag{6.68}$$

From (6.68), it is clear that k_i, k_p, δ, and β can be selected such that $\dot{V}_e(t)$ is upper bounded by a nonpositive function (i.e., for large enough k_i and k_p, for small enough δ, and even smaller β); hence,

$$\dot{V}_e(t) \le -\lambda_{3e}\left(E_b(t)+\|z(t)\|^2\right) \tag{6.69}$$

for some positive constant λ_{3e}. We can now use (6.65), (6.69), and the arguments outlined in Remark 6.2 and in the proof of Theorem 6.1 to show that all system signals remain bounded, and state a similar exponential stability result as that given by (6.10). Furthermore, by following the adaptive control derivations given in Section 6.2.4, we can redesign the control law given by (6.59) as an adaptive controller that compensates for parametric uncertainty and achieves asymptotic displacement regulation. Note that for this new adaptive controller, the regression matrix originally given in (6.47) will now be defined as follows:

$$Y(\cdot) = \left[\begin{array}{cc} -\alpha u_{xxxt}(L,t) & u_{xxx}(L,t) \end{array}\right]. \tag{6.70}$$

Compensation of Free End-Point Inertial Effects

Since we have assumed that the payload at the free end of the beam is a point-mass, the dynamic model given by (6.1) through (6.3) neglects inertial effects at the end of the beam. If it is deemed that the free end-point inertial effects cannot be neglected, we can use the following additional dynamic boundary condition:

$$Ju_{xtt}(L,t)+EIu_{xx}(L,t)=\tau(t), \tag{6.71}$$

where J denotes the inertia of the payload at the end of the beam, and $\tau(t)$ is a boundary control input torque applied to the end of the beam. To maintain the exponential stability result of Theorem 6.1, we can design the input torque $\tau(t)$ as follows:

$$\tau(t) = -Ju_{xxt}(L,t) + EIu_{xx}(L,t) - k_\tau \left(u_{xt}(L,t) + u_{xx}(L,t)\right), \qquad (6.72)$$

where k_τ is a positive control gain.

By adding the term $\frac{1}{2}J\left(u_{xt}(L,t) + u_{xx}(L,t)\right)^2$ to the function $V(t)$ of (6.13), we can easily extend the proof of Theorem 6.1 to illustrate that the additional boundary control input of (6.72) can be used in conjunction with the input force of (6.8) to compensate for free end mass and inertial effects. Note from (6.72) that the implementation of this new control strategy will require the additional measurements of the beam's end-point strain $u_{xx}(L,t)$, time derivative of the end-point strain $u_{xxt}(L,t)$, and time derivative of the end-point slope $u_{xt}(L,t)$. In a similar manner to Section 6.2.4, an adaptive control torque strategy can also be designed and combined with (6.48) to compensate for parametric uncertainty (including the unknown inertia) and produce the same asymptotic stability result of Theorem 6.2. Note, however, that since the Timoshenko beam model accounts for the rotary inertia of the beam and the deflection owing to shear as discussed in Section 6.1, it seems more reasonable to use the more complex Timoshenko model for control design when the end-point inertia cannot be neglected, as illustrated later in Section 6.3.

6.2.6 Experimental Evaluation

Experimental Setup

A schematic of the experimental setup used in the real-time implementation of the controllers is shown in Figure 6.2. A flexible beam 72 cm in length was attached to the top of a support structure. A small metal cylinder weighing 0.3 kg was attached to the free end-point of the flexible beam via a strain-gauge shear sensor to serve as the payload mass. The axis of the beam was made to coincide with the center of gravity of the mass. The displacement of the beam's end-point (i.e., $u(L,t)$) was sensed by a small laser displacement sensor mounted perpendicularly to the beam. Another laser displacement sensor was used to measure the displacement of the beam's mid-point (i.e., $u(L/2,t)$). The signal from second laser was only used to monitor the performance of the controller and was not used in the control strategy.

At the end-point of the beam, a pair of electromagnets were arranged to apply the boundary control input force on the payload mass. A custom designed software commutation strategy ensured that the desired input force

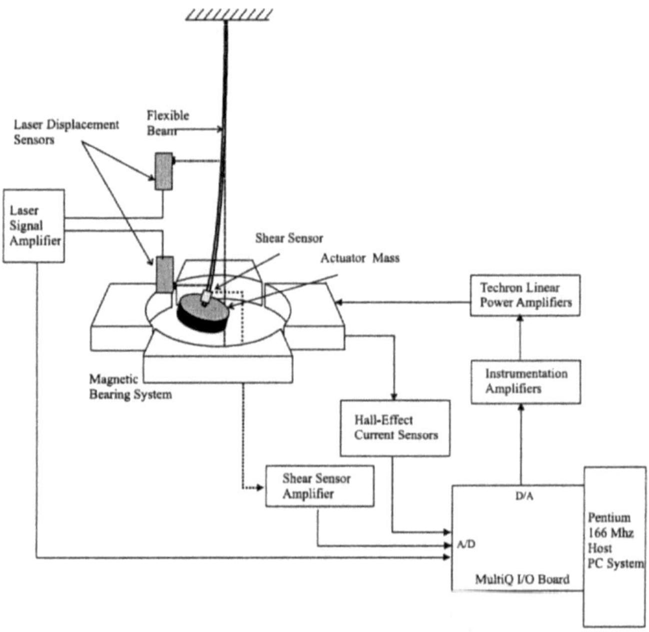

FIGURE 6.2. Schematic diagram of the Euler-Bernoulli beam experimental setup.

commanded by the control law was applied to the mass. Roughly speaking, the commutation strategy involved translation of the desired force trajectory into desired current trajectories. To ensure that the actual currents tracked the desired current trajectories, a high-gain current feedback loop was utilized to apply voltages to the two electromagnets. The two magnetizing currents were measured using hall-effect current sensors. The signal from the force sensor's bridge circuit was scaled and filtered before being read into an analog input port on a motion control board. The laser signal was sent to an analog input port on the motion control board after scaling. A Pentium 166 MHz PC running QNX hosted the control algorithm. *Qmotor* provided the environment to write the control algorithm in the C programming language. All controllers were implemented using a sampling period of 0.5 msec. The velocity of the beam's end-point (i.e., $u_t(L,t)$) and the time derivative of end-point shear (i.e., $u_{xxxt}(L,t)$) were obtained by use of a backwards difference algorithm applied to the signals $u(L,t)$ and $u_{xxx}(L,t)$, respectively, with the resulting signals being filtered by a second-order digital filter. The parameters associated with the flexible Euler-Bernoulli beam model were calculated using standard test procedures and engineering handbook tables, and were found to be $EI = 2.65$ N-m^2 and $\rho = 1.56$ kg/m.

Experimental Results under Impulse Excitation

To ensure a consistent excitation, an impulse hammer was released from a latched position and allowed to strike the beam only once and at the same point each time. First, the uncontrolled (i.e., open-loop) response of the beam's end-point and mid-point displacements when struck by the impulse hammer were recorded and are shown in Figure 6.3. Note that the uncontrolled oscillations of the beam take a long time to damp out.

Experiment 1: For comparison purposes, a standard active damper was implemented using the following control law:

$$f(t) = -k_p u_t(L,t), \tag{6.73}$$

where k_p is a positive control gain. The performance of the damper control law in terms of the beam's end-point and mid-point displacements for three values of the control gain k_p (i.e., k_p = 1, 2.5, and 6) is shown in Figure 6.4. Note that increasing the control gain k_p has a favorable effect on the damping of the low-frequency oscillations induced by the excitation; however, the high-frequency oscillations induced by the impulse excitation seem to increase as k_p is increased. It was observed that for $k_p > 6.5$, the high-frequency oscillations in the beam increased continuously even in the absence of the impulse excitation. One heuristic explanation for this behavior is that as k_p is increased, the beam dynamics began to more closely resemble a *clamped-clamped* configuration instead of the *clamped-free end* configuration described by (6.1) and (6.2).

Experiment 2: The model-based controller given by (6.59) and (6.60) was implemented with three sets of control gains:

$$k_p = 2.5, \qquad \alpha = 1.1, \qquad k_i = 0$$

$$k_p = 5, \qquad \alpha = 0.55, \qquad k_i = 0$$

$$k_p = 7.5, \qquad \alpha = 0.38, \qquad k_i = 0.$$

The performance of the controller in terms of the beam's end-point and mid-point displacements is shown in Figure 6.5. Note that the model-based controller damps out both the high-frequency and the low-frequency oscillations quite satisfactorily provided k_p is selected to be small. Several other experiments were run with k_i set to different nonzero values; however, no significant improvement in the regulation performance was observed.

Experiment 3: Finally, the adaptive version of the controller given by (6.59), (6.60), and (6.50) was implemented by assuming no a priori knowledge of the system parameters m and EI. The best performance was

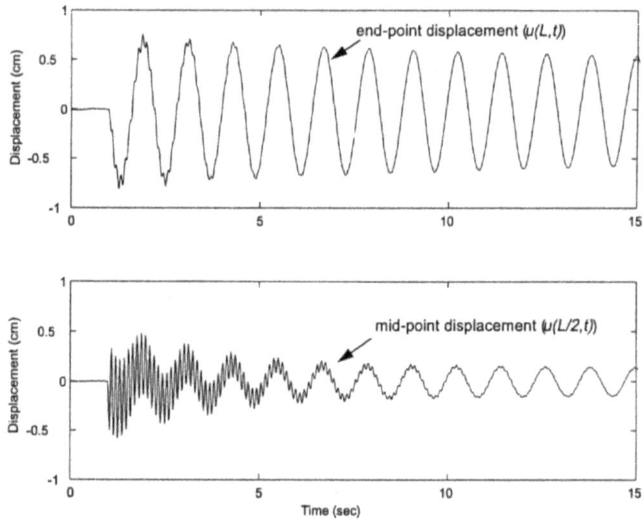

FIGURE 6.3. Uncontrolled response of the beam owing to impulse excitation.

achieved with the gains $k_p = 5.0$, $\alpha = 0.3$, $k_i = 0$, and $\Gamma = \text{diag}\{5, 5\}$. The estimates of the unknown parameters were initialized to zero and computed on-line using a standard trapezoidal rule of numerical integration. The controller performance in terms of the beam's end-point and mid-point displacements is shown in Figure 6.6. The parameter estimates are also shown in Figure 6.6.

As the above results indicate, the model-based and adaptive controllers regulate the beam displacement a little faster when compared to the active damper control of (6.73); however, the high-frequency oscillations have been significantly reduced.

Experimental Results Caused by an End-Point Offset

This set of experiments consisted of releasing the beam's end-point from an offset position via the use of a mechanical latch. The uncontrolled response of the beam's end-point when released from the end-point offset is shown in Figure 6.7. Again, note that the uncontrolled oscillations take a long time to damp out.

Experiment 1: First, the active damper control algorithm given by (6.73) was implemented. The best regulating performance was achieved with $k_p = 100$. The beam's end-point displacement is shown in Figure 6.8. The control energy (i.e., the integral of the square root of the sum of the squares of

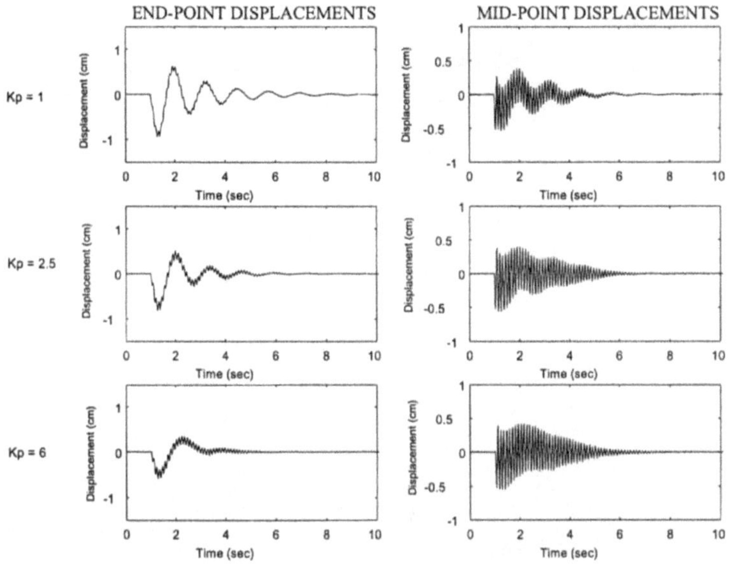

FIGURE 6.4. Impluse excitation experiment - active damper controller.

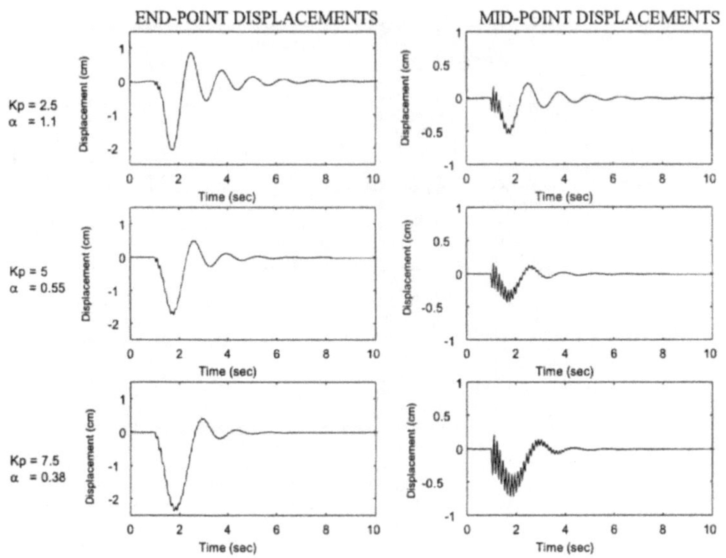

FIGURE 6.5. Impluse excitation experiment - model-based controller.

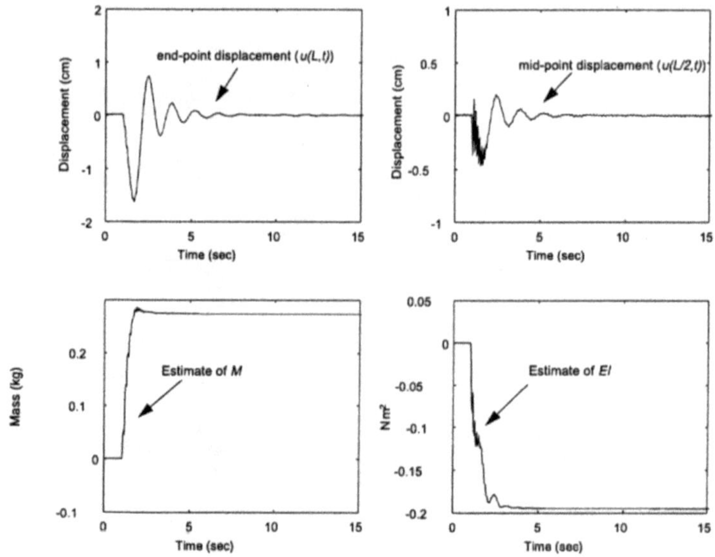

FIGURE 6.6. Impulse excitation experiment - adaptive controller.

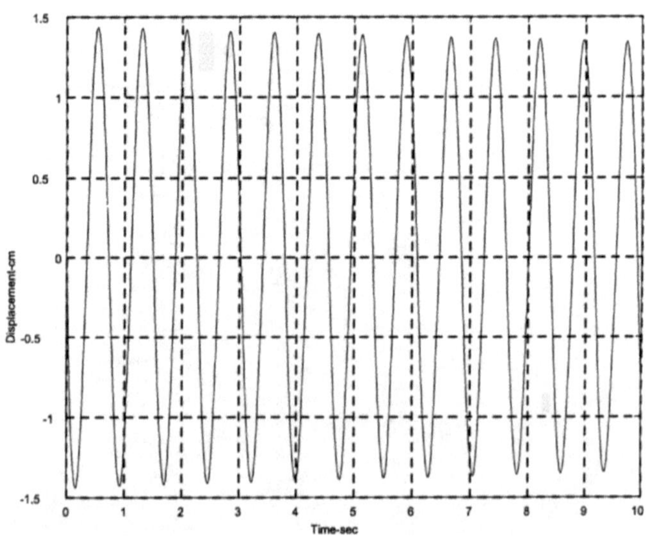

FIGURE 6.7. Uncontrolled response of the beam owing to end-point offset.

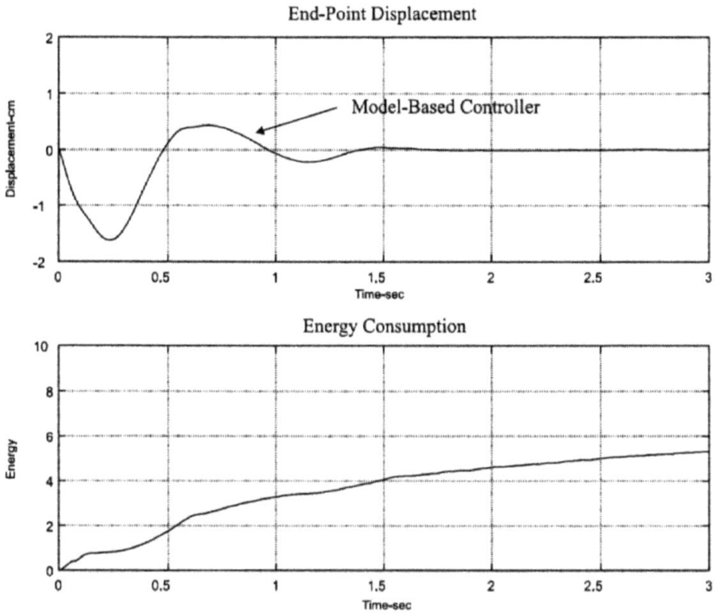

FIGURE 6.8. End-point offset experiment - active damper controller.

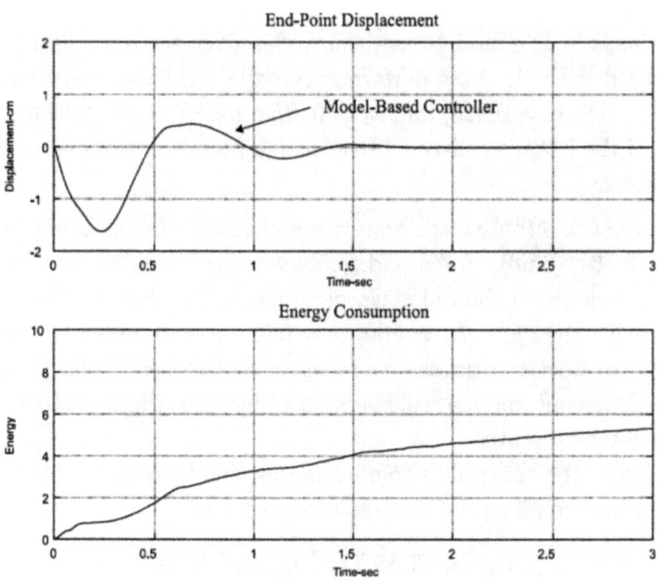

FIGURE 6.9. End-point offset experiment - model-based controller.

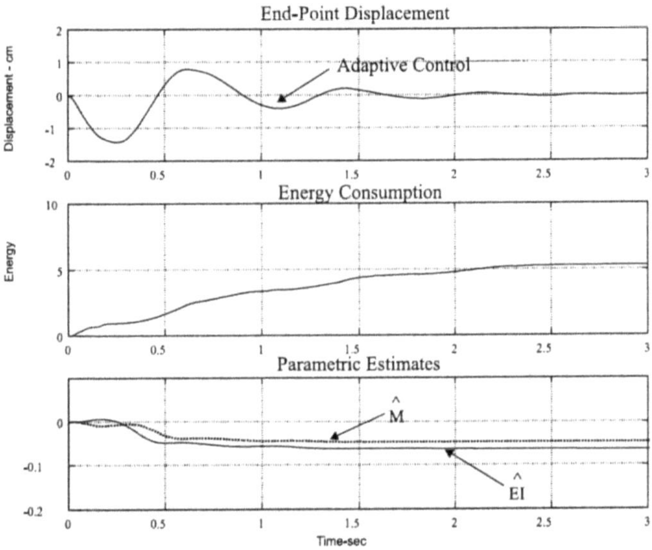

FIGURE 6.10. End-point offset experiment - adaptive controller.

the two actuator electrical currents) was computed on-line to serve as an additional performance index to compare the controllers, and is also shown in Figure 6.8.

Experiment 2: The model-based controller given by (6.59) and (6.60) was implemented with the best performance being achieved, with the control gains $k_p = 100$, $\alpha = 0.026$, and $k_i = 0$. The performance of the controller in terms of the beam's end-point displacement and control energy is shown in Figure 6.9.

Experiment 3: Finally, the adaptive controller version of the controller given by (6.59), (6.60), (6.70) and (6.50) was implemented by assuming no a priori knowledge of the unknown parameters. The best performance was achieved with the gains $k_p = 100$, $\alpha = 0.02$, $k_i = 0$, and $\Gamma = \text{diag}\{1, 1\}$, while the parameter estimates were initialized to zero. The beam's end-point displacement and control energy are shown in Figure 6.10 along with the parameter estimates.

To compare the controllers' performances, the following table is utilized to summarize the results of the above experiment:

Controller	M_p cm	t_s sec	E at 3 sec
Damper	-1.74	2.50	5.75
Model-Based	-1.63	1.35	5.39
Adaptive	-1.44	1.90	5.71

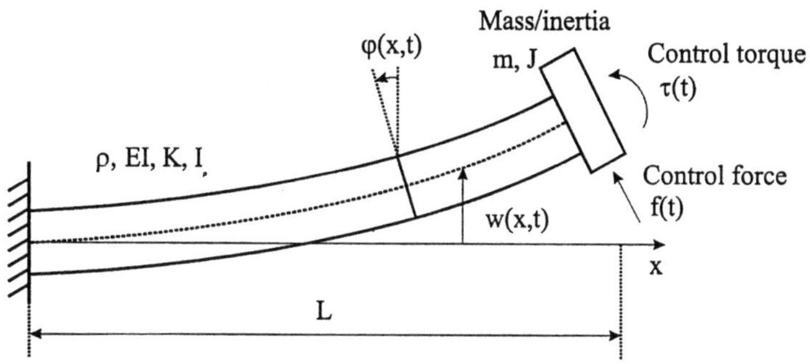

FIGURE 6.11. Schematic representation of a cantilevered Timoshenko beam with end-point mass/inertia.

where M_p denotes the maximum peak end-point displacement, t_s denotes the settling time corresponding to the time when the beam's end-point displacement is less than or equal to ± 0.8 [mm], and E denotes the control energy. While M_p and E for each controller are all relatively close in their numerical value, the model-based controller exhibited superior settling time. Furthermore, it is interesting to note that t_s of the adaptive controller is significantly better than that of damper controller, even though the adaptive controller requires no knowledge of the system parameters.

6.3 Timoshenko Beam

In this section, we consider the problem of stabilizing the distributed displacement and cross-section rotation of a cantilevered Timoshenko beam with mass and inertial dynamics at the beam's free end-point. The control strategy, which is composed of boundary force and torque inputs applied to the beam's free end-point, requires measurements of the end-point displacement, slope, rotation owing to bending, slope of the rotation owing to bending, and the time derivatives of these quantities. As in Section 6.2, a model-based controller is first developed which exponentially regulates the beam displacement and rotation. An adaptive controller is then designed to asymptotically regulate the beam displacement and rotation while compensating for parametric uncertainty.

6.3.1 System Model

The Timoshenko model for the cantilevered beam system depicted in Figure 6.11 is assumed to be described by the following PDEs [5, 19]:

$$\rho w_{tt}(x,t) + K\left(\varphi_x(x,t) - w_{xx}(x,t)\right) = 0 \tag{6.74}$$

$$I_\rho \varphi_{tt}(x,t) - EI\varphi_{xx}(x,t) + K\left(\varphi(x,t) - w_x(x,t)\right) = 0, \tag{6.75}$$

with the following boundary conditions:[6]

$$w(0,t) = \varphi(0,t) = 0, \tag{6.76}$$

$$m w_{tt}(L,t) - K\left(\varphi(L,t) - w_x(L,t)\right) = f(t), \tag{6.77}$$

and

$$J\varphi_{tt}(L,t) + EI\varphi_x(L,t) = \tau(t), \tag{6.78}$$

where $w(x,t)$ denotes the beam displacement at the position x for time t; $\varphi(x,t)$ denotes the rotation of the beam's cross-section owing to bending at the position x for time t; $f(t)$ denotes the boundary control input force; $\tau(t)$ denotes the boundary control input torque; ρ is the mass/length of the beam; $K = kGA$; k is a positive constant that depends on the shape of the beam's cross-section; G is the modulus of elasticity in shear; A is the cross-sectional area of the beam; I_ρ is the mass moment of inertia of the beam's cross-section; EI is the bending stiffness of the beam; L is the length of the beam; and m, J denote the mass and inertia of the payload/actuator, respectively, attached to the beam's end-point.

Since the control strategies will consist of relatively simple functions, we will assume the existence of a unique solution for the dynamics given in (6.74) through (6.78) under the control. Furthermore, based on the arguments outlined in Remark 5.1, we will assume that the distributed variables $w(x,t), \varphi(x,t)$ and their time derivatives $u_t(x,t), \varphi_t(x,t)$ belong to a space of functions that have the following properties.

Property 6.3: If the potential energy of the system given by (6.74) through (6.78), defined as

$$\Pi_p = \frac{1}{2}K\int_0^L \left(\varphi(\sigma,t) - w_\sigma(\sigma,t)\right)^2 d\sigma + \frac{1}{2}EI\int_0^L \varphi_\sigma^2(\sigma,t)\, d\sigma, \tag{6.79}$$

[6] Given the clamped end boundary conditons of (6.76), we also know that $w_t(0,t) = 0$ and $\varphi_t(0,t) = 0$.

is bounded, then $\frac{\partial^n}{\partial x^n} w(x,t)$ and $\frac{\partial^n}{\partial x^n} \varphi(x,t)$ are bounded for $n = 1$, 2, $\forall t \in [0, \infty)$, and $\forall x \in [0, L]$.

Property 6.4: If the kinetic energy of the system given by (6.74) through (6.78), defined as

$$\Pi_k = \frac{1}{2} \int_0^L \left(\rho w_t^2(\sigma,t) + I_\rho \varphi_t^2(\sigma,t) \right) d\sigma + \frac{1}{2} m w_t^2(L,t) + J \varphi_t^2(L,t), \tag{6.80}$$

is bounded, then $\frac{\partial^n}{\partial x^n} w_t(x,t)$ and $\frac{\partial^n}{\partial x^n} \varphi_t(x,t)$ are bounded for $n = 0$, 1, $\forall t \in [0, \infty)$, and $\forall x \in [0, L]$.

6.3.2 Problem Statement

The control objectives are to design the boundary control inputs $f(t)$ and $\tau(t)$ to drive both the beam displacement $w(x,t)$ and cross-section rotation $\varphi(x,t)$ to zero $\forall x \in [0, L]$ as $t \to \infty$. As in Section 6.2.2, in order to facilitate the control design and the subsequent Lyapunov-type stability analysis, we start by defining the auxiliary signals, denoted by $\eta_1(t)$ and $\eta_2(t)$, as follows:

$$\eta_1(t) = w_t(L,t) + w_x(L,t) - \varphi(L,t) \qquad \eta_2(t) = \varphi_t(L,t) + \varphi_x(L,t). \tag{6.81}$$

After differentiating $\eta_1(t)$ and $\eta_2(t)$ of (6.81) with respect of time, multiplying the resulting expressions by m and J, respectively, and then utilizing (6.77) and (6.78) to substitute for $m w_{tt}(L,t)$ and $J\varphi_{tt}(L,t)$, respectively, we obtain

$$m\dot{\eta}_1(t) = m w_{xt}(L,t) - m\varphi_t(L,t) + K\varphi(L,t) - K w_x(L,t) + f(t)$$

$$J\dot{\eta}_2(t) = J\varphi_{xt}(L,t) - EI\varphi_x(L,t) + \tau(t). \tag{6.82}$$

The above open-loop dynamics will dictate the design of the model-based and adaptive control laws.

6.3.3 Model-Based Control Law

Based on the form of the open-loop dynamics of (6.82), the boundary control force and torque inputs are designed as follows:

$$f(t) = -m w_{xt}(L,t) + m\varphi_t(L,t) - K\varphi(L,t) + K w_x(L,t) - k_s \eta_1(t) \tag{6.83}$$

and

$$\tau(t) = -J\varphi_{xt}(L,t) + EI\varphi_x(L,t) - k_r\eta_2(t), \tag{6.84}$$

where k_s and k_r are positive control gains. After substituting (6.83) and (6.84) into (6.82), we obtain the following closed-loop dynamics for $\eta_1(t)$ and $\eta_2(t)$:

$$m\dot{\eta}_1(t) = -k_s\eta_1(t) \qquad J\dot{\eta}_2(t) = -k_r\eta_2(t). \tag{6.85}$$

Theorem 6.3 *The model-based boundary control law given by (6.83) and (6.84) ensures that the beam displacement $w(x,t)$ and cross-section rotation $\varphi(x,t)$ are exponentially regulated in the following sense:*

$$|w(x,t)|, |\varphi(x,t)| \leq \sqrt{\frac{L\lambda_2}{\lambda_1}\kappa_o \exp\left(-\frac{\lambda_3}{\lambda_2}t\right)} \qquad \forall x \in [0,L], \tag{6.86}$$

provided the control gains k_s and k_r satisfy the following inequalities:

$$k_s > \frac{K}{2} \qquad k_r > \frac{EI}{2}, \tag{6.87}$$

where λ_1, λ_2, and λ_3 are some positive bounding constants, and the positive constant κ_o is given by

$$\begin{aligned} \kappa_o = & \int_0^L \left[\varphi_t^2(\sigma,0) + w_t^2(\sigma,0) + \varphi_\sigma^2(\sigma,0) + w_\sigma^2(\sigma,0)\right. \\ & \left.+\varphi^2(\sigma,0)\right] d\sigma + \left(w_t(L,0) + w_x(L,0) - \varphi(L,0)\right)^2 \\ & + \left(\varphi_t(L,0) + \varphi_x(L,0)\right)^2. \end{aligned} \tag{6.88}$$

Proof. The following proof utilizes the same basic steps adopted in the proof of Theorem 6.1. We begin by defining the following function:

$$V(t) = E_b(t) + \frac{1}{2}m\eta_1^2(t) + \frac{1}{2}J\eta_2^2(t) + \sum_{i=1}^{4} E_{ci}(t), \tag{6.89}$$

where

$$\begin{aligned} E_b(t) = & \frac{\rho}{2}\int_0^L w_t^2(\sigma,t)d\sigma + \frac{I_\rho}{2}\int_0^L \varphi_t^2(\sigma,t)d\sigma + \frac{EI}{2}\int_0^L \varphi_\sigma^2(\sigma,t)d\sigma \\ & +\frac{K}{2}\int_0^L \left(\varphi(\sigma,t) - w_\sigma(\sigma,t)\right)^2 d\sigma, \end{aligned} \tag{6.90}$$

$$\begin{aligned}
E_{c1}(t) &= 2\beta\rho \int_0^L w_t(\sigma,t) w_\sigma(\sigma,t)\, d\sigma, \\
E_{c2}(t) &= 2\beta I_\rho \int_0^L \sigma\varphi_t(\sigma,t)\varphi_\sigma(\sigma,t)\, d\sigma, \\
E_{c3}(t) &= -\mu\beta\rho \int_0^L w_t(\sigma,t) w(\sigma,t)\, d\sigma, \\
E_{c4}(t) &= \mu\beta I_\rho \int_0^L \varphi_t(\sigma,t)\varphi(\sigma,t)\, d\sigma,
\end{aligned} \tag{6.91}$$

and β, μ are positive weighting constants. The term $E_b(t)$ defined in (6.90) represents the beam's total energy, while the second and third terms in (6.89) are related to the kinetic energy of the end-point mass/inertia. The cross terms $E_{ci}(t)$, defined in (6.91), were originally proposed in [5] and are utilized to facilitate the stability analysis.

Notice that (6.90) can be rewritten as

$$E_b = \frac{1}{2}\int_0^L \left(\rho w_t^2 + I_\rho \varphi_t^2 + K\varphi^2 + K w_\sigma^2 + EI\varphi_\sigma^2\right) d\sigma - K\int_0^L \varphi w_\sigma d\sigma. \tag{6.92}$$

After applying (A.29) of Lemma A.13 in Appendix A to the last term on the right-hand side of (6.92), we can upper bound $E_b(t)$ as follows:

$$E_b \leq \frac{1}{2}\max\{\rho, I_\rho, EI, 2K\}\, E_n, \tag{6.93}$$

where

$$E_n = \int_0^L \left(\varphi_t^2 + w_t^2 + \varphi_\sigma^2 + w_\sigma^2 + \varphi^2\right) d\sigma. \tag{6.94}$$

After applying (A.24) of Lemma A.12 in Appendix A to the φ_σ^2 term of (6.92), we can lower bound $E_b(t)$ as shown below:

$$\frac{1}{2}\int_0^L \left(\rho w_t^2 + I_\rho\varphi_t^2 + \frac{EI}{2}\varphi_\sigma^2 + \begin{bmatrix} \varphi \\ w_\sigma \end{bmatrix}^T A \begin{bmatrix} \varphi \\ w_\sigma \end{bmatrix}\right) d\sigma \leq E_b, \tag{6.95}$$

with $A \in \mathbb{R}^{2\times 2}$ being defined as follows:

$$A = \begin{bmatrix} K + \dfrac{EI}{2L^2} & -K \\ -K & K \end{bmatrix}.$$

Since the matrix A is positive-definite, we can now see from (6.93) and (6.95) that

$$\frac{1}{2}\min\left\{\rho, I_\rho, \frac{EI}{2}, \lambda_{\min}\{A\}\right\} E_n \leq E_b \leq \frac{1}{2}\max\{\rho, I_\rho, EI, 2K\}\, E_n, \tag{6.96}$$

where $\lambda_{\min}\{\cdot\}$ denotes the minimum eigenvalue of a matrix. After utilizing (A.29) of Lemma A.13 in Appendix A, $E_{c1}(t)$ of (6.91) can be upper bounded as follows:

$$E_{c1} \leq 2\beta\rho L \int_0^L \left(w_t^2 + w_\sigma^2\right) d\sigma. \tag{6.97}$$

We can now use (6.94) and (6.97) to establish the following inequalities:

$$-2\beta\rho L E_n \leq E_{c1} \leq 2\beta\rho L E_n. \tag{6.98}$$

In a similar manner, we can use (A.24) of Lemma A.12 in Appendix A, (A.29) of Lemma A.13 in Appendix A, and (6.94) to bound $E_{c2}(t)$, $E_{c3}(t)$, and $E_{c4}(t)$ of (6.91) as follows:

$$\begin{gathered}-2\beta I_\rho L E_n \leq E_{c2} \leq 2\beta I_\rho L E_n \\ -\mu\beta\rho\max(1, L^2)E_n \leq E_{c3} \leq \mu\beta\rho\max(1, L^2)E_n \\ -\mu\beta I_\rho E_n \leq E_{c4} \leq \mu\beta I_\rho E_n.\end{gathered} \tag{6.99}$$

Now, from (6.96), (6.98), and (6.99), if β is selected as

$$\beta < \frac{\frac{1}{2}\min\left\{\rho, I_\rho, \frac{EI}{2}, \lambda_{\min}\{A\}\right\}}{2\rho L + 2I_\rho L + \mu I_\rho + \mu\rho\max(1, L^2)}, \tag{6.100}$$

then

$$\zeta_1 E_n \leq E_b + \sum_{i=1}^{4} E_{ci} \leq \zeta_2 E_n, \tag{6.101}$$

where the positive constants ζ_1 and ζ_2 are defined as follows:

$$\begin{aligned}\zeta_1 = {} & \frac{1}{2}\min\left\{\rho, I_\rho, \frac{EI}{2}, \lambda_{\min}\{A\}\right\} - 2\beta\rho L - 2\beta I_\rho L \\ & -\mu\beta I_\rho - \mu\beta\rho\max(1, L^2) \\ \zeta_2 = {} & \frac{1}{2}\max\{\rho, I_\rho, EI, 2K\} + 2\beta\rho L + 2\beta I_\rho L + \mu\beta I_\rho + \mu\beta\rho\max(1, L^2).\end{aligned} \tag{6.102}$$

Given the structure of $V(t)$ defined in (6.89) and the inequality of (6.101), it is not difficult to see that

$$\lambda_1\left(E_n(t) + \eta_1^2(t) + \eta_2^2(t)\right) \leq V(t) \leq \lambda_2\left(E_n(t) + \eta_1^2(t) + \eta_2^2(t)\right), \tag{6.103}$$

where the positive constants λ_1 and λ_2 are defined below:

$$\lambda_1 = \min\left\{\zeta_1, \frac{M}{2}, \frac{J}{2}\right\} \qquad \lambda_2 = \max\left\{\zeta_2, \frac{M}{2}, \frac{J}{2}\right\}. \tag{6.104}$$

After differentiating (6.89) with respect to time, we have

$$\dot{V} = \dot{E}_b + \sum_{i=1}^{4} \dot{E}_{ci} - k_s\eta_1^2 - k_r\eta_2^2, \tag{6.105}$$

where (6.85) has been used. To determine $\dot{E}_b(t)$ in (6.105), we differentiate (6.90) with respect to time to obtain

$$\begin{aligned}\dot{E}_b = & \int_0^L \left(Kw_t\left(w_{\sigma\sigma} - \varphi_\sigma\right) + EI\varphi_t\varphi_{\sigma\sigma}\right)d\sigma + K\int_0^L \varphi_t\left(w_\sigma - \varphi\right)d\sigma \\ & + \int_0^L \left(K\left(\varphi - w_\sigma\right)\left(\varphi_t - w_{\sigma t}\right) + EI\varphi_\sigma\varphi_{\sigma t}\right)d\sigma,\end{aligned} \tag{6.106}$$

where (6.74) and (6.75) have been utilized. If we integrate, by parts, the first integral of (6.106), we obtain

$$\begin{aligned}\dot{E}_b = & \; Kw_t\left(w_x - \varphi\right)\big|_0^L + EI\varphi_t\varphi_x\big|_0^L \\ & - \int_0^L \left(Kw_{\sigma t}\left(w_\sigma - \varphi\right) + EI\varphi_{\sigma t}\varphi_\sigma\right)d\sigma \\ & + \int_0^L \left[K\left(\varphi - w_\sigma\right)\left(\varphi_t - w_{\sigma t}\right) + EI\varphi_\sigma\varphi_{\sigma t} + \varphi_t\left(-K\varphi + Kw_\sigma\right)\right]d\sigma,\end{aligned} \tag{6.107}$$

which, after the cancellation of common terms and the application of (6.76), reduces to

$$\dot{E}_b = K\left(w_x(L) - \varphi(L)\right)w_t(L) + EI\varphi_x(L)\varphi_t(L). \tag{6.108}$$

By utilizing the definitions given in (6.81), we can rewrite (6.108) in the following advantageous form:

$$\dot{E}_b = \frac{K}{2}\eta_1^2 - \frac{K}{2}w_t^2(L) - \frac{K}{2}\left(w_x(L) - \varphi(L)\right)^2 + \frac{EI}{2}\eta_2^2 - \frac{EI}{2}\varphi_t^2(L) - \frac{EI}{2}\varphi_x^2(L). \tag{6.109}$$

To determine $\dot{E}_{c1}(t)$ in (6.105), we differentiate $E_{c1}(t)$ of (6.91) with respect to time and substitute (6.74) into the resulting equation to produce

$$\dot{E}_{c1} = 2\beta\rho\int_0^L \sigma w_t w_{\sigma t}d\sigma + 2\beta K\int_0^L x w_\sigma w_{\sigma\sigma}d\sigma - 2\beta K\int_0^L \sigma w_\sigma\varphi_\sigma d\sigma. \tag{6.110}$$

After integrating, by parts, the first two integrals of (6.110), we obtain

$$\begin{aligned}\dot{E}_{c1} = \; & \beta\rho\left(Lw_t^2(L) - \int_0^L w_t^2 d\sigma\right) + \beta K\left(Lw_x^2(L) - \int_0^L w_\sigma^2 d\sigma\right) \\ & -2\beta K \int_0^L \sigma w_\sigma \varphi_\sigma \, d\sigma.\end{aligned} \tag{6.111}$$

To determine $\dot{E}_{c2}(t)$ in (6.105), we differentiate $E_{c2}\,(t)$ of (6.91) with respect to time and substitute (6.75) into the resulting expression to yield

$$\begin{aligned}\dot{E}_{c2} = \; & 2\beta I_\rho \int_0^L \sigma\varphi_t\varphi_{\sigma t} d\sigma + 2\beta EI \int_0^L \sigma\varphi_\sigma\varphi_{\sigma\sigma} d\sigma \\ & -2\beta K \int_0^L \sigma\varphi\varphi_\sigma d\sigma + 2\beta K \int_0^L \sigma w_\sigma \varphi_\sigma \, d\sigma.\end{aligned} \tag{6.112}$$

After integrating, by parts, the first three integrals of (6.112), we have

$$\begin{aligned}\dot{E}_{c2} = \; & \beta I_\rho\left(L\varphi_t^2(L) - \int_0^L \varphi_t^2 d\sigma\right) + \beta EI\left(L\varphi_x^2(L) - \int_0^L \varphi_\sigma^2 d\sigma\right) \\ & -\beta K\left(L\varphi^2(L) - \int_0^L \varphi^2 d\sigma\right) + 2\beta K \int_0^L \sigma w_\sigma \varphi_\sigma \, d\sigma.\end{aligned} \tag{6.113}$$

In a similar manner, the time derivatives of $E_{c3}(t)$ and $E_{c4}(t)$ defined in (6.91) can be determined to be

$$\dot{E}_{c3} = -\mu\beta\rho \int_0^L w_t^2 d\sigma + \mu\beta K \int_0^L w\varphi_\sigma d\sigma - \mu\beta K\left(w(L)w_x(L) - \int_0^L w_\sigma^2 d\sigma\right) \tag{6.114}$$

and

$$\begin{aligned}\dot{E}_{c4} = \; & \mu\beta I_\rho \int_0^L \varphi_t^2 d\sigma - \mu\beta K \int_0^L \varphi^2 d\sigma + \mu\beta K \int_0^L \varphi w_\sigma d\sigma \\ & +\mu\beta EI\left(\varphi(L,t)\varphi_x(L,t) - \int_0^L \varphi_\sigma^2 d\sigma\right).\end{aligned} \tag{6.115}$$

After combining (6.111), (6.113), (6.114), and (6.115), we have

$$\begin{aligned}\sum_{i=1}^{4}\dot{E}_{ci} = {} & \beta\rho L w_t^2(L) + \beta I_\rho L\varphi_t^2(L) + \beta KLw_x^2(L) + \beta EIL\varphi_x^2(L) \\ & -\beta KL\varphi^2(L) + \mu\beta K\int_0^L \frac{d}{d\sigma}(\varphi w)\,d\sigma - \mu\beta Kw(L)w_x(L) \\ & +\mu\beta EI\varphi(L)\varphi_x(L) + \beta K(1-\mu)\left[\int_0^L \varphi^2 d\sigma\right] \\ & -\beta EI(1+\mu)\int_0^L \varphi_\sigma^2 d\sigma - \beta I_\rho(1-\mu)\int_0^L \varphi_t^2 d\sigma \\ & -\beta\rho(1+\mu)\int_0^L w_t^2 d\sigma - \beta K(1-\mu)\int_0^L w_\sigma^2 d\sigma.\end{aligned} \tag{6.116}$$

After applying (A.24) of Lemma A.12 in Appendix A to the bracketed term of (6.116), and then selecting μ such that

$$\max\left\{0, \frac{KL^2 - EI}{KL^2 + EI}\right\} < \mu < 1, \tag{6.117}$$

we can upper bound (6.116) in the following manner:

$$\begin{aligned}\sum_{i=1}^{4}\dot{E}_{ci} \le {} & \beta\rho L w_t^2(L) + \beta I_\rho L\varphi_t^2(L) + \beta EIL\varphi_x^2(L) + \mu\beta EI\left[\varphi(L)\varphi_x(L)\right] \\ & +\beta KL\left(w_x(L) - \varphi(L)\right)^2 + 2\beta KL\left[\varphi(L)\left(w_x(L) - \varphi(L)\right)\right] \\ & -\mu\beta K\left[w(L)\left(w_x(L) - \varphi(L)\right)\right] \\ & -\beta\delta_0\int_0^L \left(\varphi_t^2 + w_t^2 + \varphi_\sigma^2 + w_\sigma^2\right)d\sigma,\end{aligned} \tag{6.118}$$

where the positive constant δ_0 is defined as follows:

$$\delta_0 = \min\left\{EI(1+\mu) - KL^2(1-\mu), I_\rho(1-\mu), \rho(1+\mu), K(1-\mu)\right\}. \tag{6.119}$$

After applying (A.24) and (A.25) of Lemma A.12 in Appendix A to the last term in (6.118), and then applying (A.30) of Lemma A.13 in Appendix

A to the three bracketed terms in (6.118), we obtain

$$
\begin{aligned}
\sum_{i=1}^{4} \dot{E}_{ci} \leq\ & \beta\rho L w_t^2(L) + \beta I_\rho L \varphi_t^2(L) + \beta EIL\varphi_x^2(L) \\
& + \beta KL\left(w_x(L) - \varphi(L)\right)^2 + \mu\beta EI\left(\frac{\varphi_x^2(L)}{\delta_3} + \delta_3\varphi^2(L)\right) \\
& + 2\beta KL\left(\frac{\left(w_x(L) - \varphi(L)\right)^2}{\delta_4} + \delta_4\varphi^2(L)\right) \\
& + \mu\beta K\left(\frac{\left(w_x(L) - \varphi(L)\right)^2}{\delta_5} + \delta_5 w^2(L)\right) - \beta\delta_2 w^2(L) \\
& - \beta\delta_1\int_0^L \left(\varphi_t^2 + w_t^2 + \varphi_\sigma^2 + w_\sigma^2 + \varphi^2\right) d\sigma - \beta\delta_2\varphi^2(L),
\end{aligned} \tag{6.120}
$$

where

$$
\delta_1 = \frac{\delta_0}{3}\min\left\{1, \frac{1}{L^2}\right\}, \qquad \delta_2 = \frac{\delta_0}{3L}, \tag{6.121}
$$

and $\delta_3, \delta_4, \delta_5$ are some positive constants. After regrouping the terms in (6.120), we obtain

$$
\begin{aligned}
\sum_{i=1}^{4} \dot{E}_{ci} \leq\ & \beta\rho L w_t^2(L) + \beta I_\rho L \varphi_t^2(L) + \beta EI\left(\frac{\mu}{\delta_3} + L\right)\varphi_x^2(L) \\
& - \beta\left(\delta_2 - 2KL\delta_4 - \mu EI\delta_3\right)\varphi^2(L) - \beta\left(\delta_2 - \mu K\delta_5\right)w^2(L) \\
& + \beta K\left(\frac{\mu}{\delta_5} + \frac{2L}{\delta_4} + L\right)\left(w_x(L) - \varphi(L)\right)^2 \\
& - \beta\delta_1\int_0^L \left(\varphi_t^2 + w_t^2 + \varphi_\sigma^2 + w_\sigma^2 + \varphi^2\right) d\sigma.
\end{aligned} \tag{6.122}
$$

After substituting (6.109) and (6.122) into (6.105), we have

$$\begin{aligned}\dot{V} \leq\ & -\left(\frac{K}{2} - \beta\rho L\right) w_t^2(L) - \left(\frac{EI}{2} - \beta I_\rho L\right) \varphi_t^2(L) - \left(k_s - \frac{K}{2}\right) \eta_1^2 \\ & - \left(\frac{EI}{2} - \beta EI\left(\frac{\mu}{\delta_3} + L\right)\right) \varphi_x^2(L) - \left(k_r - \frac{EI}{2}\right) \eta_2^2 \\ & + \left(-\frac{K}{2} + \beta K\left(\frac{\mu}{\delta_5} + \frac{2L}{\delta_4} + L\right)\right) (w_x(L) - \varphi(L))^2 \\ & -\beta\left(\delta_2 - 2KL\delta_4 - \mu EI\delta_3\right) \varphi^2(L) - \beta\left(\delta_2 - \mu K\delta_5\right) w^2(L) \\ & -\beta\delta_1 \int_0^L \left(\varphi_t^2 + w_t^2 + \varphi_\sigma^2 + w_\sigma^2 + \varphi^2\right) d\sigma.\end{aligned} \tag{6.123}$$

If δ_3, δ_4, δ_5, β, k_s, and k_r are selected according to

$$\delta_3 \leq \frac{\delta_2}{2\mu EI} \qquad \delta_4 \leq \frac{\delta_2}{4KL} \qquad \delta_5 \leq \frac{\delta_2}{\mu K} \tag{6.124}$$

$$\begin{aligned}\beta <\ & \min\left\{\frac{\zeta_1}{2\rho L + 2I_\rho L + \mu I_\rho + \mu\rho \max(1, L^2)},\right. \\ & \left.\frac{K}{2\rho L}, \frac{EI}{2I_\rho L}, \frac{1}{2(\frac{\mu}{\delta_3} + L)}, \frac{1}{2(\frac{\mu}{\delta_5} + \frac{2L}{\delta_4} + L)}\right\}\end{aligned} \tag{6.125}$$

$$k_s > \frac{K}{2} \qquad k_r > \frac{EI}{2}, \tag{6.126}$$

then it is not difficult to see that $\dot{V}(t)$ can be upper bounded by

$$\dot{V}(t) \leq -\lambda_3\left(E_n(t) + \eta_1^2(t) + \eta_2^2(t)\right), \tag{6.127}$$

where the positive constant λ_3 is defined as

$$\lambda_3 = \min\left\{k_s - \frac{K}{2}, k_r - \frac{EI}{2}, \beta\delta_1\right\}, \tag{6.128}$$

and the definition of $E_n(t)$ from (6.94) has been utilized. From (6.103) and (6.127), we can obtain the following new upper bound for the time derivative of $V(t)$:

$$\dot{V}(t) \leq -\frac{\lambda_3}{\lambda_2} V(t). \tag{6.129}$$

After applying Lemma A.4 in Appendix A to (6.129), we obtain the following solution:

$$V(t) \leq V(0) \exp\left(-\frac{\lambda_3}{\lambda_2}t\right) \leq \lambda_2 \kappa_o \exp\left(-\frac{\lambda_3}{\lambda_2}t\right), \tag{6.130}$$

where (6.103) has been utilized to formulate the second inequality, and the positive constant κ_o is defined in (6.88) (note that (6.81) and (6.94) have been utilized during the construction of κ_o). In addition, we can use (A.25) of Lemma A.12 in Appendix A, (6.94), and (6.103) to derive the following inequalities:

$$\frac{1}{L}w^2(x,t) \leq \int_0^L w_\sigma^2(\sigma,t)d\sigma \leq E_n(t) \leq \frac{1}{\lambda_1}V(t) \qquad \forall x \in [0,L] \tag{6.131}$$

$$\frac{1}{L}\varphi^2(x,t) \leq \int_0^L \varphi_\sigma^2(\sigma,t)d\sigma \leq E_n(t) \leq \frac{1}{\lambda_1}V(t) \qquad \forall x \in [0,L]. \tag{6.132}$$

The result given in (6.86) is a direct consequence of the combination of (6.130), (6.131), and (6.132). □

Remark 6.3 *From (6.103) and (6.130), we can state that $E_n(t)$, $\eta_1(t)$, and $\eta_2(t)$ are bounded $\forall t \in [0,\infty)$. Since $E_n(t)$ is bounded $\forall t \in [0,\infty)$, we can use (6.93) and (A.27) of Lemma A.12 in Appendix A to show that $w(x,t)$ and $\varphi(x,t)$ are bounded $\forall t \in [0,\infty)$ and $\forall x \in [0,L]$. Since $E_n(t)$ defined in (6.94) is bounded $\forall t \in [0,\infty)$, the potential energy given by (6.79) is bounded $\forall t \in [0,\infty)$; hence, we can use Property 6.3 to state that $\frac{\partial^n}{\partial x^n}w(x,t)$ and $\frac{\partial^n}{\partial x^n}\varphi(x,t)$ are bounded for $n = 1, 2$, $\forall t \in [0,\infty)$, and $\forall x \in [0,L]$. Since $\eta_1(t)$, $\eta_2(t)$, $w_x(L,t)$, and $\varphi_x(L,t)$ are bounded $\forall t \in [0,\infty)$, we can use (6.81) to state that $w_t(L,t)$ and $\varphi_t(L,t)$ are bounded $\forall t \in [0,\infty)$; hence, it is now easy to show that the kinetic energy of the system defined in (6.80) is bounded $\forall t \in [0,\infty)$. Since the kinetic energy is bounded $\forall t \in [0,\infty)$, we can use Property 6.4 to state that $\frac{\partial^n}{\partial x^n}w_t(x,t)$ and $\frac{\partial^n}{\partial x^n}\varphi_t(x,t)$ are bounded for $n = 0, 1$, $\forall t \in [0,\infty)$, and $\forall x \in [0,L]$. From the above information, we can now state that all of the signals in the control law of (6.83) and (6.84) are bounded $\forall t \in [0,\infty)$. As a consequence, we can utilize (6.74), (6.75), (6.77), and (6.78) to illustrate that $w_{tt}(x,t)$ and $\varphi_{tt}(x,t)$ remain bounded $\forall t \in [0,\infty)$ and $\forall x \in [0,L]$.*

6.3.4 Adaptive Control Law

We now illustrate how the control law given by (6.83) and (6.84) can be redesigned to compensate for parametric uncertainty while asymptotically stabilizing the beam displacement and cross-section rotation. We begin by rewriting the open-loop dynamics of (6.82) as follows:

$$\begin{aligned} m\dot{\eta}_1(t) &= Y_1\left(w_{xt}\left(L,t\right), w_x\left(L,t\right), \varphi_t(L,t), \varphi(L,t)\right)\theta_1 + f(t) \\ J\dot{\eta}_2(t) &= Y_2\left(\varphi_{xt}(L,t), \varphi_x(L,t)\right)\theta_2 + \tau(t), \end{aligned} \tag{6.133}$$

where $Y_1(\cdot), Y_2(\cdot) \in \mathbb{R}^{1\times 2}$ are known regression matrices, and $\theta_1, \theta_2 \in \mathbb{R}^2$ are unknown, constant parameter vectors defined as follows:

$$\begin{gathered} Y_1(\cdot) = \left[\begin{array}{cc} w_{xt}(L,t) - \varphi_t(L,t) & \varphi(L,t) - w_x(L,t) \end{array}\right] \\ Y_2(\cdot) = \left[\begin{array}{cc} \varphi_{xt}(L,t) & -\varphi_x(L,t) \end{array}\right] \\ \theta_1 = \left[\begin{array}{cc} m & K \end{array}\right]^T \qquad \theta_2 = \left[\begin{array}{cc} J & EI \end{array}\right]^T. \end{gathered} \tag{6.134}$$

From the structure of (6.133), the boundary control force and torque inputs are defined as follows:

$$f(t) = -Y_1(\cdot)\hat{\theta}_1(t) - k_s\eta_1(t) \qquad \tau(t) = -Y_2(\cdot)\hat{\theta}_2(t) - k_r\eta_2(t), \tag{6.135}$$

where $\eta_1(t), \eta_2(t)$ were defined in (6.81), and $\hat{\theta}_1(t), \hat{\theta}_2(t) \in \mathbb{R}^2$ are dynamic, parameter estimate vectors defined below:

$$\hat{\theta}_1(t) = \left[\begin{array}{cc} \hat{m}(t) & \widehat{K}(t) \end{array}\right]^T \qquad \hat{\theta}_2(t) = \left[\begin{array}{cc} \hat{J}(t) & \widehat{EI}(t) \end{array}\right]^T. \tag{6.136}$$

Based on the subsequent stability analysis, the update laws for the parameter estimates are designed as follows:

$$\dot{\hat{\theta}}_1(t) = \Gamma_1 Y_1^T(\cdot)\eta_1(t) \qquad \dot{\hat{\theta}}_2(t) = \Gamma_2 Y_2^T(\cdot)\eta_2(t), \tag{6.137}$$

where $\Gamma_1, \Gamma_2 \in \mathbb{R}^{2\times 2}$ are diagonal, positive-definite, adaptation gain matrices. After defining $\tilde{\theta}_1(t) = \theta_1 - \hat{\theta}_1(t) \in \mathbb{R}^2$ and $\tilde{\theta}_2(t) = \theta_2 - \hat{\theta}_2(t) \in \mathbb{R}^2$, and then substituting (6.135) into (6.133), we obtain the following closed-loop system dynamics:

$$m\dot{\eta}_1(t) = -k_s\eta_1(t) + Y_1(\cdot)\tilde{\theta}_1(t) \qquad J\dot{\eta}_2(t) = -k_r\eta_2(t) + Y_2(\cdot)\tilde{\theta}_2(t) \tag{6.138}$$

$$\dot{\tilde{\theta}}_1(t) = -\Gamma_1 Y_1^T(\cdot)\eta_1(t) \qquad \dot{\tilde{\theta}}_2(t) = -\Gamma_2 Y_2^T(\cdot)\eta_2(t), \tag{6.139}$$

where (6.137) was utilized to form (6.139).

Theorem 6.4 *The adaptive boundary control law given by (6.135) and (6.137) ensures that the beam displacement and cross-section rotation are asymptotically regulated in the following sense:*

$$\lim_{t\to\infty} w(x,t), \varphi(x,t) = 0 \qquad \forall x \in [0, L], \tag{6.140}$$

where the control gains k_s and k_r defined in (6.135) must be selected to satisfy (6.87).

Proof. We use the following function:

$$V_a(t) = V(t) + \frac{1}{2}\sum_{i=1}^{2} \tilde{\theta}_i^T(t)\Gamma_i^{-1}\tilde{\theta}_i(t), \tag{6.141}$$

where $V(t)$ was defined in (6.89). As a straightforward extension of the arguments used in the proof of Theorem 6.3, we can select β sufficiently small to formulate the following bounds on $V_a(t)$:

$$\begin{aligned} \lambda_{1a}\left(E_n(t) + \eta_1^2(t) + \eta_2^2(t) + \sum_{i=1}^{2}\left\|\tilde{\theta}_i(t)\right\|^2\right) &\leq V_a(t) \\ \leq \lambda_{2a}\left(E_n(t) + \eta_1^2(t) + \eta_2^2(t) + \sum_{i=1}^{2}\left\|\tilde{\theta}_i(t)\right\|^2\right) \end{aligned} \tag{6.142}$$

for some positive constants λ_{1a} and λ_{2a}. After differentiating (6.141) with respect to time, and substituting from (6.74), (6.75), (6.138), and (6.139), we can proceed as in the proof of Theorem 6.3 to obtain the following upper bound for the time derivative of $V_a(t)$:

$$\begin{aligned} \dot{V}_a(t) &\leq -\lambda_3\left(E_n(t) + \eta_1^2(t) + \eta_2^2(t)\right) \\ &\leq -\lambda_4\left(E_b(t) + \eta_1^2(t) + \eta_2^2(t)\right) \triangleq -g_a(t), \end{aligned} \tag{6.143}$$

where λ_3 was defined in (6.128), λ_4 is some positive constant, and $g_a(t)$ is a non-negative function. Note that the inequality given by (6.96) was used to form the second inequality in (6.143).

After utilizing (6.142), (6.143), and the arguments outlined in Remark 6.3, we can illustrate that all system signals are bounded during closed-loop operation $\forall t \in [0, \infty)$ and $\forall x \in [0, L]$. From (6.143), the time derivative of $g_a(t)$ is given by

$$\dot{g}_a(t) = \lambda_4\left(\dot{E}_b(t) + 2\eta_1(t)\dot{\eta}_1(t) + 2\eta_2(t)\dot{\eta}_2(t)\right). \tag{6.144}$$

Since we know all of the system signals remain bounded $\forall t \in [0, \infty)$, we can use (6.109) and (6.138) to show that $\dot{E}_b(t)$, $\dot{\eta}_1(t)$, and $\dot{\eta}_2(t)$ are bounded;

hence, we can see from (6.144) that $\dot{g}_a(t)$ is also bounded $\forall t \in [0, \infty)$. We can now invoke Lemma A.6 in Appendix A to show that

$$\lim_{t \to \infty} E_b(t), \eta_1(t), \eta_2(t) = 0. \tag{6.145}$$

Finally, we can use (6.145), (6.96), (6.94), (6.131), and (6.132) to obtain the result given by (6.140). □

6.3.5 Simulation Results

To enable the simulation of the boundary control laws, Galerkin's method [10] was utilized to obtain a finite dimensional model for the beam dynamics of (6.74) and (6.75). To this end, the exact eigenvalues and eigenfunctions of the open-loop system were determined. The solutions to equations (6.74) and (6.75) have the form

$$w(x,t) = W(x)e^{i\lambda t} \qquad \varphi(x,t) = \Phi(x)e^{i\lambda t} \qquad \forall x \in [0, L], \tag{6.146}$$

where λ denotes the open-loop eigenvalue and $W(x), \Phi(x)$ are the open-loop eigenfunctions. Substitution of (6.146) into (6.74) and (6.75) yields the following form for the eigenfunctions:

$$\left.\begin{aligned} W_i(x) &= \cosh(\alpha_i x) - \cos(\beta_i x) \\ &\quad -\frac{D_{i1}}{D_{i2}}\left(\sinh(\alpha_i x) - \frac{\alpha_i'}{\beta_i'}\sin(\beta_i x)\right) \\ \Phi_i(x) &= \alpha_i' \sinh(\alpha_i x) + \beta_i' \sin(\beta_i x) \\ &\quad -\frac{D_{i1}}{D_{i2}}\left(\alpha_i' \cosh(\alpha_i x) - \alpha_i' \cos(\beta_i x)\right) \end{aligned}\right\} \text{ for } \lambda_i < \sqrt{\frac{K}{I_\rho}} \tag{6.147}$$

and

$$\left.\begin{aligned} W_i(x) &= \cos(\alpha_i x) - \cos(\beta_i x) \\ &\quad -\frac{D_{i1}}{D_{i2}}\left(\sin(\alpha_i x) + \frac{\alpha_i'}{\beta_i'}\sin(\beta_i x)\right) \\ \Phi_i(x) &= \alpha_i' \sin(\alpha_i x) + \beta_i' \sin(\beta_i x) \\ &\quad -\frac{D_{i1}}{D_{i2}}\left(-\alpha_i' \cos(\alpha_i x) + \alpha_i' \cos(\beta_i x)\right) \end{aligned}\right\} \text{ for } \lambda_i \geq \sqrt{\frac{K}{I_\rho}}, \tag{6.148}$$

where D_{i1}, D_{i2}, α_i, β_i, α_i', and β_i' are defined as follows:

$$\left.\begin{aligned} D_{i1} &= -\lambda_i^2 J\alpha_i' \sinh(\alpha_i L) - \lambda_i^2 J\beta_i' \sin(\beta_i L) \\ &\quad + EI\alpha_i\alpha_i' \cosh(\alpha_i L) + EI\beta_i\beta_i' \cos(\beta_i L) \\ D_{i2} &= -\lambda_i^2 J\alpha_i' \cosh(\alpha_i L) + \lambda_i^2 J\alpha_i' \cos(\beta_i L) \\ &\quad + EI\alpha_i\alpha_i' \sinh(\alpha_i L) + EI\beta_i\alpha_i' \sin(\beta_i L) \end{aligned}\right\} \text{ for } \lambda_i < \sqrt{\frac{K}{I_\rho}}, \tag{6.149}$$

$$\left.\begin{aligned} D_{i1} &= -\lambda_i^2 J\alpha_i' \sin(\alpha_i L) - \lambda_i^2 J\beta_i' \sin(\beta_i L) \\ &\quad + EI\alpha_i\alpha_i' \cos(\alpha_i L) + EI\beta_i\beta_i' \cos(\beta_i L) \\ D_{i2} &= \lambda_i^2 J\alpha_i' \cos(\alpha_i L) - \lambda_i^2 J\alpha_i' \cos(\beta_i L) \\ &\quad + EI\alpha_i\alpha_i' \sin(\alpha_i L) - EI\beta_i\alpha_i' \sin(\beta_i L) \end{aligned}\right\} \text{ for } \lambda_i \geq \sqrt{\frac{K}{I_\rho}}, \tag{6.150}$$

$$\left.\begin{aligned} \alpha_i^2, \beta_i^2 &= \frac{\mp\lambda_i^2(I_\rho + EI\frac{\rho}{K}) + \sqrt{\left[\lambda_i^2(I_\rho - EI\frac{\rho}{K})\right]^2 + 4EI\lambda_i^2\rho}}{2EI} \\ \alpha_i' &= \alpha_i + \frac{\lambda_i^2\rho}{K\alpha_i} \qquad \beta_i' = \beta_i - \frac{\lambda_i^2\rho}{K\beta_i} \end{aligned}\right\} \text{ for } \lambda_i < \sqrt{\frac{K}{I_\rho}}, \tag{6.151}$$

and

$$\left.\begin{aligned} \alpha_i^2, \beta_i^2 &= \frac{\lambda_i^2(I_\rho + EI\frac{\rho}{K}) \mp \sqrt{\left[\lambda_i^2(I_\rho - EI\frac{\rho}{K})\right]^2 + 4EI\lambda_i^2\rho}}{2EI} \\ \alpha_i' &= -\alpha_i + \frac{\lambda_i^2\rho}{K\alpha_i} \qquad \beta_i' = \beta_i - \frac{\lambda_i^2\rho}{K\beta_i} \end{aligned}\right\} \text{ for } \lambda_i \geq \sqrt{\frac{K}{I_\rho}}. \tag{6.152}$$

To obtain the finite dimensional, closed-loop system dynamics, we approximate the variables $w(x,t)$ and $\varphi(x,t)$ by the following finite expansions:

$$w(x,t) = \sum_{i=1}^{N} W_i(x)q_i(t) \qquad \varphi(x,t) = \sum_{i=1}^{N} \Phi_i(x)q_i(t), \tag{6.153}$$

where N denotes the number of modes used in the approximation, and $q_i(t)$ is the i-th modal coordinate. Substitution of (6.153) into (6.74), (6.75), (6.77), and (6.78) with $f(t)$ and $\tau(t)$ given by (6.135) results in the following finite dimensional, nonlinear, closed-loop system dynamics:

$$M\ddot{\mathbf{q}}(t) + B(t)\dot{\mathbf{q}}(t) + H(t)\mathbf{q}(t) = \mathbf{0}, \tag{6.154}$$

where $\mathbf{q}(t) \in \mathbb{R}^N$ is the modal coordinate vector, and the elements of the matrices $M, B(t), H(t) \in \mathbb{R}^{N\times N}$ are defined as follows:

$$\begin{aligned} M_{ij} = & \int_0^L (\rho W_i(\sigma)W_j(\sigma) + I_\rho\Phi_i(\sigma)\Phi_j(\sigma))\, d\sigma \\ & + mW_i(L)W_j(L) + J\Phi_i(L)\Phi_j(L), \end{aligned} \tag{6.155}$$

$$\begin{aligned} B_{ij}(t) = & \ \hat{m}(t)W_{ix}(L)W_j(L) + k_s W_i(L)W_j(L) + \hat{J}(t)\Phi_{ix}(L)\Phi_j(L) \\ & + (k_r - \hat{m}(t))\,\Phi_i(L)\Phi_j(L), \end{aligned} \tag{6.156}$$

and

$$\begin{aligned}H_{ij}(t) = &\int_0^L (-KW_{i\sigma\sigma}(\sigma)W_j(\sigma) - K\Phi_{i\sigma}(\sigma)W_j(\sigma) - EI\Phi_{i\sigma\sigma}(\sigma)\Phi_j(\sigma)\\ &+KW_{i\sigma}(\sigma)\Phi_j(\sigma) + K\Phi_i(\sigma)\Phi_j(\sigma))\, d\sigma\\ &+(K - \hat{K}(t) + k_s)W_{ix}(L)W_j(L)\\ &+(EI - \widehat{EI}(t) + k_r)\Phi_{ix}(L)\Phi_j(L)\\ &- \left(K - \hat{K}(t) + k_s\right)\Phi_i(L)\Phi_j(L).\end{aligned} \tag{6.157}$$

The nonlinear nature of (6.154) comes from the dependency of $B(t)$ and $H(t)$ on the parameter estimates $\hat{m}(t)$, $\hat{J}(t)$, $\widehat{K}(t)$, and $\widehat{EI}(t)$. Note that for the model-based control simulations, $\hat{m}(t) = m$, $\hat{K}(t) = K$, $\hat{J}(t) = J$, and $\widehat{EI}(t) = EI$ in (6.155) to (6.157).

The system parameters used in the simulations are given below:

$$m = 0.1 \text{ kg}, \quad J = 0.1 \text{ kg-m}^2, \quad L = 1.0 \text{ m}, \quad \rho = 1.0 \text{ kg/m},$$

$$K = 1.5 \text{ N}, \quad I_\rho = 2 \text{ kg-m}, \quad \text{and} \quad EI = 7.5 \text{ N-m}^2.$$

For $N = 4$, the following simulation results illustrate the response of the closed-loop system to a simulated impulse at the beam's free end-point. For comparison purposes, the open-loop system with $f(t) = \tau(t) = 0$ (i.e., $k_s = k_r = \hat{m}(t) = \hat{J}(t) = \hat{K}(t) = \widehat{EI}(t) = 0$ in (6.156) and (6.157)) was also simulated. The control gains used for the model-based controller were chosen as $k_s = 1.0$ and $k_r = 4.0$. The gains for the adaptive controller were $k_s = 0.75$, $k_r = 3.75$, $\Gamma_1 = \text{diag}\{4, 20\}$, and $\Gamma_2 = \text{diag}\{10, 10\}$, while the initial values of the parameter estimates were set to $\hat{m}(0) = \hat{J}(0) = \widehat{EI}(0) = 0$ and $\hat{K}(0) = 2.0$. Figure 6.12 shows the beam's end-point displacement $w(L,t)$, end-point cross-section rotation $\varphi(L,t)$, mid-point displacement $w(L/2,t)$, and mid-point cross-section rotation $\varphi(L/2,t)$ for the model-based controller in comparison to the open-loop response. Figure 6.13 depicts the boundary control force and torque inputs for the model-based controller. Figures 6.14 and 6.15 show the same plots for the adaptive controller, while Figure 6.16 presents the parameter estimates $\hat{m}(t)$, $\hat{K}(t)$, $\hat{J}(t)$, and $\widehat{EI}(t)$.

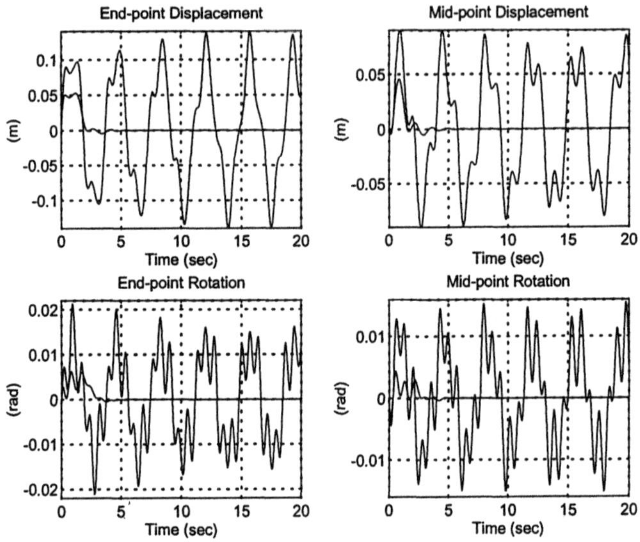

FIGURE 6.12. Simulation results for open-loop system and model-based controller.

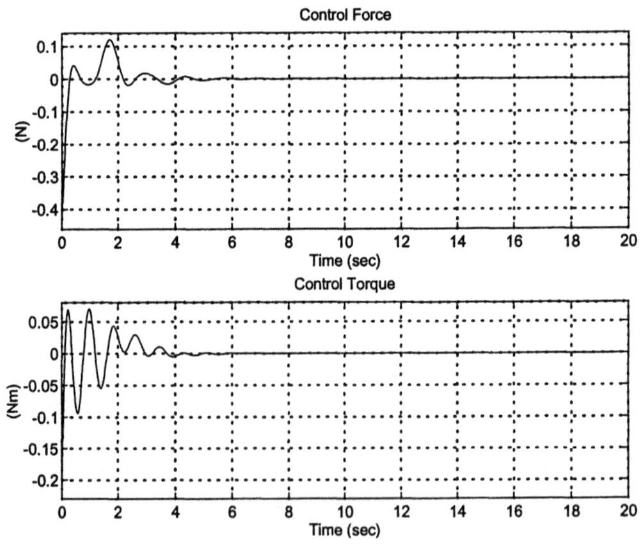

FIGURE 6.13. Control inputs for model-based controller.

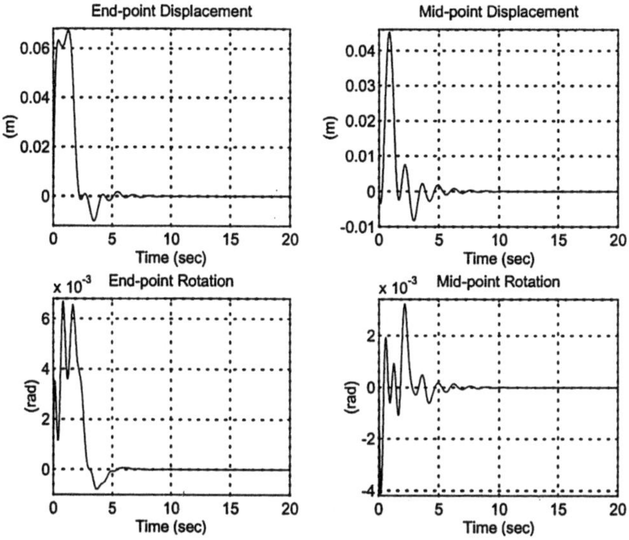

FIGURE 6.14. Simulation results for adaptive controller.

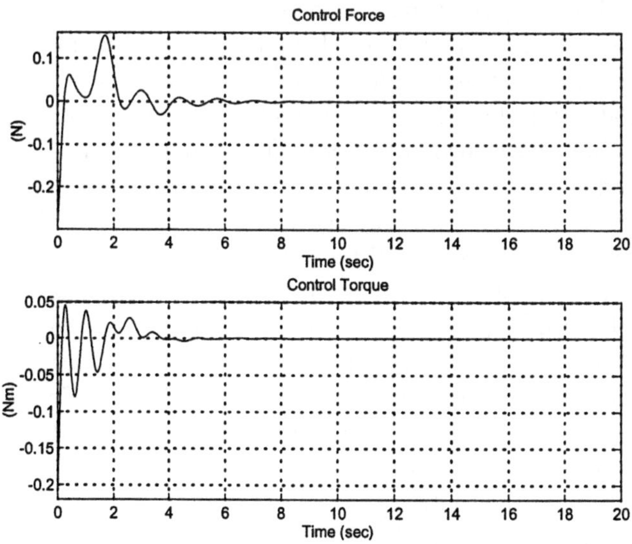

FIGURE 6.15. Control inputs for adaptive controller.

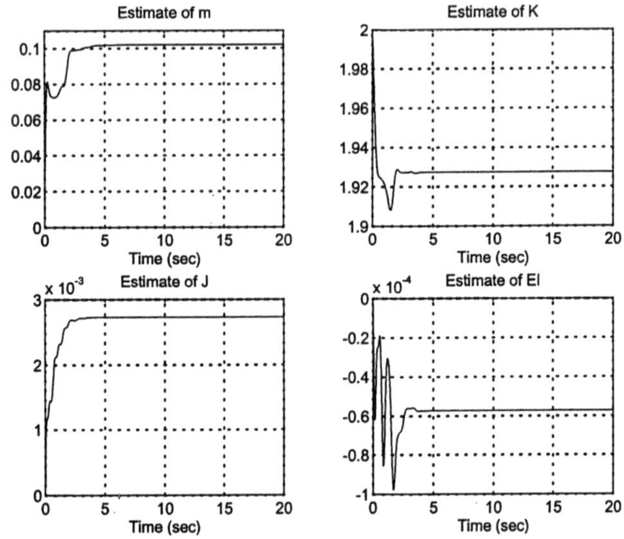

FIGURE 6.16. Parameter estimates for adaptive controller.

6.4 Notes

As stated in the Introduction, owing to its simplicity, the Euler-Bernoulli model has been largely utilized in the design of boundary controllers for beam-type systems. For example, boundary controllers for cantilevered beams were proposed in Morgül [11, 12], Chen et al. [4], Canbolat et al. [3], and Rahn and Mote [16], while boundary controllers for single flexible link robots were developed in Luo et al. [6]–[9] and Morgül [14]. Boundary controllers for rotating beam models were designed in Baillieul and Levi [2], Morgül [13], Sallet et al. [17], and Xu and Baillieul [20]. On the other hand, there seems to be very little work regarding the design of boundary controllers for the Timoshenko model. Early work was provided by Kim and Renardy [5], where the stability of a cantilevered Timoshenko beam *without* end-point mass/inertia was investigated under the influence of standard linear, force and torque control dampers. A Lyapunov-type stability analysis coupled with semigroup theory was utilized to show that the energy of the system exponentially decays after a finite initial time which depends on the control gains and the system parameters. Shi et al. [18] proposed nonlinear and linear boundary controllers for a cantilevered Timoshenko beam with end-point mass/inertia, and proved the asymptotic and exponential decay, respectively, of the energy of the system by using semigroup theory and LaSalle's invariance principle. Recently in [21], Zhang et al. proposed exponentially stabilizing boundary controllers for a cantilevered Timoshenko beam with end-point mass/inertia. In contrast to the control laws of [18, 21], the control design and analysis presented in Section 6.3 (i) contains feedforward model-based terms and (ii) illustrates how the controller can be augmented with adaptive, parameter estimates to compensate for parametric uncertainty.

References

[1] O. J. Aldraihem, R. C. Wetherhold, and T. Singh, Intelligent Beam Structures: Timoshenko Theory vs. Euler-Bernoulli Theory, *Proceedings of the IEEE Conference on Control Applications*, pp. 976-981, Dearborn, MI, Sept. 1996.

[2] J. Baillieul and M. Levi, Rotational Elastic Dynamics, *Physica*, Vol. 27(D), pp. 43-62, 1987.

[3] H. Canbolat, D. Dawson, C. Rahn, and P. Vedagarbha, Boundary Control of a Cantilevered Flexible Beam with Point-Mass Dynamics at the Free End, *Mechatronics - An International Journal*, Vol. 8, No. 3, pp. 163-186, 1998.

[4] G. Chen, M. Delfour, A. Krall, and G. Payre, Modeling, Stabilization, and Control of Serially Connected Beams, *SIAM Journal of Control Optimizations,* Vol. 25, No. 3, pp. 526–546, 1987.

[5] J. U. Kim and Y. Renardy, Boundary Control of the Timoshenko Beam, *SIAM Journal of Control and Optimization*, Vol. 25, No. 6, pp. 1417-1429, Nov. 1987.

[6] Z. Luo, Direct Strain Feedback Control of Flexible Robot Arms: New Theoretical and Experimental Results, *IEEE Transactions on Automatic Control*, Vol. 38, No. 11, pp. 1610-1622, Nov. 1993.

[7] Z. Luo and B. Guo, Further Theoretical Results on Direct Strain Feedback Control of Flexible Robot Arms, *IEEE Transactions on Automatic Control*, Vol. 40, No. 4, pp. 747-751, Apr. 1995.

[8] Z. Luo, N. Kitamura, and B. Guo, Shear Force Feedback Control of Flexible Robot Arms, *IEEE Transactions on Robotics and Automation*, Vol. 11, No. 5, pp. 760-765, Oct. 1995.

[9] Z. H. Luo and B. Z. Guo, Shear Force Feedback Control of a Single-Link Flexible Robot with a Revolute Joint, *IEEE Transactions on Automatic Control*, Vol. 42, No. 1, pp. 53-65, Jan. 1997.

[10] L. Meirovitch, *Analytical Methods in Vibrations*, New York, NY: MacMillan, 1967.

[11] O. Morgül, Orientation and Stabilization of a Flexible Beam Attached to a Rigid Body: Planar Motion, *IEEE Transactions on Automatic Control*, Vol. 36, No. 8, pp. 953-962, Aug. 1991.

[12] O. Morgül, Dynamic Boundary Control of a Euler-Bernoulli Beam, *IEEE Transactions on Automatic Control*, Vol. 37, No. 5, pp. 639-642, May 1992.

[13] O. Morgül, Control and Stabilization of a Rotating Flexible Structure, *Automatica*, Vol. 30, No. 2, pp. 351-356, Feb. 1994.

[14] O. Morgül, On Boundary Control of Single Link Flexible Robot Arms, *IFAC World Congress*, Vol. A, pp. 127-132, San Francisco, CA, July 1996.

[15] D. Oguamanam and G. Heppler, The Effect of Rotating Speed on the Flexural Vibration of a Timoshenko Beam, *Proceedings of the IEEE Conference on Robotics and Automation*, pp. 2438-2443, Minneapolis, MN, Apr. 1996.

[16] C. Rahn and C. Mote, Axial Force Stabilization of Transverse Beam Vibration, *Proceedings of the ASME Conference on Mechanical Vibration and Noise*, DE-Vol. 61, pp. 29-34, Albuquerque, NM, Sept. 1993.

[17] G. Sallet, C. Xu, and H. Laousy, Boundary Feedback Stabilization of a Rotating Body-Beam System, *Proceedings of the IEEE Conference on Decision and Control*, pp. 930-935, New Orleans, LA, Dec. 1995.

[18] D. Shi, S. Hou, and D. Feng, Feedback Stabilization of a Timoshenko Beam with an End Mass, *International Journal of Control*, Vol. 69, No. 2, pp. 285-300, Jan. 1998.

[19] S. Timoshenko, *Vibration Problems in Engineering*, New York, NY: Van Nostrand, 1955.

[20] C. Xu and J. Baillieul, Stabilizability and Stabilization of a Rotating Body-Beam System with Torque Control, *IEEE Transactions on Automatic Control*, Vol. 38, No. 12, pp. 1754-1765, Dec. 1993.

[21] F. Zhang, D. M. Dawson, M. S. de Queiroz, and P. Vedagarbha, Boundary Control of a Cantilevered Timoshenko Beam with Free-End Mass/Inertial Dynamics, *Proceedings of the IEEE Conference on Decision and Control*, pp. 245-250, San Diego, CA, Dec. 1997.

7

Boundary Control Applications

7.1 Introduction

In this chapter, we present boundary control strategies for reducing vibration in three different engineering applications of flexible mechanical systems. Two of these applications will share the interesting characteristic of having flexible and rigid subsystems, which leads to *hybrid* dynamic models containing coupled PDEs and ODEs. The design of boundary controllers and the closed-loop stability analysis will be built upon the arguments set forth in Chapters 5 and 6.

The first application will deal with the use of active vibration control for eliminating vibration-induced web breaks. This application is motivated by the fact that high-speed manufacture and transport of thin webs require tight control of the large vibration that can result from material nonuniformity, support roller motion/eccentricity, or aerodynamic excitation. Since failure of the web owing to excessive vibration wastes material and time and thereby limits the process productivity, the use of vibration control strategies is well motivated. In a typical material transport system, large rollers are often utilized at the entry and exit points to move the web-like material along the axial direction. To improve the vibration damping, much of the previous theoretical control work relies on the use of a transverse boundary control force applied at the entry or exit roller. Unfortunately, the size and weight of the entry/exit rollers and their inherent dynamic coupling with the adjacent web limits the applicability of boundary control for axially moving webs. The impracticality of boundary control seems to be the primary motivation for the current use of a *passive*, interstitial (i.e., located in between the entry and exit rollers) dancer arm to control the web vibration. To promote *active* vibration regulation of axial moving web-like materials, we present the design of a Lyapunov-based control strategy that automatically regulates the web vibration by controlling the position and orientation of a mechanical guide attached to an active interstitial actuator.

Because of the prohibitive cost of placing equipment in outer space, the use of standard, rigid link robot manipulators is impractical. As such, space-based robot manipulators are more likely to be characterized by long links manufactured from lightweight metals or composites. Unfortunately, the use of long, lightweight links greatly complicates the corresponding position control problem since the links are subject to deflection and/or vibration. The second application examines the vibration control of a robot manipulator-like device. Specifically, we design control algorithms for a nonlinear distributed parameter model of a flexible single-link robot with a payload mass at the link's free end-point. That is, the controller is based on a hybrid system model that takes into account the nonlinearities and coupling effects between the link dynamics, the actuator hub dynamics, and the payload mass dynamics.

In many engineering applications, a long rotor is utilized to transmit torque from a source to a load. As a result of shaft flexure, long rotors often exhibit dynamic misalignment or an unbalanced response. In the third application, we examine the vibration control of a flexible rotor by designing boundary controllers for a hybrid PDE-ODE model of a rotor that is free to vibrate in two dimensions. The rotor displacement is regulated via two boundary control forces applied to the free end-point of the rotor while angular velocity tracking is achieved via a boundary control torque applied to a rigid hub clamped to the other end of the rotor.

7.2 Axially Moving String System

In this section, we develop a vibration control system that uses two actuators to regulate the displacement in an axial moving string system. Specifically, we develop an active control strategy that automatically regulates the string displacement by controlling the position and orientation of a mechanical guide (see Figure 7.1). Given the dynamic model of the mechanical system, composed of a distributed parameter field equation coupled to a discrete actuator equation, a Lyapunov-type analysis is used

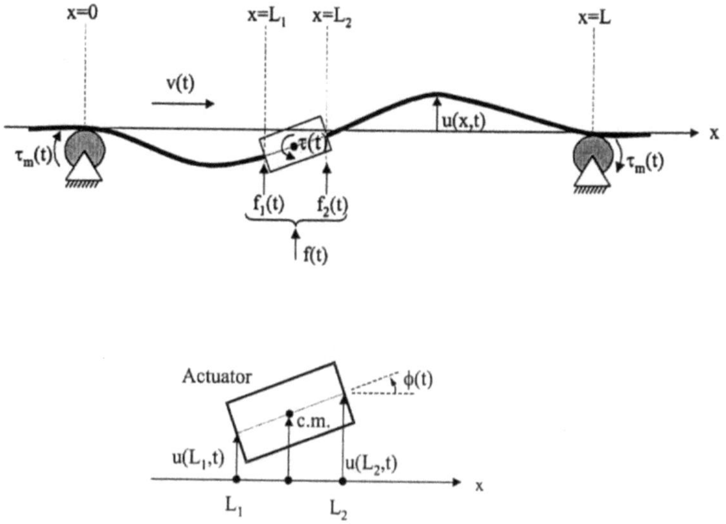

FIGURE 7.1. Schematic diagram of the axially moving string and actuator system.

to design a model-based control law that exponentially stabilizes the string displacement. We then illustrate how the model-based controller can be redesigned as an adaptive control law that asymptotically stabilizes the string displacement while compensating for parametric uncertainty.

7.2.1 System Model

For the system shown in Figure 7.1, let x and t denote the independent spatial and time variables, respectively; $u(x,t)$ denote the string displacement at the position x for time t; ρ be the mass/length of the web; P represent the constant bias tension in the web; m denote the actuator mass; J denote the actuator inertia; L be the length of the string; L_1, L_2 demark the boundaries of the actuator; v denote the constant, positive axial velocity; and $f(t), \tau(t)$ denote the control force and torque inputs applied by the interstitial actuator, respectively. To prevent buckling instabilities in the web, and to ensure that $P_o = P - \rho v^2 > 0$, we will assume that $v < \sqrt{\frac{P}{\rho}}$.

The derivation of the dynamic model of the axial moving string system uses the following assumptions: (i) the string is perfectly flexible, (ii) the string is pinned at $x = 0$ and $x = L$, (iii) the string and actuator are subject to small displacements only, i.e.,

$$\phi(t) \approx \frac{u(L_2,t) - u(L_1,t)}{L_2 - L_1}, \tag{7.1}$$

where $\phi(t)$ is defined in Figure 7.1, (iv) the rotary inertia of the string is negligible, and (v) the hollow, rigid actuator has uniform mass distribution with its center of mass located at point $\left(\frac{L_2+L_1}{2}, y(t)\right)$ where $y(t)$ is given by

$$y(t) = \frac{u(L_2,t) + u(L_1,t)}{2}. \tag{7.2}$$

Based on the above assumptions, the total kinetic energy of the system is given by

$$\begin{aligned}\Pi_k = & \frac{1}{2}\rho \int_0^{L_1} \left[u_t(\sigma,t) + vu_\sigma(\sigma,t)\right]^2 d\sigma \\ & + \frac{1}{2}\rho \int_{L_2}^{L} \left(u_t(\sigma,t) + vu_\sigma(\sigma,t)\right)^2 d\sigma + \frac{1}{2}m\dot{y}^2(t) \\ & + \frac{1}{2}J\dot{\phi}^2(t) + \frac{1}{2}\rho\left(L_2 - L_1\right)\left(\dot{y}(t) + v\phi(t)\right)^2 + \frac{1}{24}\rho\left(L_2 - L_1\right)^3 \dot{\phi}^2(t),\end{aligned} \tag{7.3}$$

where the subscripts x, t denote the partial derivatives with respect to x, t, respectively. The first two terms of (7.3) denote the kinetic energy of the axial moving string, the third and fourth terms represent the kinetic energy of the actuator, and the last two terms denote the kinetic energy of the string moving through the actuator. The total potential energy of the system is given by

$$\Pi_p = \frac{1}{2}P \int_0^{L_1} u_\sigma^2(\sigma,t)d\sigma + \frac{1}{2}P \int_{L_2}^{L} u_\sigma^2(\sigma,t)d\sigma + \frac{1}{2}P\left(L_2 - L_1\right)\phi^2(t), \tag{7.4}$$

and the total work done by the external forces is given by

$$\Pi_w = f(t)y(t) + \tau(t)\phi(t). \tag{7.5}$$

Substitution of (7.3), (7.4), and (7.5) into Hamilton's principle [15], given by

$$\int_0^t \left(\delta\Pi_k - \delta\Pi_p + \delta\Pi_w\right) dt = 0, \tag{7.6}$$

yields the following field equation:

$$\rho u_{tt}(x,t) + 2\rho v u_{xt}(x,t) - P_o u_{xx}(x,t) = 0 \qquad x \in [0, L_1] \cup [L_2, L]. \tag{7.7}$$

The boundary conditions are given by[1]

[1] The pinned boundary conditions of (7.8) imply that $u_t(0,t) = 0$ and $u_t(L,t) = 0$.

$$u(0,t) = u(L,t) = 0, \tag{7.8}$$

while the actuator dynamics are described by the following ODE:

$$Mz_{tt}(t) + N(t) = \begin{bmatrix} \dfrac{-1}{L_2 - L_1} & \dfrac{1}{2} \\ \dfrac{1}{L_2 - L_1} & \dfrac{1}{2} \end{bmatrix} \begin{bmatrix} \tau(t) \\ f(t) \end{bmatrix}, \tag{7.9}$$

where $z(t) \in \mathbb{R}^{2\times1}$, $M \in \mathbb{R}^{2\times2}$, $P_o \in \mathbb{R}$, and $N(t) \in \mathbb{R}^{2\times1}$ are defined as follows:

$$z(t) = \begin{bmatrix} u(L_1,t) \\ u(L_2,t) \end{bmatrix}, \tag{7.10}$$

$$M = \begin{bmatrix} \dfrac{m_{eq}}{4} + \dfrac{J_{eq}}{(L_2 - L_1)^2} & \dfrac{m_{eq}}{4} - \dfrac{J_{eq}}{(L_2 - L_1)^2} \\ \dfrac{m_{eq}}{4} - \dfrac{J_{eq}}{(L_2 - L_1)^2} & \dfrac{m_{eq}}{4} + \dfrac{J_{eq}}{(L_2 - L_1)^2} \end{bmatrix}, \tag{7.11}$$

$$P_o = P - \rho v^2, \qquad m_{eq} = m + \rho(L_2 - L_1), \qquad J_{eq} = J + \frac{\rho(L_2 - L_1)^3}{12}, \tag{7.12}$$

and

$$\begin{aligned} N(t) = {} & \rho v \begin{bmatrix} -1 & 1 \\ -1 & 1 \end{bmatrix} \begin{bmatrix} u_t(L_1,t) \\ u_t(L_2,t) \end{bmatrix} + P_o \begin{bmatrix} u_x(L_1,t) \\ -u_x(L_2,t) \end{bmatrix} \\ & + \frac{P_o}{L_2 - L_1} \begin{bmatrix} 1 & -1 \\ -1 & 1 \end{bmatrix} \begin{bmatrix} u(L_1,t) \\ u(L_2,t) \end{bmatrix}. \end{aligned} \tag{7.13}$$

Since the control strategies will be composed of relatively simple functions, we will assume the existence of a unique solution for the dynamics given by (7.7) through (7.9) under the control. In addition, based on the arguments outlined in Remark 5.1, we will assume that the distributed variable $u(x,t)$ and its time derivative $u_t(x,t)$ belong to a space of functions that possess the following properties.

Property 7.1: If the potential energy of the system given in (7.4) is bounded $\forall t \in [0,\infty)$, then $\dfrac{\partial^n}{\partial x^n} u(x,t)$ is bounded for $n = 1,2$, $\forall t \in [0,\infty)$, and $\forall x \in [0,L_1] \cup [L_2,L]$.

Property 7.2: If the kinetic energy of the system given in (7.3) is bounded $\forall t \in [0,\infty)$, then $\dfrac{\partial^n}{\partial x^n} u_t(x,t)$ is bounded for $n = 0,1$, $\forall t \in [0,\infty)$, and $\forall x \in [0,L_1] \cup [L_2,L]$.

7.2.2 Problem Statement

The control objective is to drive the string displacement $u(x,t)$ to zero as $t \to \infty$ for $x \in [0, L_1] \cup [L_2, L]$. To facilitate the subsequent control design and analysis, we first define an auxiliary variable, denoted by $\eta(t) \in \mathbb{R}^2$, as follows:

$$\eta(t) = \begin{bmatrix} u_t(L_1,t) + u_x(L_1,t) \\ u_t(L_2,t) - u_x(L_2,t) \end{bmatrix}. \tag{7.14}$$

After differentiating $\eta(t)$ of (7.14) with respect to time, multiplying the resulting expressions by M defined in (7.11), and then substituting from (7.9), we obtain

$$M\dot{\eta}(t) = M \begin{bmatrix} u_{xt}(L_1,t) \\ -u_{xt}(L_2,t) \end{bmatrix} - N(t) + \begin{bmatrix} \dfrac{-1}{L_2 - L_1} & \dfrac{1}{2} \\ \dfrac{1}{L_2 - L_1} & \dfrac{1}{2} \end{bmatrix} \begin{bmatrix} \tau(t) \\ f(t) \end{bmatrix}, \tag{7.15}$$

where (7.11) has been used and $\dot{\eta}(t) = \frac{d}{dt}(\eta(t))$. The above equation denotes the open-loop dynamics of the auxiliary signal $\eta(t)$ and will form the basis for the design of the model-based and adaptive control laws.

7.2.3 Model-Based Control Law

Given perfect knowledge of the system parameters, the form of (7.15), and the subsequent stability analysis, the control input vector is defined as follows:

$$\begin{bmatrix} \tau(t) \\ f(t) \end{bmatrix} = \begin{bmatrix} \dfrac{-1}{L_2 - L_1} & \dfrac{1}{2} \\ \dfrac{1}{L_2 - L_1} & \dfrac{1}{2} \end{bmatrix}^{-1} \left(-M \begin{bmatrix} u_{xt}(L_1,t) \\ -u_{xt}(L_2,t) \end{bmatrix} + N(t) - K_s \eta(t) \right), \tag{7.16}$$

where $K_s \in \mathbb{R}^{2\times 2}$ is a diagonal, positive-definite, control gain matrix. After substituting (7.16) into (7.15), we obtain the following closed-loop dynamics for $\eta(t)$:

$$M\dot{\eta}(t) = -K_s \eta(t). \tag{7.17}$$

Note that the control algorithm given in (7.16) only requires measurements of the string slope (and its time derivative) on both sides of the mechanical guide and the position and orientation (and their time derivatives) of the mechanical guide. Practically speaking, the time derivative quantities can be calculated on-line from the corresponding position or slope related quantities via a backwards difference-type algorithm; hence, implementation of the control algorithm could be performed with standard sensor hardware. That is, incremental encoders can be used to obtain position-related information while load cells mounted to the ends of the mechanical guide can be used to measure the string slope at the entry/exit points on the guide.

Theorem 7.1 *The model-based controller given by (7.16) ensures that the web displacement is exponentially regulated in the following sense:*

$$|u(x,t)| \leq \sqrt{\frac{2L_1\lambda_2}{P_o\lambda_1}\kappa_o \exp\left(-\frac{\lambda_3}{\lambda_2}t\right)} \qquad \forall x \in [0, L_1] \tag{7.18}$$

$$|u(x,t)| \leq \sqrt{\frac{2\left(L-L_2\right)\lambda_2}{P_o\lambda_1}\kappa_o \exp\left(-\frac{\lambda_3}{\lambda_2}t\right)} \qquad \forall x \in [L_2, L],$$

where λ_1, λ_2, and λ_3 are some positive bounding constants, and the positive constant κ_o is given by

$$\kappa_o = \frac{1}{2}\rho\int_0^{L_1} u_t^2(\sigma,0)\,d\sigma + \frac{1}{2}P_o\int_0^{L_1} u_\sigma^2(\sigma,0)d\sigma + \frac{1}{2}\rho\int_{L_2}^{L} u_t^2(\sigma,0)\,d\sigma$$
$$+\frac{1}{2}P_o\int_{L_2}^{L} u_\sigma^2(\sigma,0)d\sigma + \|\eta(0)\|^2, \tag{7.19}$$

with $\eta(t)$ defined in (7.14). To ensure that (7.18) is achieved, the axial velocity v and the control gain matrix K_s must be selected to satisfy the following sufficient conditions:

$$v < -2 + \sqrt{\frac{P}{\rho}+4} < \sqrt{\frac{P}{\rho}} \qquad \lambda_{\min}\left\{K_s\right\} > \frac{P_o}{2}, \tag{7.20}$$

where $\lambda_{\min}\{\cdot\}$ denotes the minimum eigenvalue of a matrix.

Proof. To prove the result given by (7.18), we define the following function:

$$V(t) = E_s(t) + E_{c1}(t) + E_{c2}(t) + \frac{1}{2}\eta^T(t)M\eta(t), \tag{7.21}$$

where

$$E_s(t) = \frac{1}{2}\rho\int_0^{L_1} u_t^2(\sigma,t)\,d\sigma + \frac{1}{2}P_o\int_0^{L_1} u_\sigma^2(\sigma,t)d\sigma$$
$$+\frac{1}{2}\rho\int_{L_2}^{L} u_t^2(\sigma,t)\,d\sigma + \frac{1}{2}P_o\int_{L_2}^{L} u_\sigma^2(\sigma,t)d\sigma, \tag{7.22}$$

$$E_{c1}(t) = 2\beta\rho\int_0^{L_1} \sigma u_\sigma(\sigma,t)\left(u_t(\sigma,t) + vu_\sigma(\sigma,t)\right)d\sigma, \tag{7.23}$$

$$E_{c2}(t) = 2\beta\rho \int_{L_2}^{L} (\sigma - L)\, u_\sigma(\sigma,t)\left(u_t(\sigma,t) + v u_\sigma(\sigma,t)\right) d\sigma, \tag{7.24}$$

and β is a positive weighting constant. To ensure that $V(t)$ defined in (7.21) is non-negative, first note that the inequality (A.29) of Lemma A.13 in Appendix A can be used to bound $E_{c1}(t)$ of (7.23) as follows:[2]

$$\begin{aligned} E_{c1} &\leq 2\beta\rho L_1 \int_0^{L_1} \left(u_t^2 + (1+v)\, u_\sigma^2\right) d\sigma \\ &\leq \frac{4\beta\rho L_1 (1+v)}{\min(\rho, P_o)} \left[\frac{1}{2}\int_0^{L_1} \left(\rho u_t^2 + P_o u_\sigma^2\right)\, d\sigma\right]. \end{aligned} \tag{7.25}$$

We can now use (7.22) and (7.25) to establish the following inequality:

$$-\frac{4\beta\rho L_1 (1+v)}{\min(\rho, P_o)} E_s \leq E_{c1} \leq \frac{4\beta\rho L_1 (1+v)}{\min(\rho, P_o)} E_s. \tag{7.26}$$

In a similar manner, we can use (A.29) of Lemma A.13 in Appendix A and (7.22) to bound $E_{c2}(t)$ of (7.24) as follows:

$$-\frac{4\beta\rho (L - L_2)(1+v)}{\min(\rho, P_o)} E_s \leq E_{c2} \leq \frac{4\beta\rho (L - L_2)(1+v)}{\min(\rho, P_o)} E_s. \tag{7.27}$$

From (7.26) and (7.27), we can see that if the weighting constant β is selected according to

$$\beta < \frac{\min(\rho, P_o)}{4\rho\,(1+v)\,(L_1 + L - L_2)}, \tag{7.28}$$

we have

$$0 \leq \zeta_1 E_s \leq E_s + E_{c1} + E_{c2} \leq \zeta_1 E_s, \tag{7.29}$$

where ζ_1 and ζ_2 are some positive constants. Given the structure of $V(t)$ defined in (7.21), the inequality given by (7.29), and the fact the matrix M defined in (7.11) is positive-definite, we can formulate the following bounds on $V(t)$:

$$\lambda_1 \left(E_s(t) + \|\eta(t)\|^2\right) \leq V(t) \leq \lambda_2 \left(E_s(t) + \|\eta(t)\|^2\right), \tag{7.30}$$

---

[2] To reduce the notational complexity in most of the following derivations, the arguments x, t will be left out of all spatial/time-dependent variables, (e.g., $u(x,t)$ will be denoted simply as u) while the argument t will be left out of the time-dependent variables (e.g., $u_x(L,t)$ will be denoted as $u_x(L)$).

where λ_1 and λ_2 are positive constants defined as follows:

$$\lambda_1 = \min\left\{1 - \frac{4\beta\rho\,(1+v)}{\min(\rho, P_o)}\,(L_1 + L - L_2)\,, \frac{\lambda_{\min}\{M\}}{2}\right\}$$
$$\lambda_2 = \max\left\{1 + \frac{4\beta\rho\,(1+v)}{\min(\rho, P_o)}\,(L_1 + L - L_2)\,, \frac{\lambda_{\max}\{M\}}{2}\right\}, \tag{7.31}$$

with $\lambda_{\max}\{\cdot\}$ denoting the maximum eigenvalue of a matrix.

After differentiating (7.21) with respect to time, we have

$$\dot{V}(t) = \dot{E}_s(t) + \dot{E}_{c1}(t) + \dot{E}_{c2}(t) - \eta^T(t)\,K_s\eta(t)\,, \tag{7.32}$$

where (7.17) has been used. To determine $\dot{E}_s(t)$ in (7.32), we differentiate (7.22) with respect to time to obtain

$$\begin{aligned}\dot{E}_s = {} & \left[\int_0^{L_1} P_o u_t u_{\sigma\sigma} d\sigma\right] + \int_0^{L_1} P_o u_\sigma u_{\sigma t} d\sigma + \left[\int_{L_2}^{L} P_o u_t u_{\sigma\sigma} d\sigma\right] \\ & + \int_{L_2}^{L} P_o u_\sigma u_{\sigma t} d\sigma - \left[\rho v \int_0^{L_1} \frac{d}{d\sigma} u_t^2 d\sigma\right] - \left[\rho v \int_{L_2}^{L} \frac{d}{d\sigma} u_t^2 d\sigma\right],\end{aligned} \tag{7.33}$$

where (7.7) has been used. After integrating, by parts, the bracketed terms on the first line of (7.33) and then integrating the bracketed terms on the second line of (7.33), we obtain

$$\dot{E}_s = P_o\left(u_t(L_1,t)u_x(L_1,t) - u_t(L_2,t)u_x(L_2,t)\right) - \rho v\left(u_t^2(L_1,t) - u_t^2(L_2,t)\right), \tag{7.34}$$

where the boundary conditions given in (7.8) have been used. After utilizing (7.14), we can rewrite (7.34) in the following advantageous form:

$$\begin{aligned}\dot{E}_s = {} & -\frac{P_o}{2}(u_t^2(L_1,t) + u_x^2(L_1,t)) - \frac{P_o}{2}(u_t^2(L_2,t) + u_x^2(L_2,t)) \\ & -\rho v\left(u_t^2(L_1,t) - u_t^2(L_2,t)\right) + \frac{P_o}{2}\eta^T\eta.\end{aligned} \tag{7.35}$$

To determine $\dot{E}_{c1}(t)$ in (7.32), we differentiate (7.23) with respect to time to obtain

$$\dot{E}_{c1} = 2\rho\beta\int_0^{L_1} \sigma u_t u_{\sigma t} d\sigma + 2\beta P_o \int_0^{L_1} \sigma u_\sigma u_{\sigma\sigma} d\sigma, \tag{7.36}$$

after using (7.7). After noting that

$$2\int_0^b \sigma u u_\sigma d\sigma = bu^2(b) - \int_0^b u^2 d\sigma \tag{7.37}$$

for some constant b, we can rewrite (7.36) as follows:

$$\dot{E}_{c1} = \rho\beta\left(L_1 u_t^2(L_1,t) - \int_0^{L_1} u_t^2 d\sigma\right) + \beta F_o\left(L_1 u_x^2(L_1,t) - \int_0^{L_1} u_\sigma^2 d\sigma\right). \tag{7.38}$$

To determine $\dot{E}_{c2}(t)$ in (7.32), we differentiate (7.24) with respect to time to obtain

$$\dot{E}_{c2} = 2\rho\beta\int_{L_2}^{L}(\sigma - L)\, u_t u_{\sigma t} d\sigma + 2\beta F_o\int_{L_2}^{L}(\sigma - L)\, u_\sigma u_{\sigma\sigma} d\sigma \tag{7.39}$$

after using (7.7). After noting that

$$2\int_a^b (\sigma - b)\, u u_\sigma d\sigma = (b-a)\, u^2(a) - \int_a^b u^2 d\sigma \tag{7.40}$$

for some constant b, we can rewrite (7.39) as

$$\begin{aligned}\dot{E}_{c2} = {} & \rho\beta\left((L - L_2)\, u_t^2(L_2,t) - \int_{L_2}^{L} u_t^2 d\sigma\right) \\ & +\beta F_o\left((L - L_2)\, u_x^2(L_2,t) - \int_{L_2}^{L} u_\sigma^2 d\sigma\right)\end{aligned} \tag{7.41}$$

After substituting (7.35), (7.38), and (7.41) into (7.32), we obtain

$$\begin{aligned}\dot{V} = {} & -\eta^T K_s \eta + \frac{P_o}{2}\eta^T\eta \\ & -\beta\rho\left(\int_0^{L_1} u_t^2 d\sigma + \int_{L_2}^{L} u_t^2 d\sigma\right) - \beta P_o\left(\int_0^{L_1} u_\sigma^2 d\sigma + \int_{L_2}^{L} u_\sigma^2 d\sigma\right) \\ & -u_t^2(L_1,t)\left(\frac{P_o}{2} + \rho v - \rho\beta L_1\right) - u_x^2(L_1,t)\left(\frac{P_o}{2} + \rho v - \beta P_o L_1\right) \\ & -u_t^2(L_2,t)\left(\frac{P_o}{2} - \rho v - \rho\beta\,(L - L_2)\right) \\ & -u_x^2(L_2,t)\left(\frac{P_o}{2} - \beta P_o\,(L - L_2)\right).\end{aligned} \tag{7.42}$$

If the axial web velocity is selected according to[3] $v < \frac{P_o}{4\rho}$, and the weighting constant β is selected to satisfy

$$\beta < \min\left\{\frac{P_o}{2\rho L_1}, \frac{1}{2L_1}, \frac{P_o}{4\rho(L-L_2)}, \frac{1}{2(L-L_2)}\right\}, \tag{7.43}$$

then we can use (7.42) to form the following upper bound[4] for $\dot{V}(t)$:

$$\begin{aligned}\dot{V} \leq & -\left(\lambda_{\min}\{K_s\} - \frac{P_o}{2}\right)\|\eta\|^2 - \beta\rho\left(\int_0^{L_1} u_t^2 d\sigma + \int_{L_2}^{L} u_t^2 d\sigma\right) \\ & -\beta P_o\left(\int_0^{L_1} u_\sigma^2 d\sigma + \int_{L_2}^{L} u_\sigma^2 d\sigma\right).\end{aligned} \tag{7.44}$$

From (7.44) and (7.22), it is clear that if the control gain matrix K_s is selected according to (7.20), then $\dot{V}(t)$ can be upper bounded by the following nonpositive function:

$$\dot{V}(t) \leq -\lambda_3\left(E_s(t) + \|\eta(t)\|^2\right), \tag{7.45}$$

where λ_3 is some positive constant.[5] From (7.30) and (7.45), we get

$$\dot{V}(t) \leq -\frac{\lambda_3}{\lambda_2}V(t), \tag{7.46}$$

whose solution, obtained by applying Lemma A.4 in Appendix A, is given by

$$V(t) \leq V(0)\exp\left(-\frac{\lambda_3}{\lambda_2}t\right) \leq \lambda_2\kappa_o\exp\left(-\frac{\lambda_3}{\lambda_2}t\right), \tag{7.47}$$

where (7.30) has been used to formulate the inequality on the right-hand side of (7.47) and the positive constant κ_o is defined in (7.19) (note that (7.14), (7.21), and (7.22) have been used during the construction of κ_o). We can now use (A.25) of Lemma A.12 in Appendix A, (7.22), and (7.30)

---

[3]The condition given on v given in (7.20) is obtained by substituting the definition for P_o from (7.12) into $v < \frac{P_o}{4\rho}$, and then solving for v.

[4]It is easy to see that the weighting constant β can be selected small enough to satisfy both (7.28) and (7.43).

[5]To ensure that λ_3 is positive, the weighting constant β, the axial web speed v, and the control gain matrix K_s must be selected to satisfy (7.43) and (7.20).

to formulate the following inequalities:

$$\frac{P_o}{2L_1}u^2(x,t) \le \frac{P_o}{2}\int_0^{L_1} u_\sigma^2(\sigma,t)d\sigma \le E_s(t) \le \frac{1}{\lambda_1}V(t) \quad \forall x \in [0, L_1]$$

$$\frac{P_o}{2(L-L_2)}u^2(x,t) \le \frac{P_o}{2}\int_{L_2}^{L} u_\sigma^2(\sigma,t)d\sigma \le E_s(t) \le \frac{1}{\lambda_1}V(t) \quad \forall x \in [L_2, L]. \tag{7.48}$$

The inequalities given in (7.18) now follow directly by combining (7.47) and (7.48). □

Remark 7.1 *From (7.21) and (7.47), we know that $E_s(t)$ and $\eta(t)$ are bounded $\forall t \in [0,\infty)$. Since $E_s(t)$ is bounded, we can use (7.22) and (A.25) of Lemma A.12 in Appendix A to show that $u(x,t)$ is bounded $\forall t \in [0,\infty)$ and $\forall x \in [0, L_1] \cup [L_2, L]$. Since $E_s(t)$ defined in (7.22) is bounded $\forall t \in [0,\infty)$, the potential energy of the system defined in (7.4) is bounded $\forall t \in [0,\infty)$; hence, we can apply Property 7.1 to show that $\frac{\partial^n}{\partial x^n}u(x,t)$ is bounded for $n = 1, 2$, $\forall t \in [0,\infty)$, and $\forall x \in [0, L_1] \cup [L_2, L]$. Since $\eta(t)$, $u_x(L_1,t)$, and $u_x(L_2,t)$ are bounded $\forall t \in [0,\infty)$, we can use (7.14) to state that $u_t(L_1,t)$ and $u_t(L_2,t)$ are bounded $\forall t \in [0,\infty)$. From the boundedness of $E_s(t)$, $u_t(L_1,t)$, and $u_t(L_2,t)$, we can see that the kinetic energy of the system defined in (7.3) is bounded $\forall t \in [0,\infty)$. Since the kinetic energy is bounded $\forall t \in [0,\infty)$, we can use Property 7.2 discussed to conclude that $\frac{\partial^n}{\partial x^n}u_t(x,t)$ is bounded for $n = 0, 1$, $\forall t \in [0,\infty)$, and $\forall x \in [0, L_1] \cup [L_2, L]$. From the above information, we can now state that all of the signals in the control law of (7.16) and in the system given by (7.7) through (7.9) remain bounded during closed-loop operation $\forall t \in [0,\infty)$.*

7.2.4 Adaptive Control Law

The control law given by (7.16) requires exact knowledge of the system parameters. We will now redesign the control law given by (7.16) to compensate for parametric uncertainty while asymptotically stabilizing the web displacement. First, note that the open-loop dynamics of (7.15) can be rewritten in the following advantageous form:

$$M\dot{\eta}(t) = Y(t)\theta + \begin{bmatrix} \frac{-1}{L_2 - L_1} & \frac{1}{2} \\ \frac{1}{L_2 - L_1} & \frac{1}{2} \end{bmatrix} \begin{bmatrix} \tau(t) \\ f(t) \end{bmatrix}, \tag{7.49}$$

where the known regression matrix $Y(t) \in \mathbb{R}^{2\times 4}$ and the unknown, constant parameter vector $\theta \in \mathbb{R}^4$ are defined as follows:

$$Y(t) = \begin{bmatrix} \dfrac{u_{xt}(L_1,t) - u_{xt}(L_2,t)}{4} & \dfrac{u_{xt}(L_1,t) + u_{xt}(L_2,t)}{(L_2 - L_1)^2} & (u_t(L_1,t) - u_t(L_2,t)) & -\dfrac{u(L_1,t) - u(L_2,t)}{L_2 - L_1} - u_x(L_1,t) \\ \dfrac{u_{xt}(L_1,t) - u_{xt}(L_2,t)}{4} & -\dfrac{u_{xt}(L_1,t) + u_{xt}(L_2,t)}{(L_2 - L_1)^2} & (u_t(L_1,t) - u_t(L_2,t)) & \dfrac{u(L_1,t) - u(L_2,t)}{L_2 - L_1} + u_x(L_2,t) \end{bmatrix} \tag{7.50}$$

and

$$\theta = \begin{bmatrix} m_{eq} & J_{eq} & \rho v & P_o \end{bmatrix}^T. \tag{7.51}$$

Based on the structure of (7.49) and the subsequent stability analysis, the control law is designed as

$$\begin{bmatrix} \tau(t) \\ f(t) \end{bmatrix} = \begin{bmatrix} \dfrac{-1}{L_2 - L_1} & \dfrac{1}{2} \\ \dfrac{1}{L_2 - L_1} & \dfrac{1}{2} \end{bmatrix}^{-1} \left(-Y(t)\hat{\theta}(t) - K_s \eta(t) \right), \tag{7.52}$$

where $\hat{\theta}(t) \in \mathbb{R}^4$ represents a dynamic, parameter estimate vector updated by

$$\dot{\hat{\theta}}(t) = \Gamma Y^T(t) \eta(t), \tag{7.53}$$

with $\Gamma \in \mathbb{R}^{4 \times 4}$ being a diagonal, positive-definite, adaptation gain matrix. After substituting (7.52) into (7.49), we can form the closed-loop dynamics for $\eta(t)$ as follows:

$$M\dot{\eta}(t) = -K_s \eta(t) + Y(t)\tilde{\theta}(t), \tag{7.54}$$

where the parameter estimation error $\tilde{\theta}(t) \in \mathbb{R}^4$ is defined as follows:

$$\tilde{\theta}(t) = \theta - \hat{\theta}(t). \tag{7.55}$$

Theorem 7.2 *The adaptive controller given by (7.52) and the parameter update law of (7.53) ensure that the web displacement is asymptotically regulated in the following sense:*

$$\lim_{t \to \infty} u(x,t) = 0 \qquad \forall x \in [0, L_1] \cup [L_2, L], \tag{7.56}$$

where the axial web velocity and the control gain matrix K_s defined in (7.52) must be selected to satisfy the sufficient conditions stated in (7.20).

Proof. The proof is based on arguments similar to those used in the proof of Theorem 7.1; hence, some of the details will not be repeated. We start by defining the following function:

$$V_a(t) = V(t) + \frac{1}{2}\tilde{\theta}^T(t)\Gamma^{-1}\tilde{\theta}(t), \tag{7.57}$$

where $V(t)$ was defined in (7.21). If β of (7.23) and (7.24) is selected according to (7.28), we can formulate the following bounds on $V_a(t)$:

$$\begin{aligned} \lambda_{1a}\left(E_s(t) + \|\eta(t)\|^2 + \left\|\tilde{\theta}(t)\right\|^2\right) &\leq V_a(t) \\ &\leq \lambda_{2a}\left(E_s(t) + \|\eta(t)\|^2 + \left\|\tilde{\theta}(t)\right\|^2\right), \end{aligned} \tag{7.58}$$

where λ_{1a} and λ_{2a} are some positive constants.

After differentiating (7.57) with respect to time, and then substituting from (7.7) and (7.54), we can follow the derivations used in Theorem 7.1 to obtain the following upper bound for the time derivative of $V_a(t)$:

$$\dot{V}_a(t) \leq -\lambda_{3a}\left(E_s(t) + \|\eta(t)\|^2\right) + \tilde{\theta}^T(t)\left(Y^T(t)\eta(t) - \Gamma^{-1}\dot{\hat{\theta}}(t)\right), \tag{7.59}$$

where λ_{3a} is some positive constant,[6] and we have used the fact that $\dot{\tilde{\theta}}(t) = -\dot{\hat{\theta}}(t)$ from (7.55). We now substitute the parameter update law of (7.53) into (7.59) to obtain the following result:

$$\dot{V}_a(t) \leq -\lambda_{3a}\left(E_s(t) + \|\eta(t)\|^2\right) \triangleq -g_a(t), \tag{7.60}$$

where $g_a(t)$ is a non-negative function. After differentiating $g_a(t)$ with respect to time, we obtain

$$\dot{g}_a(t) = \lambda_{3a}\left(\dot{E}_s(t) + 2\eta^T(t)\dot{\eta}(t)\right). \tag{7.61}$$

By utilizing (7.57), (7.58), (7.60), and arguments similar to those outlined in the proof of Theorem 7.1 and Remark 7.1, we can illustrate that all closed-loop signals are bounded $\forall t \in [0, \infty)$. Hence, we can use (7.35) and (7.54) to conclude from (7.61) that $\dot{g}_a(t)$ is also bounded $\forall t \in [0, \infty)$. We can now use Lemma A.6 in Appendix A to illustrate that

$$\lim_{t\to\infty} E_s(t), \|\eta(t)\| = 0. \tag{7.62}$$

Finally, we use (7.62) and (7.48) to obtain the result given by (7.56). □

---

[6]To ensure that λ_{3a} is positive, the weighting constant β, the axial web speed v, and the control gain matrix K_s must be selected to satisfy (7.43) and (7.20).

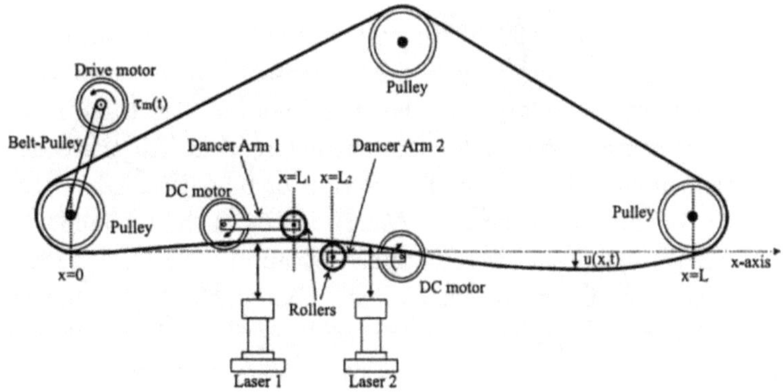

FIGURE 7.2. Schematic diagram of the experimental setup for the axially moving string and actuator system.

7.2.5 Experimental Evaluation

Experimental Setup

A schematic diagram of the experimental setup used to implement the controllers is shown in Figure 7.2. The experimental setup consists of an axially moving string driven by a brushed DC motor via a belt-pulley transmission. This string is made up of a flexible tape with negligible thickness and width of 5 mm to facilitate laser measurements. Two dancer arms mounted on brushed DC motors applied the forces $f_1(t)$ and $f_2(t)$. Encoders (1024 pulses per revolution in quadrature) mounted on the DC motors actuating the two dancer arms measured the boundary displacements $u(L_1, t)$ and $u(L_2, t)$, while two lasers (operating at a 1 kHz bandwidth), placed on either side and at a very short distance from the point of contact of the dancer arm with the string, were used to measure the slopes $u_x(L_1, t)$ and $u_x(L_2, t)$.

A Pentium 166 MHz PC running QNX hosted the control algorithm. *QMotor* provided the environment to write the control algorithm in the C programming language. The Quanser *MultiQ* I/O board [20] provided for data transfer between the computer subsystem and the electrical interface. Three A/D channels were used to sense the currents flowing through the rotor windings of the three brushed DC motors, while two A/D channels were used to read in the laser measurements. Three D/A channels delivered voltages to power the driving DC motor as well as the two DC motors that actuated the two dancer arms. These voltages went through two stages of amplification: the first stage consisted of OP07C operational amplifiers while in the second stage, Techron linear power amplifiers sourced a maximum current of 5 A at 60 V. To ensure that the actual rotor currents tracked the desired current trajectories, a high-gain current feedback loop[7] was utilized to apply voltages to the three motors. The three rotor currents were measured using hall-effect current sensors. All the controllers were implemented using a sampling period of 0.5 msec. The string velocity $v(t)$, boundary velocities $u_t(L_1, t), u_t(L_2, t)$, and boundary slope rates $u_{xt}(L_1, t), u_{xt}(L_2, t)$ were obtained by applying a backwards difference algorithm to the string displacement, boundary displacements, and boundary slopes, respectively, with the resulting signals being filtered by a second-order digital filter. The various parameters associated with the axially moving string system were calculated using standard test procedures and engineering handbook tables, and were determined to be

$$\begin{gathered}\rho = 0.02 \text{ kg/m}, \quad m = 0.03 \text{ kg}, \quad r = 0.0381 \text{ m}, \\ J_0 = 1.451 \times 10^{-5} \text{ kg-m}^2,\ J = 2.371 \times 10^{-4} \text{ kg-m}^2,\ P_0 = 12.25 \text{ N}, \\ L = 0.711 \text{ m},\ L_1 = 0.394 \text{ m}, \quad L_2 = 0.280 \text{ m}.\end{gathered} \tag{7.63}$$

Experimental Results

The vibration of the axially moving string to a consistently applied impulse was studied with all gains selected to achieve the best results possible. Since the system model presented in Section 7.2.1 is predicated on the string moving with a constant axial velocity, the DC motor/belt-pulley system was used to regulate the string axial velocity to a desired, constant velocity setpoint $v_d = 0.42$ m/s. This was accomplished by the setting the motor torque to

$$\tau_m(t) = k_d(v_d - v(t)). \tag{7.64}$$

For comparison, the "open-loop" response of the system, defined as a constant force applied by the dancer arm, was implemented with $k_d = 0.5$ and $F(t) = -[0.8228, 0.8228]^T$. Figure 7.3 shows the open-loop displacements of the string $u(L_1, t)$ and $u(L_2, t)$ measured by the laser. A second comparison was performed with a standard damper control law given by

$$F(t) = -k_s \begin{bmatrix} u_t(L_1, t) \\ u_t(L_2, t) \end{bmatrix}, \tag{7.65}$$

[7] The reader is referred to Remark 5.4 for a discussion on issues concerning the implementation of a high-gain current feedback loop.

with $k_s = 6.24$ and $k_d = 0.55$. The displacements of the string near $x = L_1$ and $x = L_2$ for the damper controller are illustrated in Figure 7.4. The model-based controller of (7.16) was then implemented with $k_s = 6.19$ and $k_d = 0.51$. The displacements of the string are shown in Figure 7.5. To illustrate the effectiveness of (7.64) in maintaining a constant axial velocity, Figure 7.6 shows the velocity setpoint error $v_d - v(t)$ during the damper and model-based control experiments.

Comparing the results of the above experiments, we observe that the string in the open-loop mode took several seconds to come back to the original displacement after the impulse disturbance was applied at approximately 8 sec. While the damper control produced some improvement in the string response, the model-based controller cleared exhibited the best transient as well as steady-state response. Note from Figure 7.5 that it took a fraction of a second for the string to return its original displacement after being disturbed by the impulse.

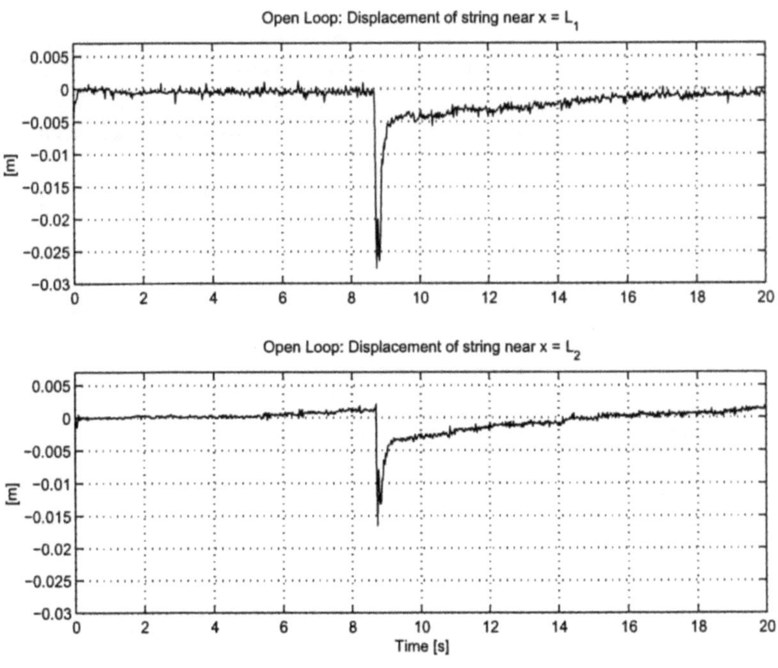

FIGURE 7.3. "Open-loop" response: (a) $u(L_1, t)$ and (b) $u(L_2, t)$.

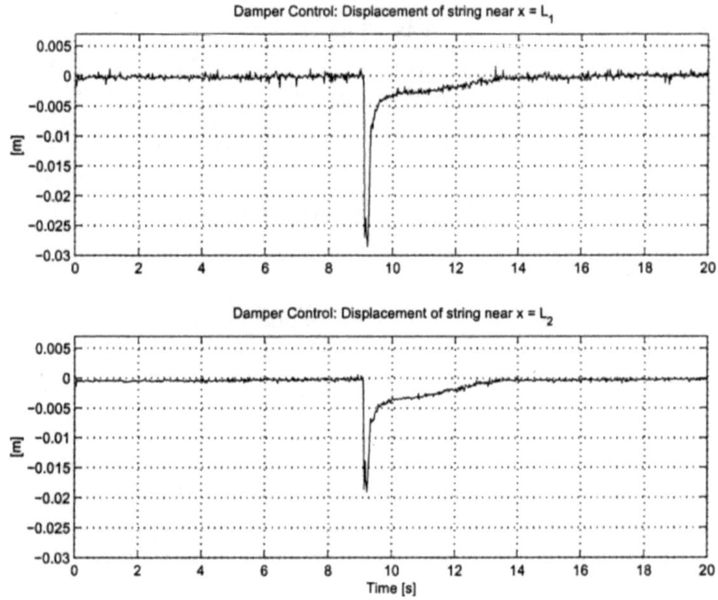

FIGURE 7.4. Damper controller: (a) $u(L_1, t)$ and (b) $u(L_2, t)$.

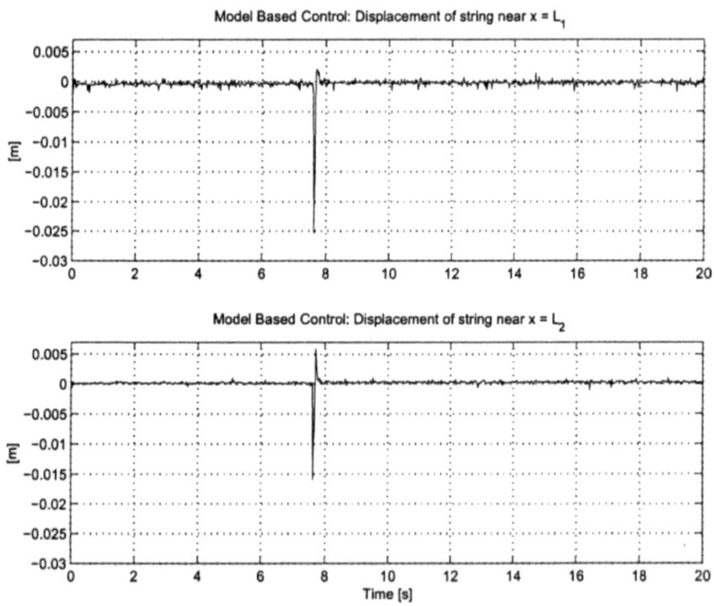

FIGURE 7.5. Model-based controller: (a) $u(L_1, t)$ and (b) $u(L_2, t)$.

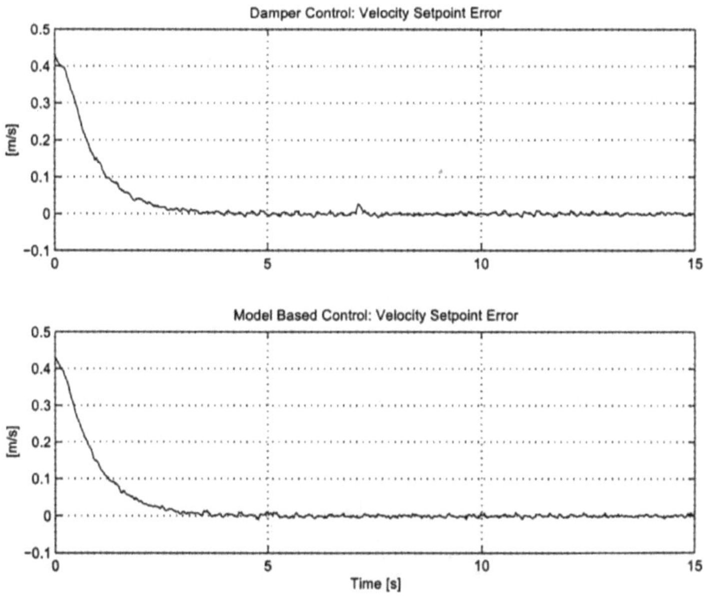

FIGURE 7.6. Velocity setpoint error: (a) damper controller and (b) model-based controller.

7.3 Flexible Link Robot Arm

In this section, we develop control strategies for a flexible link robot arm. The objectives of the control strategies are to regulate the link displacement while simultaneously driving the hub position to a desired setpoint. To this end, we first develop a model-based control law that asymptotically fulfills these control objectives. We then illustrate how the control law can be redesigned as an adaptive controller to achieve the same stability result while compensating for parametric uncertainty.

7.3.1 System Model

The robot system, illustrated in Figure 7.7, is composed of a beam clamped to a rotating, rigid actuator hub with a payload/actuator mass at its free end-point. A torque input is applied to the hub to control the system's angular position while a force input is applied to the flexible beam's end-point mass to regulate the beam displacement. The model for the single flexible-link robot arm is assumed to be described by the following equations of motion [8]:

$$\rho w_{tt}(x,t) + EIw_{xxxx}(x,t) = \rho u(x,t)\dot{q}^2(t) \tag{7.66}$$

and

$$\begin{aligned} D(t)\,\ddot{q}(t) + \frac{1}{2}\dot{D}(t)\,\dot{q}(t) + V_m(t)\,\dot{q}(t) & \\ +mu(L,t)w_t(L,t)\dot{q}(t) - EIw_{xx}(0,t) &= \tau(t), \end{aligned} \tag{7.67}$$

with the following boundary conditions:[8]

$$u(0,t) = u_x(0,t) = u_{xx}(L,t) = 0 \tag{7.68}$$

and

$$mw_{tt}(L,t) - mu(L,t)\dot{q}^2(t) - EIw_{xxx}(L,t) = f(t), \tag{7.69}$$

where $w(x,t)$ is a displacement variable defined as follows:

$$w(x,t) = u(x,t) + xq(t), \tag{7.70}$$

where $u(x,t)$ denotes the link displacement at position x for time t with respect to the (X,Y) coordinate system that rotates with the hub (see Figure 7.7); $q(t)$, $\dot{q}(t)$, $\ddot{q}(t)$ represent the angular position, velocity, and acceleration of the hub, respectively, with respect to the inertial reference direction (see Figure 7.7); ρ is the mass/length of the link; EI is the bending stiffness of the link; L is the length of the link; m represents a payload/actuator mass attached to the free end of the link; $\tau(t)$ is the control torque input applied to the hub; $f(t)$ denotes the boundary control force input applied to the mass; and the auxiliary functions $D(t)$, $\dot{D}(t)$, and $V_m(t)$ are defined as follows:

$$D(t) = J + mu^2(L,t) + \rho\int_0^L u^2(x,t)\,dx \geq J > 0, \tag{7.71}$$

$$\dot{D}(t) = \frac{d}{dt}D(t) = 2mu(L,t)u_t(L,t) + 2\rho\int_0^L u(x,t)u_t(x,t)\,dx, \tag{7.72}$$

and

$$V_m(t) = \rho\int_0^L u(x,t)w_t(x,t)dx, \tag{7.73}$$

with J denoting the hub's inertia.

The above model neglects gravitational effects, surface loads, the rotatory inertia of the link cross-sections, and the rotatory inertia of the mass at the link's free end. Furthermore, the model assumes that the motion of

---

[8] Given the clamped boundary conditions of (7.68), we also know that $u_t(0,t) = u_{xt}(0,t) = 0$.

the system is restricted to the (X, Y) plane, and that the flexible link is inextensible and straight when at rest.

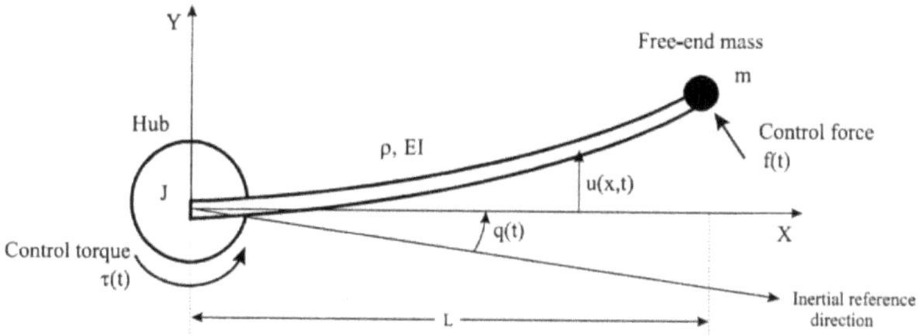

FIGURE 7.7. Flexible link robot arm.

Remark 7.2 *The variable transformation $w(x,t)$ defined in (7.70) is introduced to simplify the mathematical manipulations in the subsequent control development and stability analysis. Notice from the definition of (7.70) that the following equalities hold:*

$$\begin{aligned} w_x(x,t) &= u_x(x,t) + q(t) \quad \forall x \in [0, L] \\ \frac{\partial^n w(x,t)}{\partial x^n} &= \frac{\partial^n u(x,t)}{\partial x^n} \quad \textit{for} \quad n \geq 2 \quad \forall x \in [0, L]. \end{aligned} \tag{7.74}$$

The equalities shown in (7.74) will be exploited during the subsequent control development.

Since the control laws will consist of relatively simple functions, we will assume the existence of a unique solution for the dynamics given by (7.66) through (7.69) under the control. Based on the arguments outlined in Remark 5.1, we will assume that the link displacement variable $u(x,t)$ and its time derivative $u_t(x,t)$ belong to a space of functions that has the following properties.

Property 7.3 If the potential energy for the system given by (7.66)–(7.69), defined by

$$\Pi_p = \frac{1}{2} EI \int_0^L w_{\sigma\sigma}^2(\sigma, t) d\sigma, \tag{7.75}$$

is bounded $\forall t \in [0, \infty)$, then $\frac{\partial^n}{\partial x^n} u(x,t)$ is bounded for $n = 2, 3, 4$, $\forall t \in [0, \infty)$, and $\forall x \in [0, L]$.

Property 7.4 If the kinetic energy for the system given by (7.66)–(7.69), defined by

$$\Pi_k = \frac{1}{2}\rho \int_0^L w_t^2(\sigma,t)d\sigma + \frac{1}{2}mw_t^2(L,t) + \frac{1}{2}D(t)\,\dot{q}^2(t), \qquad (7.76)$$

is bounded $\forall t \in [0,\infty)$, then $\frac{\partial^n}{\partial x^n}u_t(x,t)$ is bounded for $n = 0, 1, 2, 3$, $\forall t \in [0,\infty)$, and $\forall x \in [0,L]$.

7.3.2 Problem Statement

The control objective is to ensure that (i) $u(x,t) \to 0\ \forall x \in [0,L]$ as $t \to \infty$ with respect to the rotating coordinate system (X,Y) attached to the hub, and (ii) $q(t) \to q_d$ as $t \to \infty$ with respect to the inertial reference direction where q_d is a desired, constant, angular position. To aid the analysis of the link displacement regulation objective, we define an auxiliary signal $\eta(t)$ as follows:

$$\eta(t) = w_t(L,t) - w_{xxx}(L,t), \qquad (7.77)$$

where $w(x,t)$ was defined in (7.70). To quantify the angular position regulation objective, we define the angular position setpoint error $e(t)$ as follows:

$$e(t) = q(t) - q_d. \qquad (7.78)$$

To formulate the open-loop dynamics for $\eta(t)$, we differentiate (7.77) with respect to time and then use (7.69) to obtain

$$m\dot{\eta}(t) = -mw_{xxxt}(L,t) + mu(L,t)\dot{q}^2(t) + EIw_{xxx}(L,t) + f(t). \qquad (7.79)$$

The open-loop dynamics for $e(t)$ are obtained by differentiating (7.78) twice with respect to time such that (7.67) can be used to produce

$$\begin{aligned} D(t)\,\ddot{e}(t) = \; & -\frac{1}{2}\dot{D}(t)\,\dot{e}(t) - V_m(t)\,\dot{q}(t) - mu(L,t)w_t'(L,t)\dot{q}(t) \\ & + EIw_{xx}(0,t) + \tau(t), \end{aligned} \qquad (7.80)$$

where we have used the fact that $\dot{e}(t) = \dot{q}(t)$.

7.3.3 Model-Based Control Law

Based on the form of (7.79) and the subsequent stability analysis, the boundary control force is designed as follows:

$$f(t) = mw_{xxxt}(L,t) - EIw_{xxx}(L,t) - k_s\eta(t), \qquad (7.81)$$

where k_s is a positive control gain. After substituting (7.81) into (7.79), we obtain the following closed-loop dynamics at the link's free end:

$$m\dot{\eta}(t) = -k_s\eta(t) + mu(L,t)\dot{q}^2(t). \tag{7.82}$$

Based on the form of (7.80) and the subsequent stability analysis, we design the hub control torque as follows:

$$\tau(t) = -k_v\dot{e}(t) - k_pe(t) + mu(L,t)\dot{q}(t)w_{xxx}(L,t) - \beta\rho L\dot{q}(t)u^2(L,t), \tag{7.83}$$

where k_v, k_p, and β are positive control gains. After substituting (7.83) into (7.80), we obtain the closed-loop hub dynamics as follows:

$$\begin{aligned} D(t)\,\ddot{e}(t) = \; & -\frac{1}{2}\dot{D}(t)\,\dot{e}(t) - mu(L,t)\dot{q}(t)\eta(t) - V_m(t)\,\dot{q}(t) + EIw_{xx}(0,t) \\ & -k_v\dot{e}(t) - k_pe(t) - \beta\rho L\dot{q}(t)u^2(L,t), \end{aligned} \tag{7.84}$$

where the definition of (7.77) has been utilized.

It is interesting to note that the boundary control force of (7.81) contains a noncollocated term in the feedback loop (i.e., $\dot{q}(t)$ appears in the definition of $\eta(t)$ through $w_t(L,t)$), while the control torque of (7.83) contains noncollocated feedforward and feedback terms (i.e., the last two terms in (7.83).

Theorem 7.3 *The model-based control law given by (7.81) and (7.83) guarantees asymptotic regulation of the link displacement and hub position setpoint error in the following sense:*

$$\lim_{t\to\infty}|u(x,t)| = 0 \qquad \forall x \in [0,L] \qquad \lim_{t\to\infty}|\dot{e}(t)| = 0 \tag{7.85}$$

$$\lim_{t\to\infty}|e(t)| = 0, \tag{7.86}$$

given that the control gains k_s, k_v, and β satisfy the following sufficient conditions:

$$k_s > \frac{EI}{2} \qquad k_v > \frac{8}{3}\beta\rho L^3 \qquad \beta < \min\left\{\frac{\min\{J,\rho,EI\}}{4\rho\max\{L^4,1\}}, \frac{EI}{2L\rho}, \frac{1}{16L^3}\right\}. \tag{7.87}$$

Proof. To prove the above result, we define the following function:

$$V(t) = E_b(t) + E_c(t) + \frac{1}{2}m\eta^2(t) + \frac{1}{2}k_pe^2(t), \tag{7.88}$$

where

$$E_b(t) = \frac{1}{2}D(t)\,\dot{e}^2(t) + \frac{1}{2}\rho\int_0^L w_t^2(\sigma,t)d\sigma + \frac{1}{2}EI\int_0^L w_{\sigma\sigma}^2(\sigma,t)d\sigma \tag{7.89}$$

and

$$E_c(t) = 2\beta\rho \int_0^L \sigma u_\sigma(\sigma,t) w_t(\sigma,t) d\sigma, \tag{7.90}$$

with β being the control gain defined in (7.83).

We now establish conditions on the control gain β to ensure that $V(t)$ defined in (7.88) is non-negative. First, note that we can use (7.71) and (7.89) to lower bound $E_b(t)$ as follows:

$$E_b \geq \frac{1}{2} \min\{J, \rho, EI\} \left(\dot{e}^2 + E_n\right) \geq 0, \tag{7.91}$$

where $E_n(t)$ is defined by

$$E_n(t) = \int_0^L \left(w_t^2(\sigma,t) + w_{\sigma\sigma}^2(\sigma,t)\right) d\sigma. \tag{7.92}$$

Now, we use (A.29) of Lemma A.13 in Appendix A to upper bound $E_c(t)$ of (7.90) as follows:

$$E_c \leq 2\beta\rho \int_0^L \left(\sigma^2 u_\sigma^2 + w_t^2\right) d\sigma. \tag{7.93}$$

After applying (A.27) of Lemma A.12 in Appendix A to the u_σ^2 term on the right-hand side of (7.93), and then utilizing (7.74) and (7.92), we have

$$E_c \leq 2\beta\rho \int_0^L \left(L^4 w_{\sigma\sigma}^2 + w_t^2\right) d\sigma \leq 2\beta\rho \max\left\{L^4, 1\right\} E_n. \tag{7.94}$$

We can now use (7.94) to establish the following inequality:

$$-2\beta\rho \max\left\{L^4, 1\right\} E_n \leq E_c \leq 2\beta\rho \max\left\{L^4, 1\right\} E_n. \tag{7.95}$$

Therefore, if β is selected according to

$$\beta < \frac{\min\{J, \rho, EI\}}{4\rho \max\{L^4, 1\}}, \tag{7.96}$$

we can use (7.91) and (7.95) to develop the following inequality:

$$\xi_1 E_n + \xi_2 \dot{e}^2 \leq E_b + E_c \tag{7.97}$$

for some positive constants ξ_1, ξ_2. Given the form of $V(t)$ defined in (7.88) and the inequality given by (7.97), we can now formulate the following inequality:

$$\lambda_1 \left(E_n(t) + \eta^2(t) + e^2(t) + \dot{e}^2(t)\right) \leq V, \tag{7.98}$$

where λ_1 is some positive constant.

After differentiating (7.88) with respect to time and substituting from (7.82), we have

$$\dot{V}(t) = \dot{E}_b(t) + \dot{E}_c(t) - k_s\eta^2(t) + mu(L,t)\dot{q}^2(t)\eta(t) + k_p e(t)\dot{e}(t). \quad (7.99)$$

To determine $\dot{E}_b(t)$ in (7.99), we differentiate (7.89) with respect to time to obtain

$$\begin{aligned}\dot{E}_b = \; & -EI\int_0^L w_t w_{\sigma\sigma\sigma\sigma} d\sigma + EI\int_0^L w_{\sigma\sigma} w_{\sigma\sigma t} d\sigma + \rho\dot{q}^2\int_0^L u w_t d\sigma \\ & +\dot{e}\left(-mu(L)\dot{q}\eta - V_m\dot{q} + EIw_{xx}(0) - k_v\dot{e} - k_p e - \beta\rho L\dot{q}u^2(L)\right),\end{aligned} \quad (7.100)$$

where (7.66) and (7.84) have been utilized. If we integrate by parts, the first two terms on the right-hand side of (7.100), we obtain

$$\dot{E}_b = -EIw_t(L)w_{xxx}(L) - mu(L)\dot{q}^2\eta - k_v\dot{e}^2 - k_p e\dot{e} - \beta\rho L\dot{q}^2u^2(L), \quad (7.101)$$

where the definition of (7.73) and the fact that $w_{xt}(0,t) = \dot{q}(t)$ (see (7.68), (7.74), and (7.78)) have been utilized to cancel common terms. We can now rewrite (7.101) into the following advantageous form:

$$\begin{aligned}\dot{E}_b = \; & -\frac{EI}{2}\left(w_t^2(L) + w_{xxx}^2(L)\right) + \frac{EI}{2}\eta^2 - mu(L)\dot{q}^2\eta \\ & -k_v\dot{e}^2 - k_p e\dot{e} - \beta\rho L\dot{q}^2u^2(L),\end{aligned} \quad (7.102)$$

upon application of (7.77).

To determine $\dot{E}_c(t)$ in (7.99), we differentiate (7.90) with respect to time, and then substitute from (7.66) to obtain

$$\dot{E}_c = A_1 + A_2 + A_3, \quad (7.103)$$

where

$$A_1 = 2\beta\rho\int_0^L \sigma u_{\sigma t} w_t d\sigma \qquad A_2 = -2\beta EI\int_0^L \sigma u_\sigma w_{\sigma\sigma\sigma\sigma} d\sigma$$
$$A_3 = 2\beta\rho\dot{q}^2\int_0^L \sigma u_\sigma u d\sigma. \quad (7.104)$$

First, note that the expression for A_1 given in (7.104) can be rewritten as follows:

$$A_1 = 2\beta\rho\int_0^L \sigma w_{\sigma t} w_t d\sigma - 2\beta\rho\dot{q}\int_0^L \sigma w_t d\sigma \quad (7.105)$$

upon the use of the time derivative of the first equation of (7.74). After integrating, by parts, the first term of (7.105), we obtain

$$A_1 = 2\beta\rho L w_t^2(L) - 2\beta\rho \int_0^L w_t^2 d\sigma - 2\beta\rho \int_0^L \sigma u_{\sigma t} w_t d\sigma - 4\beta\rho\dot{q} \int_0^L \sigma w_t d\sigma, \tag{7.106}$$

where (7.74) has been utilized. After noticing that the third term on the right-hand side of (7.106) is equal to A_1, we can rearrange (7.106) as follows:

$$A_1 = \beta\rho L w_t^2(L) - \beta\rho \int_0^L w_t^2 d\sigma - 2\beta\rho\dot{q} \int_0^L \sigma w_t d\sigma. \tag{7.107}$$

We now apply Holder's inequality [6] to the last term on the right-hand side of (7.107) to produce an upper bound on A_1 as shown below:

$$A_1 \leq \beta\rho L w_t^2(L) - \beta\rho \int_0^L w_t^2 d\sigma + 2\beta\rho\,|\dot{q}| \sqrt{\int_0^L \sigma^2 d\sigma} \sqrt{\int_0^L w_t^2 d\sigma}. \tag{7.108}$$

After applying (A.30) of Lemma A.13 in Appendix A to the last term on the right-hand side of (7.108), we obtain the final upper bound for A_1 as follows:

$$A_1 \leq \beta\rho L w_t^2(L) - \beta\rho \int_0^L w_t^2 d\sigma + 2\beta\rho\sqrt{\frac{L^3}{3}} \left(\frac{\dot{q}^2}{\delta_1} + \delta_1 \int_0^L w_t^2 d\sigma\right), \tag{7.109}$$

where δ_1 is an arbitrary positive constant.

After integrating, by parts, the expression for A_2 given in (7.104), we obtain

$$A_2 = -2\beta EI \left(L u_x(L) w_{xxx}(L) - \int_0^L u_\sigma w_{\sigma\sigma\sigma} d\sigma - \int_0^L \sigma w_{\sigma\sigma} w_{\sigma\sigma\sigma} d\sigma \right) \tag{7.110}$$

upon application of (7.68) and (7.74). After integrating, by parts, the first integral on the right-hand side of (7.110), and then applying (7.68) and (7.74), we obtain

$$A_2 = -2\beta EI \left(L u_x(L,t) w_{xxx}(L,t) + \int_0^L w_{\sigma\sigma}^2 d\sigma - \int_0^L \sigma w_{\sigma\sigma} w_{\sigma\sigma\sigma} d\sigma \right). \tag{7.111}$$

After integrating, by parts, the last integral on the right-hand side of (7.111), we obtain

$$A_2 = -2\beta EI \left(L u_x(L) w_{xxx}(L) + 2 \int_0^L w_{\sigma\sigma}^2 d\sigma + \int_0^L \sigma w_{\sigma\sigma} w_{\sigma\sigma\sigma} d\sigma \right) \tag{7.112}$$

upon application of (7.68). We can now add the expressions given in (7.111) and (7.112) to produce

$$A_2 = -2\beta EILu_x(L)w_{xxx}(L) - 3\beta EI\int_0^L w_{\sigma\sigma}^2 d\sigma. \tag{7.113}$$

Finally, we apply (A.30) of Lemma A.13 in Appendix A to the first term on the right-hand side of (7.113), and apply (7.74) and (A.28) of Lemma A.12 in Appendix A to the integral term in (7.113) to obtain an upper bound for A_2 as follows:

$$A_2 \leq 2\beta EIL\left(\delta_2 u_x^2(L) + \frac{w_{xxx}^2(L)}{\delta_2}\right) - \beta EI\frac{1}{L}u_x^2(L) - 2\beta EI\int_0^L w_{\sigma\sigma}^2 d\sigma, \tag{7.114}$$

where δ_2 is an arbitrary positive constant.

After integrating, by parts, the expression for A_3 given in (7.104), we obtain

$$A_3 = 2\beta\rho\dot{q}^2\left(Lu^2(L) - \int_0^L u^2 d\sigma\right) - 2\beta\rho\dot{q}^2\int_0^L \sigma u_\sigma u d\sigma. \tag{7.115}$$

After noting that the last term in (7.115) is equal to A_3, we can write (7.115) as follows:

$$A_3 = \beta\rho\dot{q}^2\left(Lu^2(L) - \int_0^L u^2 d\sigma\right). \tag{7.116}$$

We can now substitute (7.102), (7.103), (7.109), (7.114), and (7.116) into (7.99) to obtain the following upper bound for $\dot{V}(t)$:

$$\begin{aligned}\dot{V} \leq\ & -\beta\rho\left(1 - 2\delta_1\sqrt{\frac{L^3}{3}}\right)\int_0^L w_t^2 d\sigma - 2\beta EI\int_0^L w_{\sigma\sigma}^2 d\sigma \\ & -\beta EI\left(\frac{1}{L} - 2L\delta_2\right)u_x^2(L) - \left(\frac{EI}{2} - \beta\rho L\right)w_t^2(L) \\ & -EI\left(\frac{1}{2} - \frac{2L\beta}{\delta_2}\right)w_{xxx}^2(L) - \left(k_v - \frac{2\beta\rho}{\delta_1}\sqrt{\frac{L^3}{3}}\right)\dot{e}^2 \\ & -\left(k_s - \frac{EI}{2}\right)\eta^2 - \beta\rho\dot{q}^2\int_0^L u^2 d\sigma.\end{aligned} \tag{7.117}$$

If we let $\delta_1 = \frac{1}{4}\sqrt{\frac{3}{L^3}}$ and $\delta_2 = \frac{1}{4L^2}$, we can rewrite (7.117) as follows:

$$\begin{aligned}\dot{V} \leq & -\frac{\beta\rho}{2}\int_0^L w_t^2 d\sigma - 2\beta EI\int_0^L w_{\sigma\sigma}^2 d\sigma - \frac{\beta EI}{2L}u_x^2(L) \\ & -\left(\frac{EI}{2} - \beta\rho L\right)w_t^2(L) - EI\left(\frac{1}{2} - 8L^3\beta\right)w_{xxx}^2(L) \\ & -\left(k_v - \frac{8}{3}\beta\rho L^3\right)\dot{e}^2 - \left(k_s - \frac{EI}{2}\right)\eta^2 - \beta\rho\dot{q}^2\int_0^L u^2 d\sigma.\end{aligned} \tag{7.118}$$

If the control gains k_s and k_v are selected according to (7.87), and β is selected to satisfy

$$\beta < \min\left\{\frac{EI}{2L\rho}, \frac{1}{16L^3}\right\}, \tag{7.119}$$

we can use (7.118) to form a new upper bound for $\dot{V}(t)$ as shown below:

$$\dot{V}(t) \leq -\lambda_2\left(E_n(t) + \eta^2(t) + \dot{e}^2(t)\right), \tag{7.120}$$

where λ_2 is some positive constant.

From (7.98) and (7.120), we can state that $E_n(t)$, $e(t)$, $\dot{e}(t)$ (hence, $\dot{q}(t)$), and $\eta(t)$ are bounded $\forall t \in [0, \infty)$. Since $E_n(t)$ is bounded $\forall t \in [0, \infty)$, we can use (7.92) and (A.28) of Lemma A.12 in Appendix A to show that $u(x,t)$ is bounded $\forall t \in [0, \infty)$ and $\forall x \in [0, L]$. Since $E_n(t)$ of (7.92) is bounded $\forall t \in [0, \infty)$, the potential energy of the system given by (7.75) is bounded $\forall t \in [0, \infty)$; hence, we can use Property 7.3 to show that $\frac{\partial^n}{\partial x^n}u(x,t)$ is bounded for $n = 0, 1$, $\forall t \in [0, \infty)$, and $\forall x \in [0, L]$. Since $e(t)$ and q_d are bounded $\forall t \in [0, \infty)$, we can use (7.78) to conclude that $q(t)$ is bounded $\forall t \in [0, \infty)$. Since $u(x,t)$ and $q(t)$ are bounded, we can use (7.70) to show that $w(x,t)$ is bounded $\forall t \in [0, \infty)$ and $\forall x \in [0, L]$. Owing to the boundedness of the potential energy, we can use Property 7.3 and (7.74) to show that $\frac{\partial^n}{\partial x^n}u(x,t)$ and $\frac{\partial^n}{\partial x^n}w(x,t)$ are bounded for $n = 2, 3, 4$, $\forall t \in [0, \infty)$, and $\forall x \in [0, L]$. Since $\eta(t)$ and $w_{xxx}(L,t)$ are bounded $\forall t \in [0, \infty)$, we can use (7.77) to state that $w_t(L,t)$ is bounded $\forall t \in [0, \infty)$. Based on the above boundedness statements, we can easily show that $D(t)$ defined in (7.71) (upon application of (A.24) and (A.27) of Lemma A.12 in Appendix A) is bounded $\forall t \in [0, \infty)$. From the above information, it is easy to see that the system's kinetic energy defined in (7.76) is bounded $\forall t \in [0, \infty)$. Since the kinetic energy is bounded, we can utilize Property 7.4 to state that $\frac{\partial^n}{\partial x^n}u_t(x,t)$ is bounded for $n = 0, 1, 2, 3,$

$\forall t \in [0,\infty)$, and $\forall x \in [0,L]$. Based on the previous boundedness statements, the inequalities (A.24) and (A.27) of Lemma A.12 and (A.29) of Lemma A.13 in Appendix A can be applied to the right-hand side of (7.72) and (7.73) to illustrate that $\dot{D}(t)$ and $V_m(t)$ are bounded $\forall t \in [0,\infty)$. From the above information, we can conclude that the control inputs of (7.81) and (7.83) are bounded $\forall t \in [0,\infty)$. We can now use (7.67) and the fact that $D(t) > 0$ (see (7.71)) to conclude that $\ddot{q}(t)$ is bounded $\forall t \in [0,\infty)$. Finally, (7.66) and the second time derivative of (7.70) can be utilized to show that $w_{tt}(x,t)$ and $u_{tt}(x,t)$ are bounded $\forall t \in [0,\infty)$ and $\forall x \in [0,L]$. As a result of the boundedness of $D(t)$, we can use (7.89) and (7.92) to state that

$$E_b \leq \frac{1}{2}\max\left\{\sup_t \{D(t)\}, \rho, EI\right\}\left(\dot{q}^2 + E_n\right) < \infty. \tag{7.121}$$

From (7.120) and (7.121), we can rewrite the upper bound on $\dot{V}(t)$ as follows:

$$\dot{V}(t) \leq -\lambda_3 E_b(t) - \lambda_2 \eta^2(t) \triangleq -g(t), \tag{7.122}$$

where λ_3 is some positive constant, and $g(t)$ is a non-negative function. After differentiating $g(t)$ with respect to time, we have

$$\dot{g}(t) = \lambda_3 \dot{E}_b(t) + 2\lambda_2 \eta(t)\dot{\eta}(t). \tag{7.123}$$

Since we have illustrated that all of the system signals remain bounded, we can use (7.101) to show that $\dot{E}_b(t)$ is bounded; hence, we can see from (7.123) that $\dot{g}(t)$ is also bounded. We can now invoke Lemma A.6 in Appendix A to illustrate that

$$\lim_{t\to\infty} E_b(t), \eta(t) = 0. \tag{7.124}$$

Finally, we can utilize (7.124), the structure of $E_b(t)$ given by (7.89), (7.91), (7.74), and (A.27) of Lemma A.12 in Appendix A to prove the result given by (7.85).

To prove the result given by (7.86), we will utilize (7.84) in the following manner. Note that owing to the above boundedness statements and the result given by (7.85), we know all terms in (7.84) go to zero[9] with exception of the terms containing $\ddot{e}(t)$ and $e(t)$. Hence, if we show that $\ddot{e}(t)$ goes to zero, then it is clear that $e(t)$ will also go to zero. To this end, we first note

---

[9] From (7.74), we assume that $\lim_{t\to\infty} |u(x,t)| = 0 \ \forall x \in [0,L]$ implies that $\lim_{t\to\infty} |w_{xx}(x,t)| = 0 \ \forall x \in [0,L]$.

that $\dot{e}(t)$ can be rewritten as

$$\dot{e}(t) = \int_0^t \frac{d\dot{e}(\zeta)}{d\zeta}\, d\zeta + C, \tag{7.125}$$

where C is some constant. Since we have already shown that $\lim_{t\to\infty} \dot{e}(t) = 0$, we can use (7.125) to show that

$$\lim_{t\to\infty} \int_0^t \frac{d\dot{e}(\zeta)}{d\zeta}\, d\zeta \quad \text{exists and is finite.} \tag{7.126}$$

We now differentiate (7.84) with respect to time to obtain

$$\begin{aligned} D\dddot{e} = \; & -\frac{3}{2}\dot{D}\ddot{e} - \frac{1}{2}\ddot{D}\dot{e} - mu_t(L)\dot{q}\eta - mu(L)\ddot{q}\eta - mu(L)\dot{q}\dot{\eta} - \dot{V}_m\dot{q} \\ & -V_m\ddot{q} + EIw_{xxt}(0) - k_v\ddot{e} - k_p\dot{q} - \beta L\rho\ddot{q}u^2(L) - 2\beta L\rho\dot{q}u(L)u_t(L). \end{aligned} \tag{7.127}$$

From the previous boundedness statements, we have shown that all of the signals on the right-hand side of (7.127) are bounded $\forall t \in [0, \infty)$, except for the terms $\ddot{D}(t)$ and $\dot{V}_m(t)$. Hence, if we show $\ddot{D}(t)$ and $\dot{V}_m(t)$ are bounded $\forall t \in [0, \infty)$, we can use (7.127) and the fact that $D(t) > 0$ to conclude that $\dddot{e}(t)$ is bounded $\forall t \in [0, \infty)$. After differentiating $\dot{D}(t)$ defined in (7.72) with respect to time, we have

$$\begin{aligned} \ddot{D} = \; & 2mu_t^2(L) + 2mu(L)u_{tt}(L) + 2\rho \int_0^L u_t^2 d\sigma \\ & - \left[2EI \int_0^L uu_{\sigma\sigma\sigma\sigma}\, d\sigma\right] - \left[2\rho\ddot{q} \int_0^L \sigma u\, d\sigma\right] + 2\rho\dot{q}^2 \int_0^L u^2 d\sigma \end{aligned} \tag{7.128}$$

upon application of (7.66) and (7.74). After integrating, by parts twice, the first bracketed term in (7.128), and applying Holder's inequality [6] to the second bracketed term in (7.128), we obtain the following upper bound for $\ddot{D}(t)$:

$$\begin{aligned} \left|\ddot{D}\right| \le \; & 2mu_t^2(L) + 2m\left|u(L)\right|\left|u_{tt}(L)\right| + 2\rho \int_0^L u_t^2 d\sigma \\ & +2EI\left|u(L)\right|\left|u_{xxx}(L)\right| + 2\rho\dot{q}^2 \int_0^L u^2 d\sigma \\ & +2EI \int_0^L u_{\sigma\sigma}^2 d\sigma + 2\rho\left|\ddot{q}\right| \sqrt{\frac{L^3}{3}} \sqrt{\int_0^L u^2 d\sigma} \end{aligned} \tag{7.129}$$

upon the application of (7.68). As a result of the previous boundedness statements, we know that all the signals on the right-hand side of (7.129) are bounded; hence, it is now clear that $\ddot{D}(t)$ is also bounded $\forall t \in [0, \infty)$. A similar procedure can be applied to the time derivative of $V_m(t)$ defined in (7.73) to obtain the following upper bound for $\dot{V}_m(t)$:

$$\begin{aligned}\left|\dot{V}_m\right| \leq\ & \rho \int_0^L u_t^2 d\sigma + \rho\left|\dot{q}\right|\sqrt{\frac{L^3}{3}}\sqrt{\int_0^L u_t^2 d\sigma} \\ & + EI\left|u(L)\right|\left|u_{xxx}(L)\right| + EI\int_0^L u_{\sigma\sigma}^2 d\sigma + \rho\dot{q}^2\int_0^L u^2 d\sigma.\end{aligned} \tag{7.130}$$

From (7.130) and the previous the boundedness statements, it is easy to see that $\dot{V}_m(t)$ is bounded $\forall t \in [0, \infty)$. We now know that $\dddot{e}(t)$ is bounded $\forall t \in [0, \infty)$; hence, $\ddot{e}(t)$ is uniformly continuous from Lemma A.1 in Appendix A. Since $\ddot{e}(t)$ is uniformly continuous, we can use (7.126) along with Lemma A.2 in Appendix A to conclude that

$$\lim_{t\to\infty} \ddot{e}(t) = 0. \tag{7.131}$$

We can now apply (7.131) and (7.85) to (7.84) to obtain the result given by (7.86). □

7.3.4 Adaptive Control Law

In this section, we show how the controllers of (7.81) and (7.83) can be redesigned to compensate for parametric uncertainty while still asymptotically stabilizing the link displacement and regulating the hub to the desired setpoint. From the form of (7.79) and the fact that m and EI are now considered to be unknown, the boundary control force is designed as follows:

$$f(t) = \hat{m}w_{xxxt}(L,t) - \widehat{EI}\, w_{xxx}(L,t) - k_s\eta(t), \tag{7.132}$$

where $\hat{m}(t)$ and $\widehat{EI}(t)$ are dynamic parameter estimates for m and EI, respectively, which are updated according to

$$\begin{aligned}\dot{\hat{m}}(t) &= -\gamma_1\left(\eta(t)w_{xxxt}(L,t) + \dot{q}^2(t)u(L,t)w_{xxx}(L,t)\right) \\ \dot{\widehat{EI}}(t) &= \gamma_2\eta(t)w_{xxx}(L,t),\end{aligned} \tag{7.133}$$

with γ_1, γ_2 being positive adaptation gains. After substituting (7.132) into (7.79), we can form the closed-loop dynamics for $\eta(t)$ as

$$m\dot{\eta}(t) = -k_s\eta(t) - \tilde{m}w_{xxxt}(L,t) + \widetilde{EI}\, w_{xxx}(L,t) + mu(L,t)\dot{q}^2(t), \tag{7.134}$$

where the parameter estimation error terms $\tilde{m}(t)$ and $\overset{\sim}{EI}(t)$ are defined as follows:

$$\tilde{m}(t) = m - \hat{m}(t) \qquad \overset{\sim}{EI}(t) = EI - \overset{\wedge}{EI}(t). \tag{7.135}$$

Based on the form of (7.67), the control torque is designed as follows:

$$\tau(t) = -k_v\dot{e}(t) - k_p e(t) + \hat{m}(t)u(L,t)\dot{q}(t)w_{xxx}(L,t) - \beta\hat{\rho}(t)L\dot{q}(t)u^2(L,t), \tag{7.136}$$

where the update law for $\hat{m}(t)$ was defined in (7.133), and $\hat{\rho}(t)$ represents a dynamic parameter estimate for ρ which is updated according to

$$\dot{\hat{\rho}}(t) = \gamma_3\beta L\dot{q}^2(t)u^2(L,t), \tag{7.137}$$

with γ_3 being a positive adaptation gain. Upon the substitution of (7.136) into (7.80), the closed-loop dynamics for $e(t)$ is formed as follows:

$$\begin{aligned} D(t)\ddot{e}(t) = & -\frac{1}{2}\dot{D}(t)\dot{e}(t) - mu(L,t)\dot{q}(t)\eta(t) \\ & -\tilde{m}u(L,t)\dot{q}(t)w_{xxx}(L,t) - V_m(t)\dot{q}(t) + EIw_{xx}(0,t) \\ & -k_v\dot{e}(t) - k_p e(t) - \beta\hat{\rho}L\dot{q}(t)u^2(L,t). \end{aligned} \tag{7.138}$$

Theorem 7.4 *The adaptive boundary control law given by (7.132), (7.133), (7.137), and (7.136) ensures the same stability result given by Theorem 7.3 provided the control gains k_s, k_v, and β defined in (7.136) are selected to satisfy (7.87).*

Proof. The following proof is based on arguments similar to those used in the proof of Theorem 7.3; hence, some of the details will not be repeated. First, we define the following function:

$$V_a(t) = V(t) + \frac{1}{2}\gamma_1^{-1}\tilde{m}^2(t) + \frac{1}{2}\gamma_2^{-1}\overset{\sim}{EI}^2(t) + \frac{1}{2}\gamma_3^{-1}\tilde{\rho}^2(t), \tag{7.139}$$

where $V(t)$ was defined in (7.88) and

$$\tilde{\rho}(t) = \rho - \hat{\rho}(t). \tag{7.140}$$

As a straightforward extension of Theorem 7.3, if β defined in (7.136) is selected to be sufficiently small according to the condition in (7.87), we can formulate a lower bound for $V_a(t)$ as

$$\lambda_{1a}\left(E_n(t)+\eta^2(t)+e^2(t)+\dot{e}^2(t)+\tilde{m}^2(t)+\widetilde{EI}^2(t)+\tilde{\rho}^2(t)\right)\leq V_a(t) \tag{7.141}$$

for some positive constant λ_{1a}. After differentiating (7.139) with respect to time and substituting from (7.66), (7.134), and (7.138), we can proceed as in Theorem 7.3 to obtain the following upper bound on $\dot{V}_a(t)$:[10]

$$\begin{aligned}\dot{V}_a \leq\ & -\lambda_2\left(E_n(t)+\eta^2(t)+\dot{q}^2(t)\right)\\ & -\tilde{m}(t)\left(\eta(t)w_{xxxt}(L,t)+\dot{e}^2(t)u(L,t)w_{xxx}(L,t)+\gamma_1^{-1}\dot{\hat{m}}(t)\right)\\ & +\widetilde{EI}(t)\left(\eta(t)w_{xxx}(L,t)-\gamma_2^{-1}\dot{\widehat{EI}}(t)\right)\\ & +\tilde{\rho}(t)\left(\beta L\dot{q}^2(t)u^2(L,t)-\gamma_3^{-1}\dot{\hat{\rho}}(t)\right).\end{aligned} \tag{7.142}$$

We now substitute the parameter updates laws of (7.133) and (7.137) into (7.142) to yield the same result as that given by (7.120). The proof of Theorem 7.3 can now be followed in a similar fashion to complete the proof, and to obtain the result given by (7.85) and (7.86). □

7.3.5 Experimental Evaluation

Experimental Setup

A schematic diagram of the experimental setup used to implement the controllers is shown in Figure 7.8. The experimental setup consisted of a flexible aluminum beam attached to the shaft of a NSK Corp., model RS-0810, torque-controlled, 3-phase switched reluctance motor that was used to apply the hub control torque. To compensate for the motor friction, the actual control torque was defined as follows:

$$\tau_c(t)=\tau(t)+0.8\dot{q}(t)+4.5\mathrm{sgn}(\dot{q}(t)), \tag{7.143}$$

where $\tau_c(t)$ is the actual, applied control torque, $\tau(t)$ was given by (7.83) or (7.136), and the remaining terms in (7.143) represent the viscous and

[10]Note from (7.135) and (7.140) that $\dot{\tilde{m}}=-\dot{\hat{m}}$, $\dot{\widetilde{EI}}=-\dot{\widehat{EI}}$, and $\dot{\tilde{\rho}}=-\dot{\hat{\rho}}$.

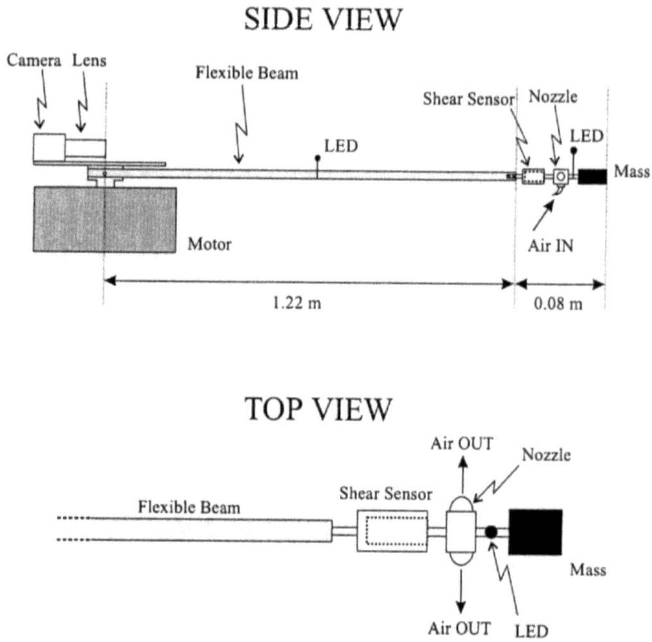

FIGURE 7.8. Schematic diagram of the flexible robot link experimental setup.

Coulomb friction compensation feedforward terms. A lightweight plastic assembly supporting two air nozzles located at the end-point of the beam was used to apply the boundary control force. The nozzles were aligned such that the outlets were in diametrically opposite directions. Compressed air at 90 psi was supplied through the lightweight, flexible air tubes controlled by high-speed proportional air valves. The algebraic relationship between the valve's output force and input voltage was obtained by clamping the beam's end-point and applying known voltages to the valves. For each applied input voltage, a Sensor Development, model 20005, shear force sensor located at the beam's end-point (see Figure 7.8) was utilized to measure a signal directly proportional to the output force. The signal from the shear sensor's bridge circuit was amplified and filtered before being read by an analog input channel on the data acquisition board. The measured relationship between input voltage and output force was then inverted to enable the application of the desired boundary control force.

A modular line scan camera mounted on the motor shaft and a high luminescence LED mounted at the beam's end-point were used to measure the beam's end-point displacement, $u(L,t)$. For monitoring purposes only, a second LED was placed at the beam's mid-point to measure the mid-point displacement, $u(L/2,t)$. The camera was a 2048 pixels, linear CCD camera with an 85 mm F1.4 Nikon lens. The camera has a resolution of 0.014 cm/pixel when the light source is one meter away, while the capture board hardware can provide sampling at a rate of 0.5 msec. The signal $u_{xxx}(L,t)$ was measured via the shear force sensor, while an incremental encoder mounted on the motor shaft was utilized to measure the hub angular position, $q(t)$. The signals $\dot{q}(t)$, $u_t(L,t)$, and $u_{xxxt}(L,t)$ were obtained by using a backwards difference algorithm and a second-order digital filter. The various parameters associated with the flexible beam model were calculated using standard test procedures and engineering handbook tables, and are given below:

$$\begin{aligned} L &= 1.22 \text{ m}, \quad m = 0.21 \text{ kg}, \\ EI &= 5.25 \text{ N-m}^2, \quad \rho = 0.35 \text{ kg/m}. \end{aligned} \tag{7.144}$$

The controllers described in the sequence were implemented via the *Qmotor* real-time control environment, and were run at a sampling period of 0.5 msec.

Experimental Results

The objective of the experiment was to regulate the hub angular position to a desired position of 20° (i.e., $q_d = 0.35$ rad) while driving the link displacement to zero. In all controllers described below, the control gains were tuned to achieve the best performance. For comparison purposes, the "open-loop" response of the system, which consisted of the following control law:

$$\tau(t) = -k_v\dot{q}(t) - k_p e(t) \qquad\qquad f(t) = 0, \tag{7.145}$$

was implemented with $k_v = 35$ and $k_p = 80$. Figure 7.9 shows the signals $u(L,t)$, $u(L/2,t)$, and $q(t)$ for the open-loop system. Next, the model-based controller given by (7.81) and (7.83) was implemented with[11] $k_s = 3.5$, $k_v = 20$, $k_p = 70$, and $\beta = 0.35$. Figure 7.10 depicts the system performance for the model-based controller. Finally, the adaptive controller defined by (7.132), (7.133), (7.136), and (7.137) was implemented by assuming no knowledge of the system parameters m, EI, and ρ (i.e., $\hat{m}(0) = \widehat{EI}(0) = \hat{\rho}(0) = 0$), and with $k_s = 7$, $k_v = 20$, $k_p = 62$, $\beta = 0.07$, $\gamma_1 = 0.001$, $\gamma_2 = 5 \times 10^{-5}$, and $\gamma_3 = 0.01$. The parameter estimates were computed

[11] Since the restrictions on the control gains k_v and β given in (7.87) are only *sufficient* conditions for achieving the asymptotic stability results, it comes as no surprise that the following results were obtained without satisfying these conditions.

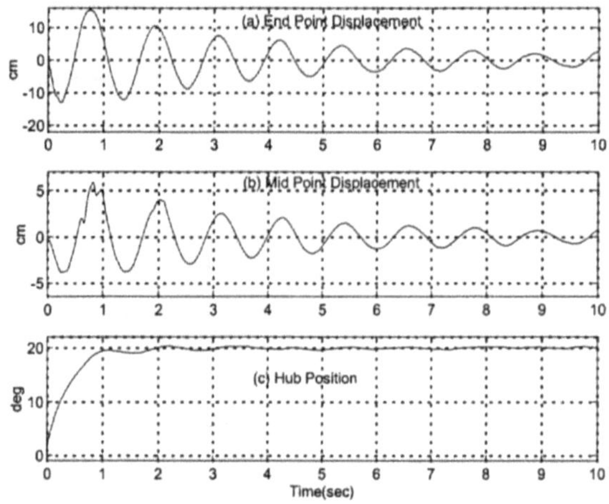

FIGURE 7.9. Open-loop system response: (a) $u(L,t)$, (b) $u(L/2,t)$, and (c) $q(t)$.

on-line using a trapezoidal numerical integration rule. The beam's end-point displacement, mid-point displacement, and the hub position for the adaptive controller are illustrated in Figure 7.11, while Figure 7.12 shows the parameter estimates.

A second experimental evaluation was performed by attaching a payload mass, consisting of a metal block weighting 0.08 kg, to the beam's end-point. The three controllers were then run *without* retuning the control gains given above (in the model-based controller, m was kept at the value given in (7.144)). The performance of the open-loop system, model-based controller, and adaptive controller are shown in Figures 7.13 through 7.16.

To compare the experimental results, Tables 7.1 and 7.2 show the maximum steady-state values of $u(L,t)$, $u(L/2,t)$, and $q(t)$ for $t \in [9,10]$ sec, and the $\mathcal{L}_2$ norm[12] of $u(L,t)$ and $u(L/2,t)$ for $t \in [0,10]$ sec for the open-loop system, model-based controller, and adaptive controller in the two experimental runs. As can be seen from the tables and the results of Figures 7.9 to 7.16, the control laws were able to damp-out the link vibrations efficiently while slightly improving the hub regulation when compared to the open-loop system. Note that although $\|u(L,t)\|_2$ and $\|u(L/2,t)\|_2$ for the adaptive controller was higher than that of the model-based controller, the adaptive controller produced smaller steady-state displacements. This can be explained by the fact that since all parameter estimates were ini-

[12]Note that the $\mathcal{L}_2$ norm of a signal $y(t)$ is defined as $\|y(t)\|_2 \triangleq \sqrt{\int_{t_o}^{t_f} y^2(\sigma)d\sigma}$, where t_o and t_f are the initial and final time, respectively.

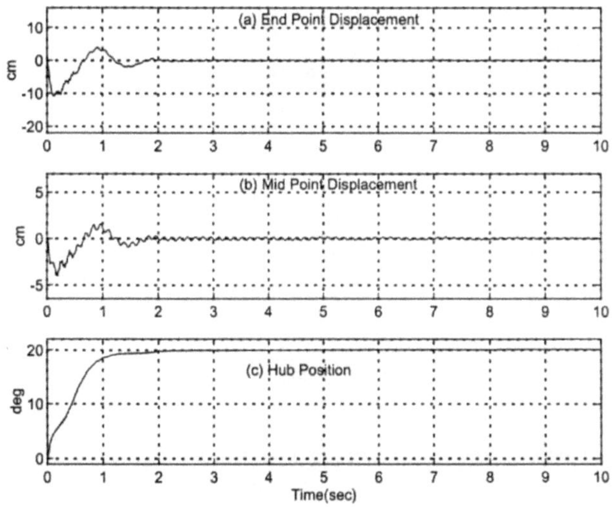

FIGURE 7.10. Model-based controller: (a) $u(L,t)$, (b) $u(L/2,t)$, and (c) $q(t)$.

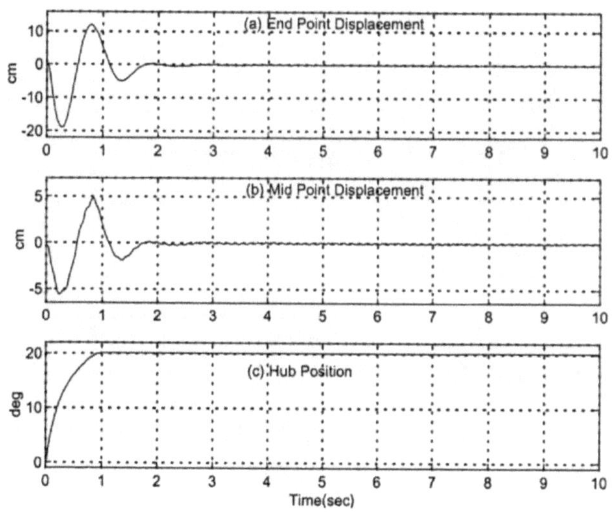

FIGURE 7.11. Adaptive controller: (a) $u(L,t)$, (b) $u(L/2,t)$, and (c) $q(t)$.

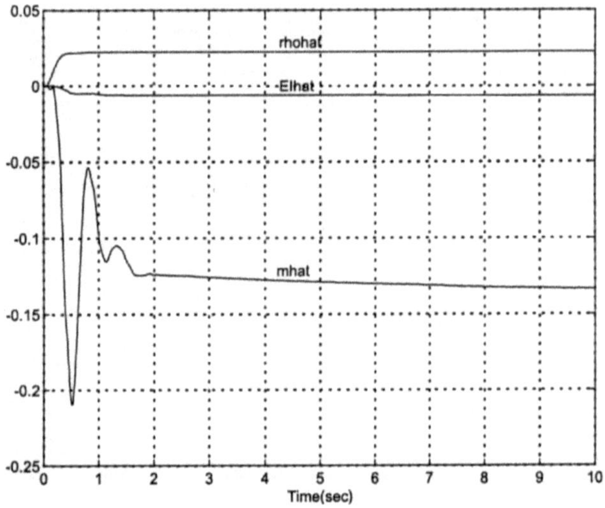

FIGURE 7.12. Parameter estimates: $\hat{m}(t)$, $\widehat{EI}\ (t)$, and $\hat{\rho}(t)$.

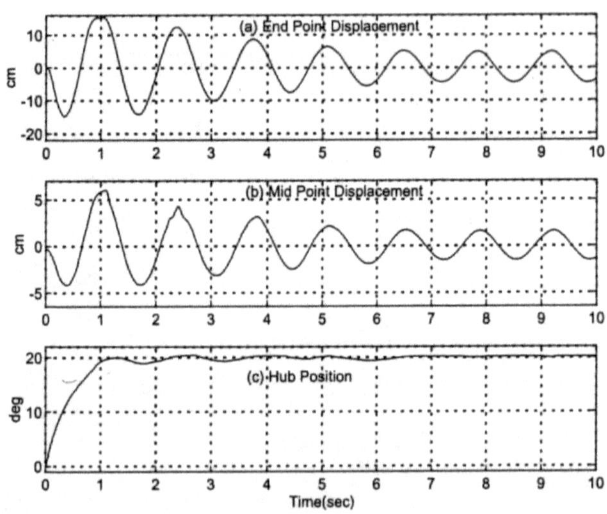

FIGURE 7.13. Open-loop response with payload mass: (a) $u(L,t)$, (b) $u(L/2,t)$, and (c) $q(t)$.

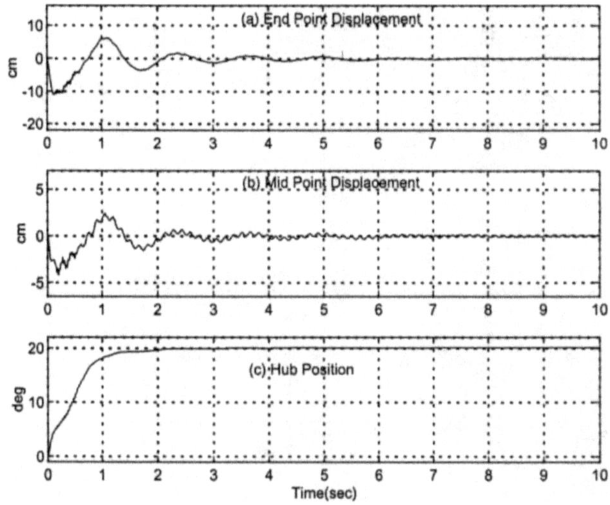

FIGURE 7.14. Model-based controller with payload mass: (a) $u(L,t)$, (b) $u(L/2,t)$, and (c) $q(t)$.

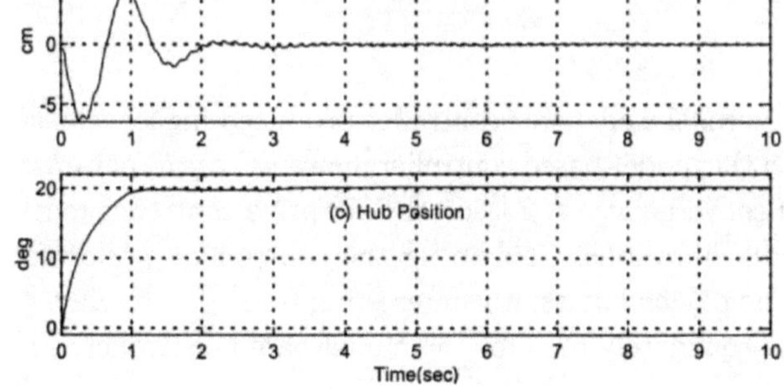

FIGURE 7.15. Adaptive controller with payload mass: (a) $u(L,t)$, (b) $u(L/2,t)$, and (c) $q(t)$.

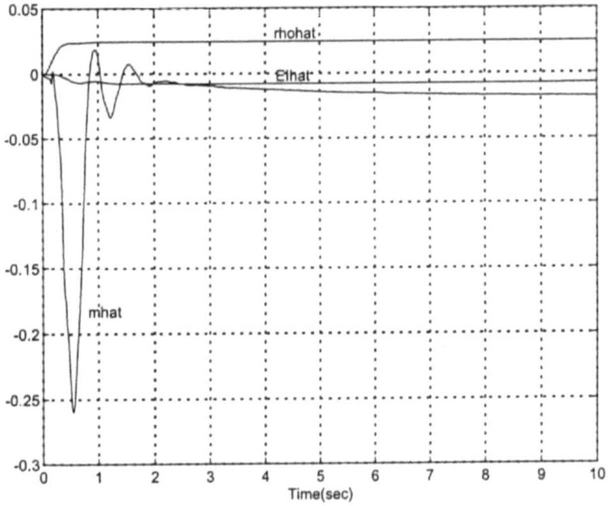

FIGURE 7.16. Parameter estimates with payload mass: $\hat{m}(t)$, $\hat{EI}\ (t)$, and $\hat{\rho}(t)$.

TABLE 7.1. System Performance without Payload Mass

	Open-Loop system	Model-based controller	Adaptive controller
$\lvert u(L,t)\rvert_{\max,ss}$	2.5 [cm]	0.2 [cm]	0.148 [cm]
$\lvert u(L/2,t)\rvert_{\max,ss}$	0.86 [cm]	0.11 [cm]	0.08 [cm]
$\lvert q(t)\rvert_{\max,ss}$	20.2°	20.07°	20.13°
$\lVert u(L,t)\rVert_2$	0.0680	0.0244	0.0476
$\lVert u(L/2,t)\rVert_2$	0.0230	0.0083	0.0155

tialized to zero, the adaptive controller produced higher transient displacements than the model-based controller; however, as the parameter estimates started to converge at $t \approx 2$ [sec], the adaptive controller quickly damped-out the link vibrations. Comparing the controllers' performance without and with the payload mass, it is interesting to note that while $\lVert u(L,t)\rVert_2$ increased approximately 52% for the model-based controller, $\lVert u(L,t)\rVert_2$ only increased approximately 20% for the adaptive controller, which indicates its expected ability to cope with unknown parameters. The $\lvert u(L,t)\rvert_{\max,ss}$ and $\lvert u(L/2,t)\rvert_{\max,ss}$ measures of performance also indicate the ability of the adaptive controller to quickly damp-out the link vibrations in spite of payload variation.

TABLE 7.2. System Performance with Payload Mass

	Open-Loop system	Model-based controller	Adaptive controller
$\lvert u(L,t)\rvert_{\max,ss}$	4.87 [cm]	0.164 [cm]	0.1 [cm]
$\lvert u(L/2,t)\rvert_{\max,ss}$	1.61 [cm]	0.141 [cm]	0.05 [cm]
$\lvert q(t)\rvert_{\max,ss}$	20.17°	20.10°	20.09°
$\lVert u(L,t)\rVert_2$	0.0850	0.0370	0.0573
$\lVert u(L/2,t)\rVert_2$	0.0277	0.0105	0.0188

7.4 Flexible Rotor System

In this section, we construct boundary control laws for the flexible rotor system depicted in Figure 7.17. Specifically, we seek to regulate the rotor displacement and provide hub angular velocity tracking. A pair of perpendicular, boundary control forces are applied to the rotor's free end-point mass with the objective of stabilizing the rotor displacement while a control torque is applied to the hub to rotate the entire rotor system.

7.4.1 System Model

The rotor system under consideration is modeled as a Euler-Bernoulli beam clamped to a rotating, rigid hub with a payload/actuator mass attached to its free end. Based on the standard Euler-Bernoulli beam modeling assumptions, Hamilton's principle [8] can be used to show that the field equation for the rotor system is given by

$$\begin{aligned} &\rho\Big(q_{tt}(x,t)+2\dot{\theta}(t)\,Sq_t(x,t) \\ &+\ddot{\theta}(t)\,Sq(x,t)-\dot{\theta}^2(t)\,q(x,t)\Big)+EIq_{xxxx}(x,t) = 0, \end{aligned} \tag{7.146}$$

where ρ is the mass/length of the rotor, EI is the bending stiffness of the rotor, $\theta(t), \dot{\theta}(t), \ddot{\theta}(t)$ represent the hub's angular position, velocity, and acceleration about the $x-$axis (see Figure 7.17), respectively, $S \in \mathbb{R}^{2\times 2}$ denotes the following skew-symmetric matrix:

$$S = \begin{bmatrix} 0 & -1 \\ 1 & 0 \end{bmatrix}, \tag{7.147}$$

and $q(x,t) \in \mathbb{R}^2$ denotes the rotor's composite displacement vector defined as follows:

$$q(x,t) = \begin{bmatrix} u(x,t) & v(x,t) \end{bmatrix}^T, \tag{7.148}$$

with $u(x,t), v(x,t)$ denoting the displacement of the rotor in the u, v directions, respectively (note that the $u-v$ coordinate system in Figure 7.17 is a rotating coordinate system fixed to the rotor). The boundary conditions for the rotor system are given by[13]

$$q(0,t) = q_x(0,t) = q_{xx}(L,t) = 0 \tag{7.149}$$

and

$$\begin{aligned} &m\left[q_{tt}(L,t) + 2\dot{\theta}(t)\,Sq_t(L,t) + \ddot{\theta}(t)\,Sq(L,t)\right. \\ &\left.-\dot{\theta}^2(t)\,q(L,t)\right] - EIq_{xxx}(L,t) = F(t), \end{aligned} \tag{7.150}$$

where m represents the payload/actuator mass located at the free end of the rotor, L denotes the length of the rotor, $F(t) \in \mathbb{R}^2$ is the control force input vector defined as follows:

$$F(t) = \begin{bmatrix} f_1(t) & f_2(t) \end{bmatrix}^T, \tag{7.151}$$

with $f_1(t), f_2(t)$ being the individual control force inputs applied at the free end of the rotor along the u, v directions, respectively. The dynamics of the rotating hub at the rotor's clamped end are given by

$$J\ddot{\theta}(t) = \tau(t), \tag{7.152}$$

where $\tau(t)$ denotes the control torque input applied to the hub, and J represents the hub's inertia.

The rotor model described above is based on the following assumptions: (i) the gravitational effects, surface loads, and the rotatory inertia of the rotor cross-sections are assumed to be negligible; (ii) the flexible rotor is in a straight position when the system is at rest; (iii) the flexible rotor is inextensible and has infinite torsional stiffness; and (iv) the inertia of the rotor's free end mass is negligible.

Since the control strategies will consist of relatively simple functions, we will assume the existence of a unique solution for the dynamics given by (7.146), (7.150), and (7.152) under the control. Based on the arguments outlined in Remark 5.1, we will assume that the distributed variable $q(x,t)$ and its time derivative $q_t(x,t)$ belong to a space of functions that has the following properties.

---

[13] Given the clamped boundary conditions of (7.149), we also know that $q_t(0,t) = q_{xt}(0,t) = 0$.

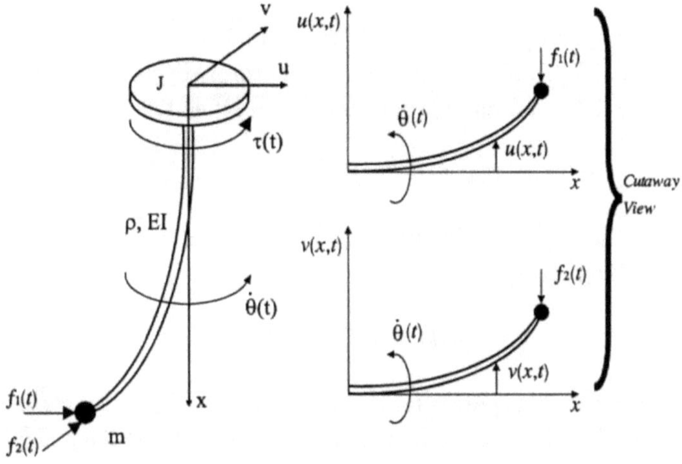

FIGURE 7.17. Schematic diagram of the flexible rotor system.

Property 7.5 If the potential energy of the system given by (7.146) through (7.150), defined as

$$\Pi_p = \frac{1}{2} EI \int_0^L q_{\sigma\sigma}^T(\sigma, t) q_{\sigma\sigma}(\sigma, t) d\sigma, \tag{7.153}$$

is bounded $\forall t \in [0, \infty)$, then $\frac{\partial^n}{\partial x^n} q(x,t)$ is bounded for $n = 2, 3, 4$, $\forall t \in [0, \infty)$, and $\forall x \in [0, L]$.

Property 7.6 If the kinetic energy given by (7.146) through (7.150), defined as

$$\begin{aligned}\Pi_k = {} & \frac{\rho}{2} \int_0^L \left(q_t(\sigma,t) + \dot{\theta}(t) S q(\sigma,t) \right)^T \left(q_t(\sigma,t) + \dot{\theta}(t) S q(\sigma,t) \right) d\sigma \\ & + \frac{m}{2} \left(q_t(L,t) + \dot{\theta}(t) S q(L,t) \right)^T \left(q_t(L,t) + \dot{\theta}(t) S q(L,t) \right) \\ & + \frac{J}{2} \dot{\theta}^2(t),\end{aligned} \tag{7.154}$$

is bounded $\forall t \in [0, \infty)$, then $\frac{\partial^n}{\partial x^n} q_t(x,t)$ is bounded for $n = 0, 1, 2, 3$, $\forall t \in [0, \infty)$, and $\forall x \in [0, L]$.

7.4.2 Problem Statement

The control objective is to ensure that (i) $q(x,t) \to 0 \; \forall x \in [0, L]$ as $t \to \infty$, and (ii) $\dot{\theta}(t) \to \omega_d(t)$ as $t \to \infty$ where $\omega_d(t)$ denotes the desired angular velocity trajectory for the rotor (we assume that $\omega_d(t)$ and $\dot{\omega}_d(t)$ are bounded signals). To aid the analysis of the rotor displacement regulation objective, we define an auxiliary signal $\eta(t) \in \mathbb{R}^2$ as follows:

$$\eta(t) = q_t(L,t) + \dot{\theta}(t) Sq(L,t) - q_{xxx}(L,t). \tag{7.155}$$

To quantify the angular velocity tracking objective, we define the angular velocity tracking error $e(t)$ as

$$e(t) = \dot{\theta}(t) - \omega_d(t). \tag{7.156}$$

With the intent of preparing for the design of the subsequent control laws, we will now obtain the open-loop dynamics for the auxiliary signal $\eta(t)$ and angular velocity tracking error $e(t)$. To this end, we first differentiate (7.155) with respect to time, and then multiply the resulting expression by m. After substituting for $mq_{tt}(L,t)$ from (7.150), we obtain the following open-loop dynamics for $\eta(t)$:

$$\begin{aligned} m\dot{\eta}(t) = & -m\dot{\theta}(t) Sq_t(L,t) + m\dot{\theta}^2(t) q(L,t) - mq_{xxxt}(L,t) \\ & + EIq_{xxx}(L,t) + F(t). \end{aligned} \tag{7.157}$$

After differentiating (7.156) with respect to time, multiplying the resulting expression by J, and substituting for $J\ddot{\theta}(t)$ from (7.152), we can obtain the open-loop dynamics for $e(t)$ as follows:

$$J\dot{e}(t) = -J\dot{\omega}_d(t) + \tau(t). \tag{7.158}$$

7.4.3 Model-Based Control Law

In this section, we present the design of a model-based boundary control law. Based on the structure of (7.157) and the subsequent stability analysis, the control force input is designed as follows:

$$\begin{aligned} F(t) = & -k_s\eta(t) - EIq_{xxx}(L,t) \\ & -m\left(\dot{\theta}^2(t) q(L,t) - \dot{\theta}(t) Sq_t(L,t) - q_{xxxt}(L,t)\right), \end{aligned} \tag{7.159}$$

where k_s is a positive control gain. After substituting (7.159) into (7.157), we can formulate the closed-loop dynamics for $\eta(t)$ as follows:

$$m\dot{\eta}(t) = -k_s\eta(t). \tag{7.160}$$

Based on the form of (7.158) and the subsequent stability analysis, the control torque input is defined as follows:

$$\tau(t) = J\dot{\omega}_d(t) - k_r e(t), \tag{7.161}$$

where k_r is a positive control gain. After substituting (7.161) into (7.158), we obtain the closed-loop dynamics for $e(t)$, as shown below:

$$J\dot{e}(t) = -k_r e(t). \tag{7.162}$$

Theorem 7.5 *The model-based control law given by (7.159) and (7.161) ensures that the rotor displacement and the angular velocity tracking error are exponentially regulated in the following sense:*

$$\begin{aligned} \|q(x,t)\| &\leq \sqrt{\frac{2\lambda_2 L^3}{\lambda_1 EI}\kappa_0 \exp\left(-\frac{\lambda_3}{\lambda_2}t\right)} \qquad \forall x \in [0,L] \\ |e(t)| &\leq \sqrt{\frac{\lambda_2}{\lambda_1}\kappa_0 \exp\left(-\frac{\lambda_3}{\lambda_2}t\right)}, \end{aligned} \tag{7.163}$$

where λ_1, λ_2, and λ_3 are some positive bounding constants, and κ_0 is positive constant defined as

$$\begin{aligned} \kappa_0 = {} & \frac{1}{2}\rho\int_0^L \left(q_t(\sigma,0) + \dot{\theta}(0) Sq(\sigma,0)\right)^T \left(q_t(\sigma,0) + \dot{\theta}(0) Sq(\sigma,0)\right) d\sigma \\ & + \frac{1}{2}EI\int_0^L q_{\sigma\sigma}^T(\sigma,0)\, q_{\sigma\sigma}(\sigma,0)\, d\sigma + \|\eta(0)\|^2 + e^2(0). \end{aligned} \tag{7.164}$$

To ensure that the result given in (7.163) is valid, the control gain k_s must be selected to satisfy the following sufficient condition:

$$k_s > \frac{EI}{2}. \tag{7.165}$$

Proof. To examine the stability of the closed-loop system, we define the following function:

$$V(t) = E_b(t) + E_c(t) + \frac{1}{2}m\eta^T(t)\eta(t) + \frac{1}{2}Je^2(t), \tag{7.166}$$

where the energy related term $E_b(t)$ is defined as follows:

$$\begin{aligned} E_b(t) = {} & \frac{1}{2}\rho\int_0^L \left(q_t(\sigma,t) + \dot{\theta}(t) Sq(\sigma,t)\right)^T \left(q_t(\sigma,t) + \dot{\theta}(t) Sq(\sigma,t)\right) d\sigma \\ & + \frac{1}{2}EI\int_0^L q_{\sigma\sigma}^T(\sigma,t)\, q_{\sigma\sigma}(\sigma,t)\, d\sigma, \end{aligned} \tag{7.167}$$

and the "cross" term $E_c(t)$ is defined as

$$E_c(t) = 2\beta\rho \int_0^L \sigma q_\sigma^T(\sigma,t) \left(q_t(\sigma,t) + \dot{\theta}(t)\, Sq(\sigma,t) \right) d\sigma, \tag{7.168}$$

with β being a positive weighting constant.

To ensure that $V(t)$ defined in (7.166) is non-negative, first note that the inequality (A.29) of Lemma A.13 in Appendix A can be used to upper bound $E_c(t)$ of (7.168) as follows:

$$E_c(t) \leq 2\beta\rho L \int_0^L \left[q_\sigma^T q_\sigma + \left(q_t + \dot{\theta} Sq \right)^T \left(q_t + \dot{\theta} Sq \right) \right] d\sigma. \tag{7.169}$$

We can also lower bound $E_b(t)$ of (7.167) as follows:

$$E_b(t) \geq \frac{1}{2} \min\{\rho, EI\} \int_0^L \left[q_{\sigma\sigma}^T q_{\sigma\sigma} + \left(q_t + \dot{\theta} Sq \right)^T \left(q_t + \dot{\theta} Sq \right) \right] d\sigma. \tag{7.170}$$

After applying the inequalities given in (A.27) of Lemma A.12 in Appendix A to (7.169), we can use the inequality of (7.170) to obtain the following bounds for $E_c(t)$:

$$-\frac{4\beta\rho L \max\{1, L^2\}}{\min\{\rho, EI\}} E_b(t) \leq E_c(t) \leq \frac{4\beta\rho L \max\{1, L^2\}}{\min\{\rho, EI\}} E_b(t). \tag{7.171}$$

Provided β is selected to satisfy the following inequality:

$$\beta < \frac{\min\{\rho, EI\}}{4\rho L \max\{1, L^2\}}, \tag{7.172}$$

we can use (7.171) to state that

$$0 \leq \xi_1 E_b(t) \leq E_b(t) + E_c(t) \leq \xi_2 E_b(t), \tag{7.173}$$

where ξ_1, ξ_2 are some positive constants. Based on the structure of (7.166) and (7.173), we can formulate the following for $V(t)$:

$$\lambda_1 \left(\|\eta(t)\|^2 + e^2(t) + E_b(t) \right) \leq V(t) \leq \lambda_2 \left(\|\eta(t)\|^2 + e^2(t) + E_b(t) \right), \tag{7.174}$$

where λ_1 and λ_2 are defined as follows:

$$\begin{aligned} \lambda_1 &= \min\left\{ \frac{m}{2}, \frac{J}{2}, 1 - \frac{4\beta\rho L \max\{1, L^2\}}{\min\{\rho, EI\}} \right\} > 0 \\ \lambda_2 &= \max\left\{ \frac{m}{2}, \frac{J}{2}, 1 + \frac{4\beta\rho L \max\{1, L^2\}}{\min\{\rho, EI\}} \right\} > 0, \end{aligned} \tag{7.175}$$

with β satisfying (7.172).

After differentiating (7.166) with respect to time, and then substituting from (7.162) and (7.160), we have

$$\dot{V}(t) = -k_s \eta^T(t)\eta(t) - k_r e^2(t) + \dot{E}_b(t) + \dot{E}_c(t). \tag{7.176}$$

To determine $\dot{E}_b(t)$ in (7.176), we differentiate (7.167) with respect to time, substitute the field equation of (7.146), and cancel common terms to obtain

$$\dot{E}_b = EI\left[\int_0^L q_{\sigma\sigma}^T q_{\sigma\sigma t} d\sigma - \int_0^L q_t^T q_{\sigma\sigma\sigma\sigma} d\sigma\right] + \dot{\theta}EI\left[\int_0^L q^T S q_{\sigma\sigma\sigma\sigma} d\sigma\right] \tag{7.177}$$

After integrating, by parts once, the first bracketed term in (7.177), integrating, by parts twice, the second bracketed term in (7.177), applying the boundary conditions given in (7.149), and then canceling common terms, we obtain

$$\dot{E}_b = -EIq_{xxx}^T(L)\left(q_t(L) + \dot{\theta}Sq(L)\right), \tag{7.178}$$

which can be rewritten as

$$\begin{aligned}\dot{E}_b = & -\frac{EI}{2}\left(q_t(L) + \dot{\theta}Sq(L)\right)^T\left(q_t(L) + \dot{\theta}Sq(L)\right) \\ & -\frac{EI}{2}q_{xxx}^T(L)\,q_{xxx}(L) + \frac{EI}{2}\eta^T\eta\end{aligned} \tag{7.179}$$

upon application of (7.155).

To determine $\dot{E}_c(t)$ in (7.176), we first differentiate (7.168) with respect to time to obtain

$$\dot{E}_c = 2\beta\rho\int_0^L \sigma q_{\sigma t}^T\left(q_t + \dot{\theta}Sq\right)d\sigma + 2\beta\rho\int_0^L \sigma q_\sigma^T\left(q_{tt} + \ddot{\theta}Sq + \dot{\theta}Sq_t\right)d\sigma. \tag{7.180}$$

After substituting (7.146) into (7.180) for ρq_{tt}, we obtain

$$\dot{E}_c = A_1 + A_2 + A_3 + A_4 + A_5, \tag{7.181}$$

where the auxiliary terms A_i are defined as follows:

$$\begin{aligned} A_1 &= 2\beta\rho\int_0^L \sigma q_t^T q_{\sigma t} d\sigma, \qquad A_2 = 2\beta\rho\dot{\theta}\int_0^L \sigma q_{\sigma t}^T Sq d\sigma, \\ A_3 &= -2\beta EI\int_0^L \sigma q_\sigma^T q_{\sigma\sigma\sigma\sigma} d\sigma, \quad A_4 = 2\beta\rho\dot{\theta}^2\int_0^L \sigma q^T q_\sigma d\sigma, \\ A_5 &= 2\beta\rho\dot{\theta}\int_0^L \sigma q_t^T S q_\sigma d\sigma. \end{aligned} \tag{7.182}$$

After integrating, by parts, the A_1, A_2, and A_3 terms in (7.182), and then applying the boundary conditions of (7.149), we have

$$A_1 = \beta\rho \left(Lq_t^T(L)\, q_t(L) - \int_0^L q_t^T q_t d\sigma \right),$$

$$A_2 = 2\beta\rho\dot{\theta} \left(Lq_t^T(L)\, Sq(L) + \int_0^L \left(q^T Sq_t + \sigma q_\sigma^T Sq_t\right) d\sigma \right), \tag{7.183}$$

and

$$A_3 = -2\beta EI \left(Lq_x^T(L)\, q_{xxx}(L) - \int_0^L q_\sigma^T q_{\sigma\sigma\sigma} d\sigma - \int_0^L \sigma q_{\sigma\sigma}^T q_{\sigma\sigma\sigma} d\sigma \right). \tag{7.184}$$

After integrating, by parts, the first integral in (7.184), we obtain the following expression:

$$A_3 = -2\beta EI \left(Lq_x^T(L)\, q_{xxx}(L) + \int_0^L q_{\sigma\sigma}^T q_{\sigma\sigma} d\sigma - \int_0^L \sigma q_{\sigma\sigma}^T q_{\sigma\sigma\sigma} d\sigma \right) \tag{7.185}$$

upon application of (7.149). After integrating, by parts, the second integral in (7.185), we have

$$A_3 = -2\beta EI \left(Lq_x^T(L)\, q_{xxx}(L) + 2\int_0^L q_{\sigma\sigma}^T q_{\sigma\sigma} d\sigma + \int_0^L \sigma q_{\sigma\sigma}^T q_{\sigma\sigma\sigma} d\sigma \right) \tag{7.186}$$

upon use of (7.149). After adding (7.185) and (7.186), we obtain

$$A_3 = -\beta EI \left(2Lq_x^T(L)\, q_{xxx}(L) + 3\int_0^L q_{\sigma\sigma}^T q_{\sigma\sigma} d\sigma \right). \tag{7.187}$$

Finally, we apply (A.30) of Lemma A.13 and (A.28) of Lemma A.12 in Appendix A to the first and second terms on the right-hand side of (7.187), respectively, to formulate the following upper bound for A_3:

$$A_3 \le 2\beta EIL \left(\delta q_x^T(L)\, q_x(L) + \frac{1}{\delta} q_{xxx}^T(L)\, q_{xxx}(L) \right) - \frac{\beta EI}{L} q_x^T(L)\, q_x(L) - 2\beta EI \int_0^L q_{\sigma\sigma}^T q_{\sigma\sigma} d\sigma, \tag{7.188}$$

where δ is an arbitrary positive constant. After integrating, by parts, the expression for A_4 in (7.182), we have

$$A_4 = 2\beta\rho\dot{\theta}^2 \left(Lq^T(L)\, q(L) - \int_0^L q^T q d\sigma \right) - 2\beta\rho\dot{\theta}^2 \int_0^L \sigma q^T q_\sigma d\sigma. \tag{7.189}$$

Since the last term on the right-hand side of (7.189) is equal to A_4, we can rewrite the expression for A_4 as follows:

$$A_4 = \beta\rho\dot{\theta}^2 \left(Lq^T(L)\,q(L) - \int_0^L q^T q d\sigma \right). \tag{7.190}$$

We can now substitute (7.179), (7.181), (7.183), (7.188), (7.190), and the expression for A_5 given in (7.182) into (7.176) to formulate the following upper bound for $\dot{V}(t)$:

$$\begin{aligned}
\dot{V} \leq\ & -\left(k_s - \frac{EI}{2}\right)\|\eta\|^2 - k_r e^2 \\
& -\left(\frac{EI}{2} - \beta\rho L\right)\left(q_t(L) + \dot{\theta}Sq(L)\right)^T \left(q_t(L) + \dot{\theta}Sq(L)\right) \\
& -\beta\rho \int_0^L \left(q_t + \dot{\theta}Sq\right)^T \left(q_t + \dot{\theta}Sq\right) d\sigma - 2\beta EI \int_0^L q_{\sigma\sigma}^T q_{\sigma\sigma} d\sigma \\
& -EI\left(\frac{1}{2} - \frac{2\beta L}{\delta}\right) q_{xxx}^T(L)\, q_{xxx}(L) \\
& -\beta EI\left(\frac{1}{L} - 2L\delta\right) q_x^T(L)\, q_x(L).
\end{aligned} \tag{7.191}$$

If we select the positive constant δ such that

$$\delta < \frac{1}{2L^2}, \tag{7.192}$$

and, in addition, restrict the control gain k_s and the weighting constant β as follows:

$$k_s > \frac{EI}{2} \qquad \beta < \min\left\{\frac{EI}{2\rho L}, \frac{1}{8L^3}\right\}, \tag{7.193}$$

then $\dot{V}(t)$ can be further upper bounded as shown below:

$$\begin{aligned}
\dot{V} \leq\ & -\left(k_s - \frac{EI}{2}\right)\|\eta\|^2 - k_r e^2 \\
& -2\beta\left(\frac{1}{2}\rho \int_0^L \left(q_t + \dot{\theta}Sq\right)^T \left(q_t + \dot{\theta}Sq\right) d\sigma + \frac{1}{2}EI \int_0^L q_{\sigma\sigma}^T q_{\sigma\sigma} d\sigma\right).
\end{aligned} \tag{7.194}$$

Finally, by applying the definition of $E_b(t)$ given in (7.167) to the right-hand side of (7.194), $\dot{V}(t)$ can be upper bounded as follows:

$$\dot{V}(t) \leq -\lambda_3 \left(\|\eta(t)\|^2 + e^2(t) + E_b(t)\right), \tag{7.195}$$

where λ_3 is defined as follows:[14]

$$\lambda_3 = \min\left\{k_s - \frac{EI}{2}, k_r, 2\beta\right\}. \tag{7.196}$$

Based on the structure of (7.174) and (7.195), we can invoke Lemma A.4 in Appendix A to obtain the following solution for (7.195):

$$\begin{aligned} V(t) &\leq V(0)\exp\left(-\frac{\lambda_3}{\lambda_2}t\right) \\ &\leq \lambda_2\left(E_b(0) + \|\eta(0)\|^2 + e^2(0)\right)\exp\left(-\frac{\lambda_3}{\lambda_2}t\right) \\ &= \lambda_2\kappa_0\exp\left(-\frac{\lambda_3}{\lambda_2}t\right), \end{aligned} \tag{7.197}$$

where (7.167) was used to define the constant κ_0. In addition, we can use (7.174), (7.167), and (A.28) of Lemma A.12 in Appendix A to formulate the following inequalities:

$$\begin{aligned} \frac{EI}{2}\frac{1}{L^3}q^T(x,t)\,q(x,t) &\leq \frac{EI}{2}\int_0^L q_{\sigma\sigma}^T(\sigma,t)\,q_{\sigma\sigma}(\sigma,t)\,d\sigma \\ &\leq E_b(t) \leq \frac{1}{\lambda_1}V(t) \quad \forall x \in [0,L] \end{aligned} \tag{7.198}$$

and

$$e^2(t) \leq \frac{1}{\lambda_1}V(t). \tag{7.199}$$

The result given by (7.163) now follows directly from (7.197), (7.198), and (7.199). □

Remark 7.3 *Note that the proof of Theorem 7.5 requires the weighting constant β to be selected sufficiently small to satisfy the conditions of (7.172) and (7.193). Thus, the overall condition on β can be written as follows:*

$$0 < \beta < \min\left\{\frac{\min\{\rho, EI\}}{4\rho L\max\{L^2,1\}}, \frac{EI}{2\rho L}, \frac{1}{8L^3}\right\}. \tag{7.200}$$

Remark 7.4 *From (7.174) and (7.195), we can state that $E_b(t)$, $e(t)$, and $\eta(t)$ are bounded $\forall t \in [0,\infty)$. Since $E_b(t)$ is bounded $\forall t \in [0,\infty)$, we can use (7.167) and (A.28) of Lemma A.12 in Appendix A to show that $q(x,t)$*

[14] Note that λ_3 is positive as a result of the first condition given in (7.193).

is bounded $\forall t \in [0,\infty)$ *and* $\forall x \in [0,L]$. *Since* $E_b(t)$ *is bounded* $\forall t \in [0,\infty)$, *the potential energy given by (7.153) is bounded* $\forall t \in [0,\infty)$*; hence, we can use Property 7.5 to show that* $\frac{\partial^n}{\partial x^n}q(x,t)$ *is bounded for* $n = 2, 3, 4$, $\forall t \in [0,\infty)$, *and* $\forall x \in [0,L]$. *Since* $e(t)$ *and* $\omega_d(t)$ *are bounded* $\forall t \in [0,\infty)$, *we can use (7.156) to state that* $\dot{\theta}(t)$ *is bounded* $\forall t \in [0,\infty)$. *Since* $\eta(t)$, $\dot{\theta}(t)$, $q(L,t)$, *and* $q_{xxx}(L,t)$ *are all bounded* $\forall t \in [0,\infty)$, *we can use (7.155) to state that* $q_t(L,t)$ *is bounded* $\forall t \in [0,\infty)$*; hence, the kinetic energy of the mechanical system defined in (7.154) is bounded* $\forall t \in [0,\infty)$. *Since the kinetic energy is bounded, we can use Property 7.6 to state that* $\frac{\partial^n}{\partial x^n}q_t(x,t)$ *is bounded for* $n = 0, 1, 2, 3$, $\forall t \in [0,\infty)$, *and* $\forall x \in [0,L]$. *From the above information, we can state that the control inputs of (7.159) and (7.161) are all bounded; hence, we can utilize (7.146), (7.150), and (7.152) to show that* $q_{tt}(x,t)$ *and* $\ddot{\theta}(t)$ *remain bounded* $\forall t \in [0,\infty)$ *and* $\forall x \in [0,L]$.

7.4.4 Adaptive Control Law

In this section, we illustrate how the model-based control law can be redesigned to compensate for constant parametric uncertainty while asymptotically regulating the rotor displacement and the angular velocity tracking error. Specifically, based on the subsequent stability analysis, the control force is defined as follows:

$$F(t) = -k_s\eta(t) - \widehat{EI}(t)\,\Omega_e(t) - \hat{m}(t)\,\Omega_m(t), \tag{7.201}$$

where $\eta(t)$ was defined in (7.155), $\widehat{EI}(t)$ is a dynamic estimate of the rotor's bending stiffness, $\hat{m}(t)$ is a dynamic estimate of the mass, and the auxiliary functions $\Omega_e(t), \Omega_m(t) \in \mathbb{R}^2$ are defined as follows:

$$\Omega_e(t) = q_{xxx}(L,t) \quad \Omega_m(t) = \dot{\theta}^2(t)\,q(L,t) - \dot{\theta}(t)\,Sq_t(L,t) - q_{xxxt}(L,t). \tag{7.202}$$

Motivated by the subsequent stability analysis, the bending stiffness and mass estimates are updated according to

$$\dot{\widehat{EI}}(t) = \gamma_e\Omega_e^T(t)\,\eta(t) \qquad \dot{\hat{m}}(t) = \gamma_m\Omega_m^T(t)\,\eta(t), \tag{7.203}$$

where γ_e, γ_m are positive adaptation gains. After substituting (7.201) into the open-loop dynamics of (7.157), we obtain the following closed-loop dynamics for $\eta(t)$:

$$m\dot{\eta}(t) = -k_s\eta(t) + \widetilde{EI}(t)\,\Omega_e(t) + \tilde{m}(t)\,\Omega_m(t), \tag{7.204}$$

where (7.202) has been utilized, and $\widetilde{EI}(t), \tilde{m}(t)$ denote the parameter estimation errors defined as follows:

$$\widetilde{EI}(t) = EI - \widehat{EI}(t) \qquad \tilde{m}(t) = m - \hat{m}(t). \tag{7.205}$$

Based on the subsequent stability analysis, the control torque input is defined as

$$\tau(t) = \hat{J}(t)\dot{\omega}_d(t) - k_r e(t), \tag{7.206}$$

where $\hat{J}(t)$ denotes the estimate of the hub inertia that is updated by

$$\dot{\hat{J}}(t) = -\gamma_j \dot{\omega}_d(t) e(t), \tag{7.207}$$

with γ_j being a positive adaptation gain. After substituting (7.206) into the open-loop dynamics of (7.158), we obtain the closed-loop dynamics for $e(t)$ as follows:

$$J\dot{e}(t) = -k_r e(t) - \tilde{J}(t)\dot{\omega}_d(t), \tag{7.208}$$

where the inertia estimation error $\tilde{J}(t)$ is defined as follows:

$$\tilde{J}(t) = J - \hat{J}(t). \tag{7.209}$$

Theorem 7.6 *The adaptive control law given by (7.201), (7.203), (7.206), and (7.207) ensures that the rotor displacement and the angular velocity tracking error are asymptotically regulated in the following sense:*

$$\lim_{t\to\infty} \|q(x,t)\|, |e(t)| = 0 \qquad \forall x \in [0, L], \tag{7.210}$$

provided the control gain k_s is selected to satisfy (7.165).

Proof. The following proof is based on arguments similar to those used in the proof of Theorem 7.5; hence, some of the details will not be repeated. To prove the result given by (7.210), we define the following function:

$$V_a(t) = V(t) + \frac{1}{2}\gamma_e^{-1}\widetilde{EI}^2(t) + \frac{1}{2}\gamma_m^{-1}\tilde{m}^2(t) + \frac{1}{2}\gamma_j^{-1}\tilde{J}^2(t), \tag{7.211}$$

where $V(t)$ was previously defined in (7.166), $\widetilde{EI}(t), \tilde{m}(t)$ were defined in (7.205), and $\tilde{J}(t)$ was defined in (7.209). If β in (7.168) is selected sufficiently small according to the condition given in (7.172), we can follow the arguments given in the proof of Theorem 7.5 to formulate bounds on $V_a(t)$ as shown below:

$$\begin{aligned} \lambda_{1a}\left(\|\eta\|^2 + e^2 + E_b + \widetilde{EI}^2 + \tilde{m}^2 + \tilde{J}^2\right) &\le V_a(t) \\ &\le \lambda_{2a}\left(\|\eta\|^2 + e^2 + E_b + \widetilde{EI}^2 + \tilde{m}^2 + \tilde{J}^2\right), \end{aligned} \tag{7.212}$$

where $\lambda_{1a}, \lambda_{2a}$ are some positive bounding constants.

After differentiating (7.211) with respect to time, and then substituting from (7.146), (7.204), and (7.208), we can follow the derivations used in Theorem 7.5 to obtain the following upper bound for the time derivative of $V_a(t)$:

$$\begin{aligned}\dot{V}_a(t) \leq \; & -\lambda_3\left(E_b + \|\eta\|^2 + e^2\right) + \widetilde{EI}\left(\Omega_e^T\eta + \gamma_e^{-1}\dot{\widetilde{EI}}\right) \\ & +\tilde{m}\left(\Omega_m^T\eta + \gamma_m^{-1}\dot{\tilde{m}}\right) + \tilde{J}\left(\dot{\omega}_d e + \gamma_j^{-1}\dot{\tilde{J}}\right),\end{aligned} \tag{7.213}$$

where λ_3 is the positive constant defined in (7.196). We can now substitute the parameter update laws given by (7.203) and (7.207) into (7.213), and then proceed in a similar manner as in the proof of Theorem 7.5 to obtain a simplified upper bound on the time derivative of $V_a(t)$

$$\dot{V}_a(t) \leq -\lambda_3\left(\eta^T(t)\,\eta(t) + e^2(t) + E_b(t)\right) \triangleq -g_a(t). \tag{7.214}$$

We can now use (7.212), (7.214), and arguments similar to those outlined in Remark 7.4 to state that all signals remain bounded during closed-loop operation. From (7.214), the time derivative of $g_a(t)$ can be found to be

$$\dot{g}_a(t) = \lambda_3\left(2\eta^T(t)\,\dot{\eta}(t) + 2e(t)\,\dot{e}(t) + \dot{E}_b(t)\right). \tag{7.215}$$

Since we know that all system signals remain bounded $\forall t \in [0, \infty)$, we can use (7.179), (7.204), and (7.208) to show that $\dot{E}_b(t)$, $\dot{\eta}(t)$, and $\dot{e}(t)$ are bounded $\forall t \in [0, \infty)$; hence, we can see from (7.215) that the time derivative of the right-hand side of (7.214) is also bounded $\forall t \in [0, \infty)$. We can now apply Lemma A.6 in Appendix A to show that

$$\lim_{t\to\infty} E_b(t),\, \|\eta(t)\|,\, |e(t)| = 0. \tag{7.216}$$

Finally, we can use (7.216) and the inequality-type bound developed in (7.198) to obtain the result given by (7.210). □

7.4.5 Experimental Evaluation

Experimental Setup

A schematic diagram of the experimental setup used to implement the controllers is shown in Figure 7.18. The experimental setup consisted of a flexible rotor (hollow PVC pipe 0.016 m in diameter) rotated by a brushed DC motor via a belt-pulley transmission. A circular disk, which served as the payload mass, was made of magneto-ferrous material and was mounted at the free end of the rotor. Two linear CCD cameras, placed 90 deg apart, measured the free end-point position of the rotor $q(L,t)$ via an LED attached to the center of the free end payload. The ultra-high-intensity (5000 mcd) LED with a viewing angle of 30 steradians was detected by the cameras, which were placed at a distance of 0.86 m below the LED. The data provided by the camera was sampled at 2 kHz, and went through several stages of hardware and software decoding before the data was available as the rotor end-point position in meters. A high-sensitivity slip-ring was placed at the clamped end of the rotor to provide the necessary signals from the rotating shear sensor and LED. At the end of the rotor, two pairs of electromagnets[15] were arranged to apply the control forces $f_1(t)$ and $f_2(t)$ on the payload mass at the end-point. A custom designed software commutation strategy ensured that the desired force commanded by the control law was applied to the payload. Roughly speaking, the commutation strategy involved translation of the desired force trajectory into desired current trajectories. To ensure that the actual magnetizing currents track the desired current trajectories, a high-gain current feedback loop was utilized to apply voltages to the four electromagnets and the brushed DC motor.[16] The four magnetizing currents were measured using hall-effect current sensors. A two-axis shear sensor, attached just above the payload mass, was used to measure the shear at the rotor's free end (i.e., $q_{xxx}(L,t)$) along the u and v axes.

The controllers were implemented via the *Qmotor* control environment. See Appendix D for a description of the C code used to implement the control algorithm. All the controllers were implemented using a sampling period of 0.5 msec. The data acquisition system consisted of four A/D channels to sense the currents flowing through the coils of the electromagnets, two more A/D channels to read the shear sensor signals, and five D/A channels to output the voltages that powered the four electromagnets and the DC motor. These voltages went through two stages of amplification; the first stage consists of AD-624 instrumentation amplifiers, while in the second stage, Techron linear power amplifiers connected in the master-slave configuration sourced a maximum current of 20 A at 200 V. The rotor's end-point velocity, the hub's angular velocity, and the time derivative of the end-point shear were obtained by use of a backwards difference algorithm applied to the end-point position, angular position, and end-point shear,

[15] The 2 kg electromagnets had a 10 cm airgap and a bandwidth of 2 kHz.

[16] The reader is referred to Remark 5.4 for a discussion on issues concerning the implementation of a high-gain current feedback loop.

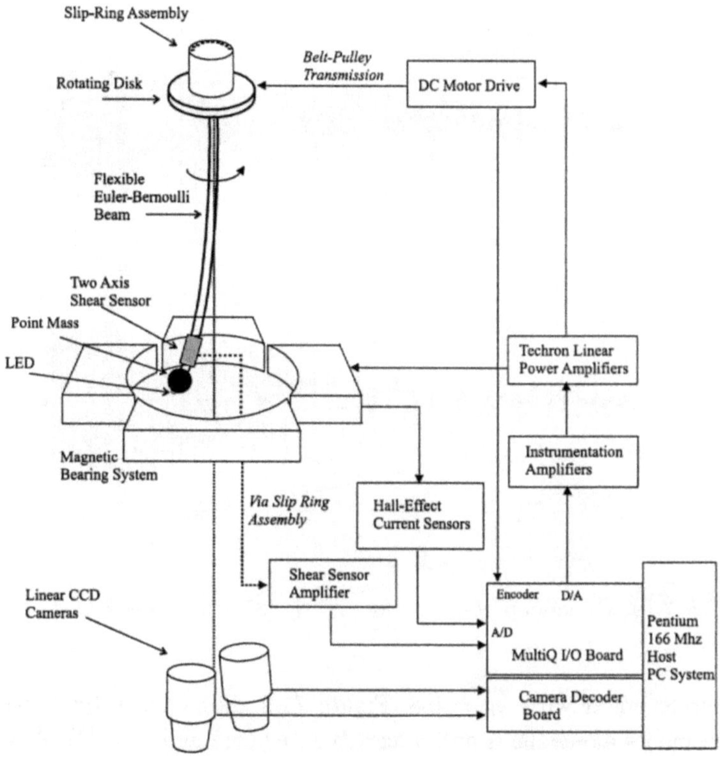

FIGURE 7.18. Schematic diagram of the flexible rotor experimental setup.

respectively, with the resulting signals being filtered by a second-order digital filter. The various parameters associated with the flexible rotor system were calculated using standard test procedures and engineering handbook tables, and are given below:

$$\begin{aligned} EI &= 3.8 \times 10^{-7}\ \text{N-m}^2, & m &= 0.25\ \text{kg}, \\ J &= 3.6 \times 10^{-5}\ \text{kg-m}^2, & \rho &= 0.3\ \text{kg/m}. \end{aligned} \qquad (7.217)$$

Experimental Results

The objective of the experiment was to regulate the angular velocity of the rotor to a desired setpoint $\omega_d = 380$ rpm while driving the rotor displacement in the u and v axes to zero. For comparison, the "open-loop" response of the system, which consisted of the following control law:

$$\tau(t) = -k_r e(t) \qquad F(t) = 0, \qquad (7.218)$$

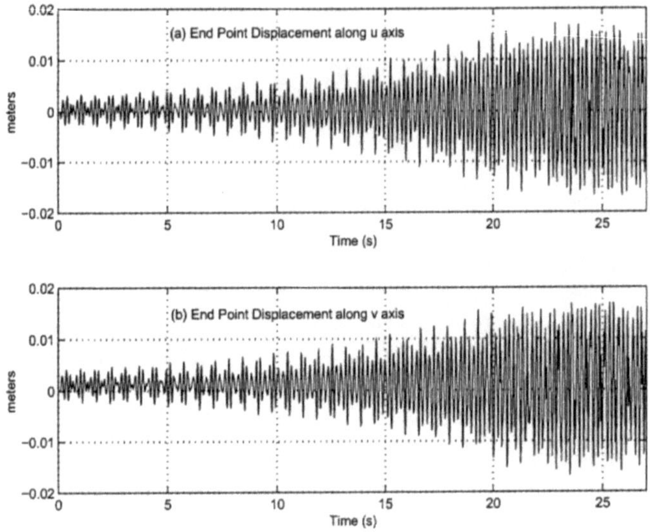

FIGURE 7.19. Open-loop response for $\omega_d = 380$ [rpm]: (a) $u(L,t)$ and (b) $v(L,t)$.

was implemented with $k_r = 0.8$. Figure 7.19 shows the rotor's free end displacements along the u and v axes (i.e., $u(L,t)$ and $v(L,t)$). A second comparison was performed with a standard damper controller formulated as shown below:

$$\tau(t) = -k_r e(t) \qquad F(t) = -k_d q_t(L,t). \tag{7.219}$$

The control gains $k_r = 0.9$ and $k_d = 10.2$ resulted in the best controller performance. The free end displacements of the rotor along the u and v axes are illustrated in Figure 7.20. Next, the model-based controller of (7.159) and (7.161) was implemented. The set of control gains that resulted in the best performance was determined to be $k_s = 1.24$ and $k_r = 0.92$. The rotor's free end displacements along the u and v axes, and the angular velocity setpoint error are illustrated in Figures 7.21 and 7.22, respectively.

Finally, the adaptive controller given by (7.201), (7.203), and (7.161) was implemented. The dynamic estimates $\hat{m}(t)$ and $\widehat{EI}(t)$ were initialized to approximately 80% of their nominal values, while the estimate of the parameter J was not required for the angular velocity setpoint problem (i.e., $\dot{\omega}_d = 0$). A projection algorithm was utilized in the adaptive update laws to ensure that parameter estimates remained between known lower and upper bounds (for details on the use of adaptive projection algorithms see [24]). The parameter estimates were computed on-line using a standard, trapezoidal numerical integration rule. The best performance of the

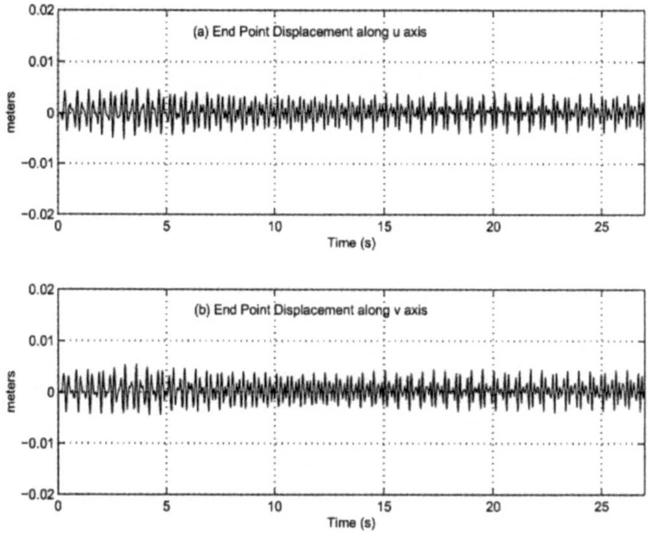

FIGURE 7.20. Damper controller for $\omega_d = 380$ [rpm]: (a) $u(L,t)$ and (b) $v(L,t)$.

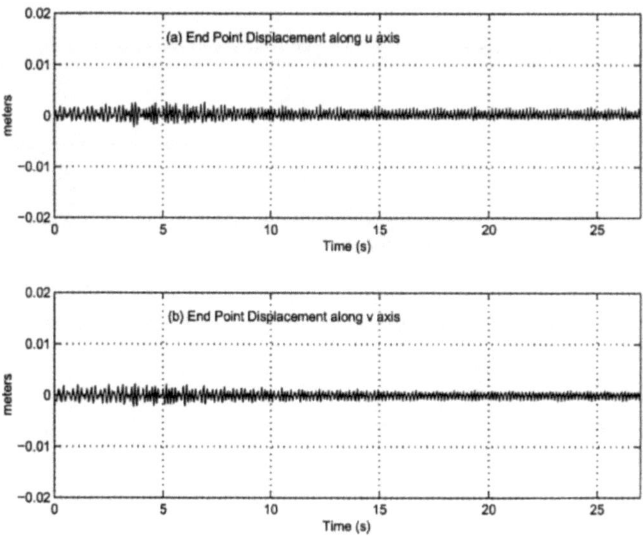

FIGURE 7.21. Model-based controller for $\omega_d = 380$ [rpm]: (a) $u(L,t)$ and (b) $v(L,t)$.

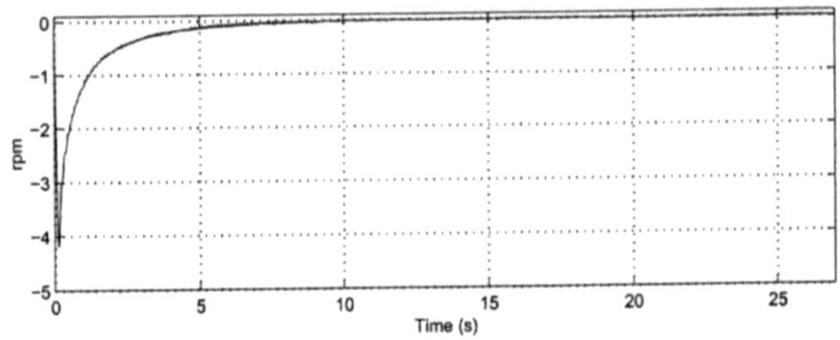

FIGURE 7.22. Model-based controller for $\omega_d = 380$ [rpm]: $e(t)$.

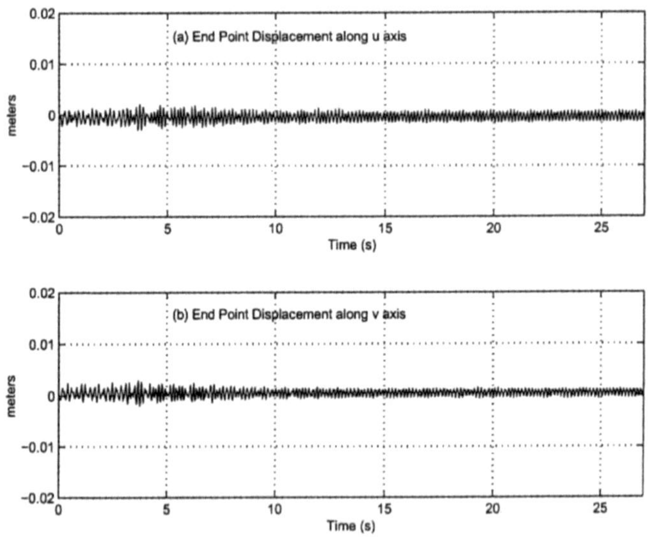

FIGURE 7.23. Adaptive controller for $\omega_d = 380$ [rpm]: (a) $u(L,t)$ and (b) $v(L,t)$.

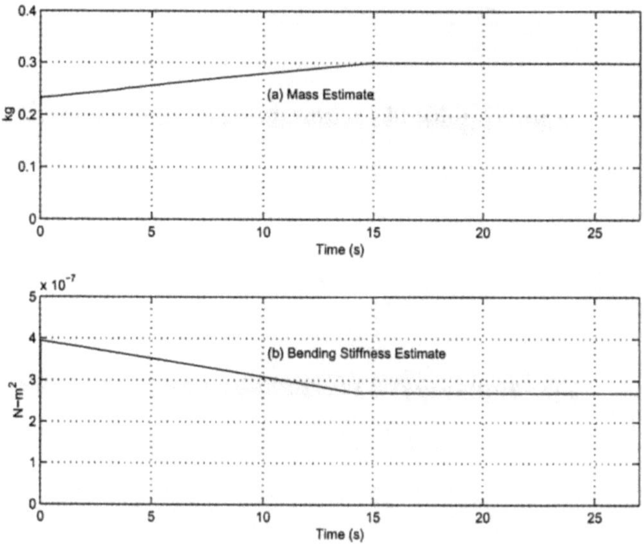

FIGURE 7.24. Adaptive controller for $\omega_d = 380$ [rpm]: (a) $\hat{m}(t)$ and (b) $\widehat{EI}(t)$.

adaptive controller was achieved with the following set of gains: $k_s = 1.27$, $k_r = 0.89$, $\gamma_e = 0.3 \times 10^{-3}$, and $\gamma_m = 7 \times 10^{-3}$. Figure 7.23 shows the free end displacements while Figure 7.24 depicts the dynamic parameter estimates. The angular velocity setpoint error for the adaptive controller was very similar to the result obtained for the model-based controller (see Figure 7.22) and hence is omitted. Comparing the results of the above experiments, we first observe that, while the peak steady-state, free-end rotor displacement of the damper controller was approximately 16.7% of the peak open-loop displacement, the peak steady-state, free-end rotor displacement of the model-based controller was only about 4.7% of the peak open-loop displacement. The peak steady-state, free-end rotor displacement for the adaptive controller settled to approximately 6.1% of the open-loop displacement. We also note that the angular velocity setpoint error was approximately 0.01% of the desired angular speed in the damper, model-based and adaptive controllers.

The model-based controller of (7.159) and (7.161) was also implemented with the desired angular velocity of the rotor ω_d set to 1000 rpm. Figure 7.25 shows the free end displacement of the rotor along the u and v axes. At such high speeds, any impact of the actuator mass with the magnetic bearings can cause damage to the experimental setup; hence, the system was not run in the open-loop mode at this desired speed.

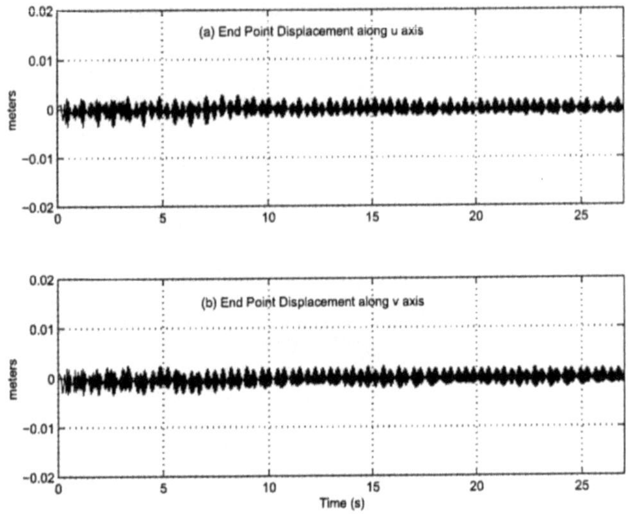

FIGURE 7.25. Model-based controller for $\omega_d = 1000$ rpm: (a) $u(L,t)$ and (b) $v(L,t)$.

To evaluate the system performance when subjected to a desired, time-varying angular velocity, the model-based control law defined in (7.159) and (7.161) was implemented with the desired angular velocity trajectory set to

$$\omega_d(t) = 900\left(1 + 0.15\sin(0.75t)\right)\left(1 - \exp(-3t)\right) \text{ rpm.} \tag{7.220}$$

Figures 7.26 and 7.27 show the desired angular velocity trajectory, the angular velocity tracking error and the free end displacements of the rotor along the u and v axes. Again, at such high speeds, it was deemed unsafe to run the system in the open-loop mode. It can be seen from Figures 7.25 and 7.27 that the performance of the controllers for the setpoint and tracking objectives are comparable at high speeds around 1000 rpm.

7.5 Notes

A number of researchers have investigated the vibration control for axially moving materials such as webs. For example, Ulsoy [25] demonstrated how control and observation spillover can destabilize the vibration of an axially moving string under state feedback control. Specifically, the control design of [25] was based on a reduced-order, discretized version of the infinite dimensional, axially moving string model. Unfortunately, point actuators

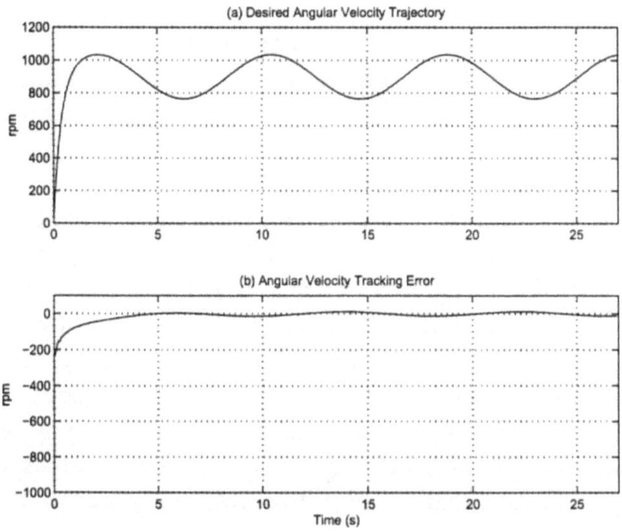

FIGURE 7.26. Model-based controller for angular velocity tracking: (a) $\omega_d(t)$ and (b) $e(t)$.

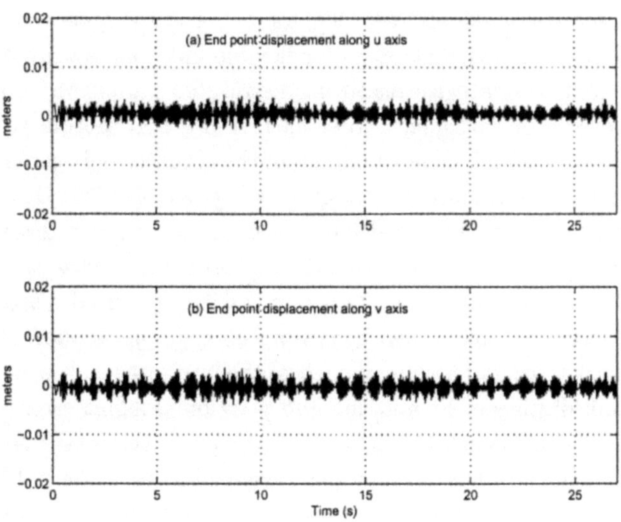

FIGURE 7.27. Model-based controller for angular velocity tracking: (a) $u(L,t)$ and (b) $v(L,t)$.

can force high-frequency modes that have not been included in the reduced-order, discrete model, hence making the closed-loop system unstable. To avoid spillover instabilities [3], Yang and Mote [28] used a transfer function approach to develop a class of asymptotically stabilizing controllers for distributed parameter models of axially moving strings and beams. Using a carefully positioned point sensor and point actuator, a quasi-passive controller ensured that all of the vibration modes decay to zero. In [23], Renshaw et al. illustrated that Lyapunov theory can be applied to axially moving systems without using the expected material derivative for time differentiation of the Lyapunov function. In [10], Lee and Mote developed Lyapunov-based, boundary control laws that asymptotically stabilize the vibration of an axially moving string and minimize the energy reflected from the boundaries. A number of other researchers have developed controllers for distributed parameter systems similar to the axially moving web. For example, Morgül [16] introduced dynamic, boundary feedback controllers for the wave equation that included proportional and strictly positive real derivative feedback. In [7], Joshi and Rahn developed and experimentally implemented boundary controllers for a linear gantry crane model with a flexible cable, while in [1], Baicu et al. experimentally demonstrated how a similar boundary control law can be used to stabilize the out-of-plane vibration of a flexible cable.

For the flexible link robots, most control design work has been directed towards discretized models (see [29] for a literature review of finite dimensional model-based controllers for flexible link robots); however, some boundary control strategies based on distributed parameter models have been proposed. For example, Luo et al. [11]–[14] and Morgül [17] utilized linear PDE models (the ODE dynamics of the actuator hub were neglected) to develop vibration control strategies to regulate the link displacement. The control inputs in [17] consisted of a boundary force and torque both applied to the link's free end instead of a torque applied to the actuator hub as in the work of Luo et al. Later, in [18] Morgül, used a linear hybrid model to design a control law, consisting of a torque applied to the hub and boundary force and torque applied to the link's free end, which ensured that the link displacement and the hub position setpoint error were both asymptotically regulated. Ge et al. [5] augmented the linear hybrid model of [18] with a linear ODE dynamics of a payload mass at the link's free end to design a nonlinear strain feedback control law which guaranteed only closed-loop stability (i.e., the asymptotic or exponential regulation of the link displacement and hub position were not explicitly proven). Recently in [21], Queiroz et al. designed a boundary control strategy for a nonlinear hybrid model of flexible link robots that asymptotically regulated the link displacement and hub position.

The vibration control of flexible rotor systems is another application that has received some focused attention. For example, Baillieul and Levi [2] introduced an idealized model for studying the dynamics and control of rotating Euler-Bernoulli beam-like structures. Specifically, the rotor system studied in [2] consisted of an Euler-Bernoulli beam perpendicularly attached to the center of a rigid rotating hub. However, the vibration of the beam was confined to a single plane that was fixed with respect to the rotating hub (i.e., one-dimensional displacement, flexible rotor system). For this simplified flexible rotor model, Baillieul and Levi [2] and later Bloch and Titi [4] illustrated the effects of structural and viscous damping, respectively, on the open-loop system response. Subsequently, Xu and Baillieul [26, 27], Morgül [19], and Laousy et al. [9] proposed boundary controllers to achieve regulation of the angular velocity and beam displacement provided the desired angular velocity setpoint was sufficiently small. In [22], Queiroz et al. augmented the nonlinear rotor model used in [9] to include a payload/actuator mass at the free end of the rotor. An adaptive boundary controller was then proposed which asymptotically stabilized the rotor displacement and angular velocity with the restriction on the desired angular velocity setpoint still in place.

References

[1] C. F. Baicu, C. D. Rahn, and B. D. Nibali, Active Boundary Control of Elastic Cables: Theory and Experiment, *Journal of Sound and Vibration*, Vol. 198, No. 1, pp. 17-26, Jan. 1996.

[2] J. Baillieul and M. Levi, Rotational Elastic Dynamics, *Physica*, Vol. 27(D), pp. 43-62, 1987.

[3] M. Balas, Active Control of Flexible Systems, *Journal of Optimization Theory and Applications*, Vol. 25, No. 3, pp. 415-436, July 1978.

[4] A. M. Bloch and E. S. Titi, On the Dynamics of Rotating Elastic Beams, *Proceedings of the Conference on New Trends in System Theory*, Conte, Perdon, and Wyman, Eds., Birkhäuser, Genoa, Italy, July 1990.

[5] S. S. Ge, T. H. Lee, and G. Zhu, Improving Regulation of a Single-Link Flexible Manipulator with Strain Feedback, *IEEE Transactions on Robotics and Automation*, Vol. 14, No. 1, pp. 179-185, Feb. 1998.

[6] G. H. Hardy, J. E. Littlewood, and G. Polya, *Inequalities*, Cambridge University Press, 1959.

[7] S. Joshi and C. D. Rahn, Position Control of a Flexible Cable Gantry Crane: Theory and Experiment, *Proceedings of the American Control Conference*, pp. 2820-2824, Seattle, WA, June 1995.

[8] J. L. Junkins and Y. Kim, *Introduction to Dynamics and Control of Flexible Structures*, Washington, DC: AIAA Education Series, 1993.

[9] H. Laousy, C. Z. Xu, and G. Sallet, Boundary Feedback Stabilization of a Rotating Body-Beam System, *IEEE Transactions on Automatic Control*, Vol. 41, No. 2, pp. 241-244, Feb. 1996.

[10] S. Y. Lee and C. D. Mote, Jr., Vibration Control of an Axially Moving String by Boundary Control, *ASME Journal of Dynamic Systems, Measurement, and Control*, Vol. 118, pp. 66–74, Mar. 1996.

[11] Z. Luo, Direct Strain Feedback Control of Flexible Robot Arms: New Theoretical and Experimental Results, *IEEE Transactions on Automatic Control*, Vol. 38, No. 11, pp. 1610-1622, Nov. 1993.

[12] Z. Luo and B. Guo, Further Theoretical Results on Direct Strain Feedback Control of Flexible Robot Arms, *IEEE Transactions on Automatic Control*, Vol. 40, No. 4, pp. 747-751, Apr. 1995.

[13] Z. Luo, N. Kitamura, and B. Guo, Shear Force Feedback Control of Flexible Robot Arms, *IEEE Transactions on Robotics and Automation*, Vol. 11, No. 5, pp. 760-765, Oct. 1995.

[14] Z. H. Luo and B. Z. Guo, Shear Force Feedback Control of a Single-Link Flexible Robot with a Revolute Joint, *IEEE Transactions on Automatic Control*, Vol. 42, No. 1, pp. 53-65, Jan. 1997.

[15] L. Meirovitch, *Analytical Methods in Vibrations*, New York, NY: Macmillan, 1967.

[16] O. Morgül, A Dynamic Control Law for the Wave Equation, *Automatica*, Vol. 30, No. 11, pp. 1785–1792, Nov. 1994.

[17] O. Morgül, Dynamic Boundary Control of a Euler-Bernoulli Beam, *IEEE Transactions on Automatic Control*, Vol. 37, No. 5, pp. 639-642, May 1992.

[18] O. Morgül, On Boundary Control of Single Link Flexible Robot Arms, *IFAC World Congress*, Vol. A, pp. 127-132, San Francisco, CA, July 1996.

[19] O. Morgül, Control and Stabilization of a Rotating Flexible Structure, *Automatica*, Vol. 30, No. 2, pp. 351-356, Feb. 1994.

[20] http://www.quanser.com.

[21] M. S. de Queiroz, D. M. Dawson, M. Agarwal, and F. Zhang, Adaptive Nonlinear Boundary Control of a Flexible Link Robot Arm, *IEEE Transactions on Robotics and Automation*, Vol. 15, No. 4, pp. 779-787, Aug. 1999.

[22] M. S. de Queiroz, D. M. Dawson, and F. Zhang, Boundary Control of a Rotating Flexible Body-Beam System, *Proceedings of the IEEE Conference on Control Applications*, pp. 812-817, Hartford, CT, Oct. 1997.

[23] A. A. Renshaw, C. D. Rahn, J. A. Wickert, and C. D. Mote, Jr., Energy and Conserved Functionals for Axially Moving Materials, *ASME Journal of Vibration and Acoustics*, to appear.

[24] S. Sastry and M. Bodson, *Adaptive Control: Stability, Convergence, and Robustness*, Englewoods Cliff, NJ: Prentice Hall Co., 1989.

[25] A. G. Ulsoy, Vibration Control in Rotating or Translating Elastic Systems, *ASME Journal of Dynamic Systems, Measurement, and Control*, Vol. 106, No. 1, pp. 6-14, Jan. 1984.

[26] C. Xu and J. Baillieul, Stabilizability and Stabilization of a Rotating Body-Beam System with Torque Control, *IEEE Transactions on Automatic Control*, Vol. 38, No. 12, pp. 1745-1765, Dec. 1993.

[27] C. Xu and G. Sallet, Boundary Stabilization of Rotating Flexible Systems, *Lecture Notes in Control and Information Sciences 185*, R. Curtain, A. Bensoussan, and J. Lions, Eds., New York, NY: Springer-Verlag, pp. 347-365, 1992.

[28] B. Yang and C. D. Mote, Jr., Active Vibration Control of the Axially Moving String in the S Domain, *ASME Journal of Applied Mechanics*, Vol. 58, pp. 189-196, 1991.

[29] J. H. Yang, F. L. Lian, L. C. Fu, Nonlinear Adaptive Control for Flexible-Link Manipulators, *IEEE Transactions on Robotics and Automation*, Vol. 13, No. 1, pp. 140-148, Feb. 1997.

Appendix A

Mathematical Background

In this appendix, we present several mathematical tools in the form of definitions and lemmas that aid the control designs and closed-loop stability analyses presented in the book. The proofs of most of the following lemmas are omitted, but can be found in the cited references.

Definition A.1 *[7]*

Let $f(t) \in \mathbb{R}$ be a function of time on $[0, \infty)$. Let the 2-norm (denoted by $\|\cdot\|_2$) of $f(t)$ be defined as

$$\|f(t)\|_2 = \sqrt{\int_0^\infty f^2(\tau)\, d\tau}. \tag{A.1}$$

If $\|f(t)\|_2 < \infty$, then we say that the function $f(t)$ belongs to the subspace $\mathcal{L}_2$ of the space of all possible functions (i.e., $f(t) \in \mathcal{L}_2$). Let the ∞-norm (denoted by $\|\cdot\|_\infty$) of $f(t)$ be defined as

$$\|f(t)\|_\infty = \sup_t |f(t)|. \tag{A.2}$$

If $\|f(t)\|_\infty < \infty$, then we say that the function $f(t)$ belongs to the subspace $\mathcal{L}_\infty$ of the space of all possible functions (i.e., $f(t) \in \mathcal{L}_\infty$).

Definition A.2 *[9]*

Let $f(t) \in \mathbb{R}$ be a function of time on $[0, \infty)$. The function $f(t)$ is uniformly continuous if for each positive number ϵ_o, there exists a positive number δ_o such that

$$|f(t) - f(t_1)| < \epsilon_o \quad \text{for} \quad \max\{0, t_1 - \delta_o\} < t < t_1 + \delta_o, \tag{A.3}$$

where t_1 is a specific instant of time.

Lemma A.1 *[8]*

Let $f(t) \in \mathbb{R}$ be a function of time on $[0, \infty)$. If $\dot{f}(t) \triangleq \frac{d}{dt} f(t)$ is bounded for $t \in [0, \infty)$, then $f(t)$ is uniformly continuous for $t \in [0, \infty)$.

Lemma A.2 *[5]*

Let $f(t) \in \mathbb{R}$ be a function of time on $[0, \infty)$. If $f(t)$ is uniformly continuous and if the integral

$$\lim_{t \to \infty} \int_0^t |f(\tau)| \, d\tau \tag{A.4}$$

exists and is finite, then

$$\lim_{t \to \infty} |f(t)| = 0. \tag{A.5}$$

Lemma A.3 *[7]*

Let $f(t) \in \mathbb{R}$ be a function of time on $[0, \infty)$. If $f(t) \in \mathcal{L}_\infty$, $\dot{f}(t) \in \mathcal{L}_\infty$, and $f(t) \in \mathcal{L}_2$, then

$$\lim_{t \to \infty} f(t) = 0. \tag{A.6}$$

Lemma A.4 *[1]*

Let $V(t) \in \mathbb{R}$ be a non-negative function of time on $[0, \infty)$ that satisfies the differential inequality

$$\dot{V}(t) \leq -\gamma V(t), \tag{A.7}$$

where γ is a positive constant. Given (A.7), then

$$V(t) \leq V(0) \exp(-\gamma t) \qquad \forall t \in [0, \infty), \tag{A.8}$$

where $\exp(\cdot)$ denotes the base of the natural logarithm.

Lemma A.5 *[1]*

Let $V(t) \in \mathbb{R}$ be a non-negative function of time on $[0, \infty)$ that satisfies the differential inequality

$$\dot{V} \leq -\gamma V + \epsilon, \tag{A.9}$$

where γ and ϵ are positive constants. Given (A.9), then

$$V(t) \leq V(0) \exp(-\gamma t) + \frac{\epsilon}{\gamma}(1 - \exp(-\gamma t)) \qquad \forall t \in [0, \infty). \tag{A.10}$$

Lemma A.6

Let $V(t) \in \mathbb{R}$ be a non-negative function of time on $[0, \infty)$. If (i) $\dot{V}(t) \leq -f(t)$ where $f(t)$ is a non-negative function, and (ii) $f(t)$ is uniformly continuous (or if $\dot{f}(t) \in \mathcal{L}_\infty$), then

$$\lim_{t \to \infty} f(t) = 0. \tag{A.11}$$

Proof. First, we define the following function:

$$V_n(t) = V(t) - \int_0^t \left(\dot{V}(\tau) + f(\tau)\right) d\tau, \tag{A.12}$$

which is lower bounded by zero since $V(t) \geq 0$ and $\dot{V}(t) \leq -f(t)$. If we differentiate (A.12) with respect to time, we obtain

$$\dot{V}_n(t) = -f(t). \tag{A.13}$$

We now apply a lemma from page 125 of [8] that states that if (i) $V_n(t) \in \mathbb{R}$ is a non-negative function of time on $[0, \infty)$, (ii) $\dot{V}_n(t) = -f(t)$ where $f(t)$ is a non-negative function, and (iii) $f(t)$ is uniformly continuous, then

$$\lim_{t \to \infty} f(t) = 0. \tag{A.14}$$

Application of the above lemma to (A.12) and (A.13) yields the result given by (A.11). □

Lemma A.7 *[1]*

Let $r(t), e(t) \in \mathbb{R}$ be functions of time on $[0, \infty)$. Given the differential equation

$$r(t) = \dot{e}(t) + \alpha e(t), \tag{A.15}$$

if $r(t)$ is exponentially stable in the sense that

$$|r(t)| \leq \beta_0 \exp(-\beta_1 t), \tag{A.16}$$

where β_0, β_1 are positive constants, then $e(t)$ and $\dot{e}(t)$ are exponentially stable in the sense that

$$|e(t)| \leq \exp(-\alpha t) |e(0)| + \frac{\beta_0}{\alpha - \beta_1} (\exp(-\beta_1 t) - \exp(-\alpha t)) \tag{A.17}$$

$$\begin{aligned} |\dot{e}(t)| \leq \; & \alpha \exp(-\alpha t) |e(0)| + \beta_0 \exp(-\beta_1 t) \\ & + \frac{\alpha \beta_0}{\alpha - \beta_1} (\exp(-\beta_1 t) - \exp(-\alpha t)). \end{aligned} \tag{A.18}$$

Lemma A.8 *[2]*

Let the transfer function matrix $H(s) \in \mathbb{R}^{\kappa \times \kappa}(\sim)$ be exponentially stable and strictly proper, and let $h(t)$ be the corresponding impulse response (obtained by evaluating the inverse Laplace transform of $H(s)$), then

(i) if $u \in \mathcal{L}_2$, then $y = h * u \in \mathcal{L}_2 \cap \mathcal{L}_\infty$, $\dot{y} \in \mathcal{L}_2$, y is continuous, and $y(t) \to 0$ as $t \to \infty$;

(ii) if $u \in \mathcal{L}_\infty$, then $y = h * u \in \mathcal{L}_\infty$, $\dot{y} \in \mathcal{L}_\infty$, and y is uniformly continuous;

(iii) if $u \in \mathcal{L}_p$ and $1 < p < \infty$, then $y = h * u \in \mathcal{L}_p$ and $\dot{y} \in \mathcal{L}_p$.

Lemma A.9 *[4]*

Let $A \in \mathbb{R}^{\kappa \times \kappa}$ be a real, symmetric, positive-definite matrix; therefore, all of the eigenvalues of A are real and positive. Let $\lambda_{\min}\{A\}$ and $\lambda_{\max}\{A\}$ denote the minimum and maximum eigenvalues of A, respectively; then for $\forall x \in \mathbb{R}^{\kappa}$

$$\lambda_{\min}\{A\} \|x\|^2 \leq x^T A x \leq \lambda_{\max}\{A\} \|x\|^2, \tag{A.19}$$

where $\|\cdot\|$ denotes the standard Euclidean norm. This lemma is often referred to as the Rayleigh-Ritz theorem.

Lemma A.10 *[1, 6]*

If a function $N_d(x, y) \in \mathbb{R}$ is given by

$$N_d = \Omega(x)xy - k_n \Omega^2(x)x^2, \tag{A.20}$$

where $x, y \in \mathbb{R}$, $\Omega(x) \in \mathbb{R}$ is a function dependent only on x, and k_n is a positive constant, then $N_d(x, y)$ can be upper bounded as follows:

$$N_d \leq \frac{y^2}{k_n}. \tag{A.21}$$

The bounding of $N_d(x, y)$ in the above manner is often referred to as *nonlinear damping* [6] since a nonlinear control function, $k_n \Omega^2(x)x$, can be used to "damp-out" an unmeasurable quantity (e.g., y) multiplied by a known, measurable nonlinear function, $\Omega(x)$.

Lemma A.11 *[1]*

Let $x(t) \in \mathbb{R}$ be a function of time on $[0, \infty)$. If the time derivative of $x(t)$ satisfies the following relationship:

$$|\dot{x}(t)| \leq \beta_0 \exp(-\beta_1 t), \tag{A.22}$$

where β_0, β_1 are positive constants, then $x(t) \in \mathcal{L}_\infty$ (note that it is assumed that $\dot{x}(t) \in \mathcal{L}_\infty$ since $x(t)$ is differentiable).

Lemma A.12 *[3]*

Let $\phi(x, t) \in \mathbb{R}$ be a function defined on $x \in [0, L]$ and $t \in [0, \infty)$ that satisfies the *boundary* condition

$$\phi(0, t) = 0 \qquad \forall t \in [0, \infty); \tag{A.23}$$

then the following inequalities hold:

$$\int_0^L \phi^2(\sigma,t)d\sigma \leq L^2 \int_0^L \phi_\sigma^2(\sigma,t)d\sigma \tag{A.24}$$

$$\phi^2(x,t) \leq L \int_0^L \phi_\sigma^2(\sigma,t)d\sigma \qquad \forall x \in [0,L]\,. \tag{A.25}$$

If in addition to (A.23), the function $\phi(x,t)$ satisfies the boundary condition

$$\phi_x(0,t) = 0 \qquad \forall t \in [0,\infty), \tag{A.26}$$

then the following inequalities also hold:

$$\int_0^L \phi_\sigma^2(\sigma,t)d\sigma \leq L^2 \int_0^L \phi_{\sigma\sigma}^2(\sigma,t)d\sigma \tag{A.27}$$

$$\phi^2(x,t) \leq L^3 \int_0^L \phi_{\sigma\sigma}^2(\sigma,t)d\sigma, \;\; \phi_x^2(x,t) \leq L \int_0^L \phi_{\sigma\sigma}^2(\sigma,t)d\sigma \;\; \forall x \in [0,L]\,. \tag{A.28}$$

Lemma A.13

Let $f(t), g(t) \in \mathbb{R}$ be functions of time on $[0,\infty)$, then the following inequalities hold:

$$f^2(t) + g^2(t) \geq |f(t)g(t)| \tag{A.29}$$

$$\frac{f^2(t)}{\delta} + \delta g^2(t) \geq |f(t)g(t)| \qquad \forall \delta > 0. \tag{A.30}$$

References

[1] D. M. Dawson, J. Hu, and T. C. Burg, *Nonlinear Control of Electric Machinery*, New York, NY: Marcel Dekker, 1998

[2] C. A. Desoer and M. Vidyasagar, *Feedback Systems: Input-Output Properties*, Academic Press, 1975.

[3] G. H. Hardy, J. E. Littlewood, and G. Polya, *Inequalities*, Cambridge, MA: Cambridge University Press, 1959.

[4] R. Horn and C. Johnson, *Matrix Analysis*, Cambridge, MA: Cambridge University Press, 1985.

[5] H. Khalil, *Nonlinear Systems*, Upper Saddle River, NJ: Prentice Hall, 1996.

[6] M. Krstić, I. Kanellakopoulos, and P. Kokotović, *Nonlinear and Adaptive Control Design,* New York, NY: John Wiley & Sons, 1995.

[7] S. Sastry and M. Bodson, *Adaptive Control,* Englewood Cliffs, NJ: Prentice Hall, 1989.

[8] J. J. Slotine and W. Li, *Applied Nonlinear Control,* Englewood Cliffs, NJ: Prentice Hall, 1991.

[9] G. Thomas and R. Finney, *Calculus and Analytic Geometry*, Reading, MA: Addison Wesley, 1982.

Appendix B

Bounds for General Rigid Mechanical System

In this appendix, the derivation of the upper bound for the variable $\chi(\cdot)$ given in (4.68) is illustrated. The triangle inequality is first used to find an upper bound for $\|\chi\|$ based on the definition of $\chi(\cdot)$ in (4.63) as follows:

$$\|\chi\| \leq \left\|\tilde{Y}\right\| + \|M(q)\eta\| + 2\|M(q)e_f\| + \|V_m(q,\dot{q})e_f\| + \|V_m(q,\dot{q})e\|. \quad \text{(B.1)}$$

In [1, 2], it has been shown that

$$\left\|\tilde{Y}\right\| \leq \zeta_0\|e\| + \zeta_1\|\dot{e}\|^2 + \zeta_2\|e\|^2 + \zeta_3\|\dot{e}\|, \quad \text{(B.2)}$$

where the ζ_i's are some positive bounding constants that depend on the physical parameters of the mechanical system and the bounds on desired motion trajectory. This inequality combined with induced matrix norms for the other terms on the right-hand side of (B.1) allows for the following new bound on $\|\chi\|$:

$$\begin{aligned}\|\chi\| \leq\ & \zeta_0\|e\| + \zeta_1\|\dot{e}\|^2 + \zeta_2\|e\|^2 + \zeta_3\|\dot{e}\| + \|M(q)\|_{i2}\|\eta\| \\ & +2\|M(q)\|_{i2}\|e_f\| + \|V_m(q,\dot{q})\|_{i\infty}\|e_f\| + \|V_m(q,\dot{q})\|_{i\infty}\|e\|.\end{aligned} \quad \text{(B.3)}$$

Properties 3.1 and 4.2 can be used to bound the occurrences of $\|M(q)\|_{i2}$ and $\|V_m(q,\dot{q})\|_{i\infty}$, respectively, in (B.3) to yield

$$\begin{aligned}\|\chi\| \leq\ & \zeta_0\|e\| + \zeta_1\|\dot{e}\|^2 + \zeta_2\|e\|^2 + \zeta_3\|\dot{e}\| + m_2\|\eta\| \\ & +2m_2\|e_f\| + \zeta_{c1}\|\dot{q}\|\|e_f\| + \zeta_{c1}\|\dot{q}\|\|e\|.\end{aligned} \quad \text{(B.4)}$$

Substitutions are now made in (B.4) using $\dot{q}(t) = \dot{q}_d(t) - \dot{e}(t)$, the error system dynamics for $\dot{e}(t)$ in (4.58), and the assumed bound on $\dot{q}_d(t)$ from (4.4) to find a new upper bound. The result of these substitutions is the following inequality, which contains known constants and the variables $e(t)$, $e_f(t)$, and $\eta(t)$:

$$\begin{aligned} \|\chi\| \leq\ & \zeta_0 \|e\| + \zeta_1 \|e\|^2 + \zeta_1 \|e_f\|^2 + \zeta_1 \|\eta\|^2 + 2\zeta_1 \|e\| \|e_f\| \\ & +2\zeta_1 \|e\| \|\eta\| + 2\zeta_1 \|e_f\| \|\eta\| + \zeta_2 \|e\|^2 + \zeta_3 \|e_f\| + \zeta_3 \|e\| \\ & +\zeta_3 \|\eta\| + m_2 \|\eta\| + 2m_2 \|e_f\| + \zeta_{c1}\zeta_{d2} \|e_f\| + \zeta_{c1} \|e\| \|e_f\| \\ & +\zeta_{c1} \|e_f\|^2 + \zeta_{c1} \|\eta\| \|e_f\| + \zeta_{c1}\zeta_{d2} \|e\| + \zeta_{c1} \|e\|^2 \\ & +\zeta_{c1} \|e_f\| \|e\| + \zeta_{c1} \|\eta\| \|e\| . \end{aligned} \tag{B.5}$$

Now, note from (4.69) that $\|x\| \geq \|e_f\|, \|e\|, \|\eta\|$; hence, after collecting terms, a final upper bound can be placed on (B.5) as shown below:

$$\|\chi\| \leq \alpha_o \|x\| + \alpha_1 \|x\|^2 , \tag{B.6}$$

where α_o, α_1 are positive bounding constants defined as

$$\alpha_o = \zeta_0 + 2\zeta_{c1}\zeta_{d2} + 3\zeta_3 + 3m_2 \qquad \alpha_1 = 9\zeta_1 + \zeta_2 + 6\zeta_{c1}. \tag{B.7}$$

References

[1] F. L. Lewis, C. T. Abdallah, and D. M. Dawson, *Control of Robot Manipulators*, New York, NY: Macmillan Publishing Co., 1993.

[2] N. Sadegh and R. Horowitz, Stability and Robustness Analysis of a Class of Adaptive Controllers for Robotic Manipulators, *International Journal of Robotic Research*, Vol. 9, No. 9, pp. 74-92, June 1990.

Appendix C
Bounds for the Puma Robot

To prove the existence of the bounds given in (4.105), we will make use of the inequalities given in (4.104). First, note that the matrices in (4.105) contain mismatches *only* in the position variable. For general, revolute joint mechanical systems, these mismatches can be arranged to have one of the following general forms:

$$\left|\cos\left(\sum_{i=j}^{k}\xi_i\right)-\cos\left(\sum_{i=j}^{k}\nu_i\right)\right|, \quad \left|\sin\left(\sum_{i=j}^{k}\xi_i\right)-\sin\left(\sum_{i=j}^{k}\nu_i\right)\right|, \tag{C.1}$$

where $\xi_i(t), \nu_i(t)$ are the i-th elements of $\forall\xi(t), \nu(t) \in \mathbb{R}^n$, respectively, $j \in \{1, 2, ..., n\}$, $k \in \{j, j+1, ..., n\}$, and n represents the system's total degrees-of-freedom (DOF). That is, all the elements of the matrices in (4.105) will contain terms similar to those given by (C.1). For example, the most complicated elements of the centripetal-Coriolis matrix (i.e., $V_{m53}(q, \dot{q})$) for the six-DOF Puma Robot Manipulator has the following form [1]:

$$\begin{aligned} V_{m53} = \; & 0.0025\cos(q_2+q_3)\sin(q_4)\sin(q_5)\,\dot{q}_1 \\ & +0.00064\cos(q_2+q_3)\sin(q_4)\,\dot{q}_1, \end{aligned} \tag{C.2}$$

where $q_i(t)$ denotes the i-th element of the link position vector $q(t) \in \mathbb{R}^6$. If we define

$$\tilde{V} \triangleq V_m(\xi, \dot{q}) - V_m(\nu, \dot{q}), \tag{C.3}$$

then, according to (C.2), the mismatch for element $V_{m53}(q,\dot{q})$ becomes

$$\tilde{V}_{53} = \Omega_1(\xi,\nu,\dot{q}) + \Omega_2(\xi,\nu,\dot{q}), \tag{C.4}$$

where $\Omega_1(\xi,\nu,\dot{q})$ and $\Omega_2(\xi,\nu,\dot{q})$ are scalar functions defined as

$$\Omega_1 = 0.00064\dot{q}_1\left[\cos\left(\xi_2+\xi_3\right)\sin\left(\xi_4\right) - \cos\left(\nu_2+\nu_3\right)\sin\left(\nu_4\right)\right] \tag{C.5}$$

and

$$\begin{aligned} \Omega_2 = &\ 0.0025\dot{q}_1\cos\left(\xi_2+\xi_3\right)\sin\left(\xi_4\right)\sin\left(\xi_5\right) \\ &-0.0025\dot{q}_1\cos\left(\nu_2+\nu_3\right)\sin\left(\nu_4\right)\sin\left(\nu_5\right). \end{aligned} \tag{C.6}$$

Note that (C.5) can be rewritten as

$$\begin{aligned} \Omega_1 = &\ 0.00064\dot{q}_1\left[\cos\left(\xi_2+\xi_3\right) - \cos\left(\nu_2+\nu_3\right)\right]\sin\left(\xi_4\right) \\ &+0.00064\dot{q}_1\cos\left(\nu_2+\nu_3\right)\left[\sin\left(\xi_4\right) - \sin\left(\nu_4\right)\right]; \end{aligned} \tag{C.7}$$

hence, an upper bound can be placed on $|\Omega_1|$ as follows:

$$|\Omega_1| \le 0.00064\left|\dot{q}_1\right|\left[\left|\cos\left(\xi_2+\xi_3\right) - \cos\left(\nu_2+\nu_3\right)\right| + \left|\sin\left(\xi_4\right) - \sin\left(\nu_4\right)\right|\right]. \tag{C.8}$$

Upon the application of (4.104) to (C.8), we have the following new upper bound:

$$|\Omega_1| \le 0.00512\left|\dot{q}_1\right|\left[\left|\tanh\left(\xi_2-\nu_2\right)\right| + \left|\tanh\left(\xi_3-\nu_3\right)\right| + \left|\tanh\left(\xi_4-\nu_4\right)\right|\right]. \tag{C.9}$$

Similar arguments can be applied to (C.6) to show that

$$\begin{aligned} |\Omega_2| \le &\ 0.02\left|\dot{q}_1\right|\left[\left|\tanh\left(\xi_2-\nu_2\right)\right| + \left|\tanh\left(\xi_3-\nu_3\right)\right|\right. \\ &\left. + \left|\tanh\left(\xi_4-\nu_4\right)\right| + \left|\tanh\left(\xi_5-\nu_5\right)\right|\right]. \end{aligned} \tag{C.10}$$

From (C.9) and (C.10), it is clear that $\tilde{V}_{53}(\cdot)$ of (C.4) can be upper bounded as follows:

$$\begin{aligned} \left|\tilde{V}_{53}\right| \le &\ \zeta_{53}\left|\dot{q}_1\right|\left[\left|\tanh\left(\xi_2-\nu_2\right)\right| + \left|\tanh\left(\xi_3-\nu_3\right)\right|\right. \\ &\left. + \left|\tanh\left(\xi_4-\nu_4\right)\right| + \left|\tanh\left(\xi_5-\nu_5\right)\right|\right], \end{aligned} \tag{C.11}$$

where ζ_{53} is some positive bounding constant. We can now use the fact that

$$\|x\| = \sqrt{\sum_{i=1}^{n}\left|x_i\right|^2} \ge \left|x_i\right| \qquad \forall x \in \mathbb{R}^n, \tag{C.12}$$

where x_i is the i-th element of x, to show that

$$\left|\tilde{V}_{53}\right| \leq \bar{\zeta}_{53} \|\dot{q}\| \|\text{Tanh}(\xi - \nu)\|, \tag{C.13}$$

where $\bar{\zeta}_{53}$ is some positive bounding constant.

Since all of the elements of $\tilde{V}(\cdot)$, defined in (C.3), can be upper bounded in a similar fashion as shown in (C.13), it is now easy to see that $\left\|\tilde{V}\right\|_{i\infty}$ can be upper bounded as in the third inequality of (4.105). Similar arguments can be followed to prove the other two inequalities given in (4.105).

References

[1] F. L. Lewis, C. T. Abdallah, and D. M. Dawson, *Control of Robot Manipulators*, New York, NY: MacMillan Publishing Co., 1993.

Appendix D

Control Programs

In this appendix, we present the C program codes written for the real-time implementation of the DCAL controller of Section 3.5 in Chapter 3 and the flexible rotor boundary controller of Section 7.4 in Chapter 7. While the DCAL controller was implemented via the *WinMotor* control environment, the boundary controller used the *Qmotor* environment. See Section 1.4 in Chapter 1 for a description of these real-time control environments.

The control program for both *WinMotor* and *Qmotor,* shown in the following, contain the subroutines described below:

(i) *init_control()*: Initializes the hardware boards and allocates memory for the second-order filters.
(ii) *input()*: Reads in data from the A/D, encoder channels, and digital inputs; uses a backwards difference/filter to find the approximate time derivative of signals; provides safety features to shut the system off if the inputs go out of their safe range.
(iii) *control()*: The heart of the program that contains the control algorithm and generates the value for the applied control input (usually a voltage).
(iv) *output()*: Sends out the signals via the D/A and digital outputs; contains safety features to shut the system off if the outputs go beyond their safe range.
(v) *end_control()*: Resets the hardware and de-allocates the memory reserved for the filters.

It should be noted that while the *init_ control()* and *end_ control()* subroutines are executed only once at the start and completion of the control algorithm, the *input()*, *control()* and *output()* subroutines are executed continuously at a specified sampling period. In the following codes, the variables containing the prefix "*VF_* " denote variables which the user can interactively change their values from the GUI without the need to recompile the program. The variables containing the prefix "*VD_* " represent variables which are available for on-line graphing or storage.

D.1 DCAL Controller

```
/*=======================================================
Note: Positive joint directions are as follows:
Base          Clockwise
Elbow         Clockwise
=======================================================*/

#include ''pc.h''
#include ''wmresour.h''
#include <math.h>

#define PI 3.1415

/*These are predefined variables*/
float state[10], statedot[10];
int system_order=4;

extern float tsamp;
extern int home_state;
int Safety;  /*1 no error  0 problem has occured*/
float VD_Base_Pos, VD_Elbow_Pos;
float VD_Base_Vel, VD_Elbow_Vel;
float err1, err2;
float u1, u2;
float VD_Time,VF_init;

float  I1  = 0.267,     /*robot parameters*/
       M1  = 73.0,
       I2  = 0.334,
       M2  = 9.78,
       I3  = 0.0075,
       M3  = 14.0,
```

```
      I3c = 0.040,
      M4  = 5.0685,  /*these must be calculated for the new link*/
      I4  = 0.096,
      Mp  = 0.0,
      Ip  = 0.0,
      L1  = 0.359,
      L2  = 0.30,
      L3  = 0.136,
      L4  = 0.133,
      Fd1 = 5.3,
      Fd2 = 1.1;

float P1, P2, P3, M11, M12, M21, M22,
      MI11, MI12, MI21, MI22, detMinv;

float VD_Error1, VD_Error2;
float cosq2, sinq2,r1,r2,VD_Edot1,VD_Edot2;
float Yd11, Yd12, Yd13, Yd14, Yd15, Yd21, Yd22, Yd23, Yd24, Yd25;
float dphi1hat[2], dphi2hat[2];
float dphi3hat[2], dphi4hat[2], dphi5hat[2];
float phi1hat, phi2hat, phi3hat, phi4hat, phi5hat;
float VD_phi1hat,VD_phi2hat,VD_phi3hat,VD_phi4hat,VD_phi5hat;
float VF_Gamma1,VF_Gamma2,VF_Gamma3,VF_Gamma4,VF_Gamma5;
float VD_q1d,VD_q1dotd,VD_q1ddotd,VD_q1tdotd;
float VD_q2d,VD_q2dotd,VD_q2ddotd,VD_q2tdotd;
float oldpos1,oldpos2,VF_K1,VF_K2,VF_Amp1,VF_Amp2;
float VF_qdfreq,VF_qdpeak,VF_qdtau,VF_Alpha1,VF_Alpha2;
float VD_Temp1,VD_Temp2;

/*--------- CONTROL ROUTINE ---------*/
control()
{
int i;
float tt[8], texp, tsin, tcos, tausq, freqsq;
float qd, qdotd, qddotd, qtdotd;
float acc1, acc2;

/*-------- Desired Trajectory --------*/
tt[1] = VD_Time;
for (i=2; i < 7; i++)
  tt[i] = tt[i-1] * VD_Time;
texp = exp(-VF_qdtau*tt[3]);
tsin = VF_qdpeak*sin(VF_qdfreq*VD_Time);
tcos = VF_qdpeak*cos(VF_qdfreq*VD_Time);
```

```
tausq = VF_qdtau*VF_qdtau;
freqsq= VF_qdfreq*VF_qdfreq;

qd = tsin*(1.0-texp);
qdotd = (3.0*tt[2]*VF_qdtau*tsin-VF_qdfreq*tcos)
   *texp+VF_qdfreq*tcos;
qddotd = (6*tt[2]*VF_qdfreq*VF_qdtau*tcos
   +(freqsq-9*tt[4]*tausq+6*tt[1]*VF_qdtau)*tsin)
   *texp-freqsq*tsin;
qtdotd = (VF_qdfreq*(freqsq-27*tt[4]*tausq
   +18*VD_Time*VF_qdtau)*tcos
   -3*VF_qdtau*(3*tt[2]*freqsq
   -(9*tt[6]*tausq-18*tt[3]*VF_qdtau+2))*tsin)*texp
   -freqsq*VF_qdfreq*tcos;

VD_q1d=VF_Amp1*qd;
VD_q1dotd=VF_Amp1*qdotd;
VD_q1ddotd=VF_Amp1*qddotd;
VD_q1tdotd=VF_Amp1*qtdotd;
VD_q2d=VF_Amp2*qd;
VD_q2dotd=VF_Amp2*qdotd;
VD_q2ddotd=VF_Amp2*qddotd;
VD_q2tdotd=VF_Amp2*qtdotd;

VD_Error1=VD_q1d-VD_Base_Pos;
VD_Error2=VD_q2d-VD_Elbow_Pos;

VD_Edot1=VD_q1dotd-VD_Base_Vel;
VD_Edot2=VD_q2dotd-VD_Elbow_Vel;

r1=VF_Alpha1*VD_Error1+VD_Edot1;
r2=VF_Alpha2*VD_Error2+VD_Edot2;

sinq2=sin(VD_q2d);
cosq2=cos(VD_q2d);

/*--- Regression Terms ---*/
Yd11=VD_q1ddotd;
Yd12=VD_q2ddotd;
Yd13=2*cosq2*VD_q1ddotd
     +cosq2*VD_q2ddotd
     -sinq2*VD_q1dotd*VD_q2dotd
     -sinq2*(VD_q1dotd+VD_q2dotd)*VD_q2dotd;
Yd14=VD_q1dotd;
```

```
Yd15=0.0;

Yd21=0.0;
Yd22=VD_q1ddotd+VD_q2ddotd;
Yd23=cosq2*VD_q1ddotd+sinq2*VD_q1dotd*VD_q1dotd;
Yd24=0.0;
Yd25=VD_q2dotd;

/*--- Adaptive Update Laws ---*/
dphi1hat[1]=VF_Gamma1*(Yd11*r1+Yd21*r2);
dphi2hat[1]=VF_Gamma2*(Yd12*r1+Yd22*r2);
dphi3hat[1]=VF_Gamma3*(Yd13*r1+Yd23*r2);
dphi4hat[1]=VF_Gamma4*(Yd14*r1+Yd24*r2);
dphi5hat[1]=VF_Gamma5*(Yd15*r1+Yd25*r2);

/* trapezoidal integration algorithm */
phi1hat=phi1hat+0.5*tsamp*(dphi1hat[0]+dphi1hat[0]);
dphi1hat[0]=dphi1hat[1];

phi2hat=phi2hat+0.5*tsamp*(dphi2hat[0]+dphi2hat[0]);
dphi2hat[0]=dphi2hat[1];

phi3hat=phi3hat+0.5*tsamp*(dphi3hat[0]+dphi3hat[0]);
dphi3hat[0]=dphi3hat[1];

phi4hat=phi4hat+0.5*tsamp*(dphi4hat[0]+dphi4hat[0]);
dphi4hat[0]=dphi4hat[1];

phi5hat=phi5hat+0.5*tsamp*(dphi5hat[0]+dphi5hat[0]);
dphi5hat[0]=dphi5hat[1];

/*------- Control Law ------- */
u1=Yd11*phi1hat+Yd12*phi2hat+Yd13*phi3hat
   +Yd14*phi4hat+Yd15*phi5hat
   +VF_K1*r1+VD_Error1;
u2=Yd21*phi1hat+Yd22*phi2hat+Yd23*phi3hat
   +Yd24*phi4hat+Yd25*phi5hat
   +VF_K2*r2+VD_Error2;

VD_Temp1=u1;
VD_Temp2=u2;
VD_phi1hat=phi1hat;
VD_phi2hat=phi2hat;
VD_phi3hat=phi3hat;
```

```
VD_phi4hat=phi4hat;
VD_phi5hat=phi5hat;

/*--- Error System (Display Only) in Degrees ---*/
VD_Error1=VD_Error1*180/PI;
VD_Error2=VD_Error2*180/PI;
VD_Base_Pos=VD_Base_Pos*180/PI;
VD_Elbow_Pos=VD_Elbow_Pos*180/PI;
}


/*--------- INIT_CONTROL ROUTINE ---------*/
init_control()
{
int i;

VD_Base_Pos=0.0;
VD_Base_Vel=0.0;
VD_Elbow_Pos=0.0;
VD_Elbow_Vel=0.0;

P1=I1+I2+I3c+I3+I4+Ip + (M3+M4+Mp)*L1*L1
   + M2*L3*L3 + M4*L4*L4 + Mp*L2*L2;
P2=I3+I4+Ip + M4*L4*L4 + Mp*L2*L2;
P3=M4*L1*L4 + Mp*L1*L2;

dphi1hat[0]=0.0;
dphi2hat[0]=0.0;
dphi3hat[0]=0.0;
dphi4hat[0]=0.0;
dphi5hat[0]=0.0;

phi1hat=3.473*VF_init;
phi2hat=0.193*VF_init;
phi3hat=0.242*VF_init;
phi4hat=5.3*VF_init;
phi5hat=1.1*VF_init;

oldpos1=0.0;
oldpos2=0.0;

VD_Time=-1*tsamp;
Reset_HCTL(0,2);
```

```
Dac_Out(0, 0, 0);
Dac_Out(0, 1, 0);
Dac_Switch(0, 1, 1);

*PC_PARM7=0x24;
*PC_PARM6=0x00;

Safety=1;
}

/*--------- INPUT ROUTINE ---------*/
input()
{
int limit;
long pos1, pos2, temp;

limit=*PC_PARM7;  /*check for limit switch problems*/
limit=limit&0xE7;
if (limit!=0x24)
   {
   Safety=0;      /*indicate a problem has occured*/
   if (limit & 0x80)
      *C30_PARM1=DDR_BS_CW_LMT;  /*Base CW Limit*/
   if (limit & 0x40)
      *C30_PARM1=DDR_EB_CCW_LMT;  /*Elbow CCW Limit*/
   if (!(limit & 0x20))
      *C30_PARM1=DDR_BS_EMERG;  /*Base Emergency*/
   if (!(limit & 0x04))
      *C30_PARM1=DDR_EB_EMERG;  /*Elbow Emergency*/
   if (limit & 0x02)
      *C30_PARM1=DDR_BS_CCW_LMT;  /*Base CCW Limit*/
   if (limit & 0x01)
      *C30_PARM1=DDR_EB_CW_LMT;  /*Elbow CW Limit*/
   *C30_COMMAND=C30_ERROR;
   }

/*--- Get the joint positions ---*/
Get_Position(0, &pos1, &pos2);

VD_Base_Pos=(float)pos1/153600*2*PI;
VD_Elbow_Pos=(float)pos2/153600*2*PI;

VD_Base_Vel=(VD_Base_Pos-oldpos1)/tsamp;
VD_Elbow_Vel=(VD_Elbow_Pos-oldpos2)/tsamp;
```

```
oldpos1=VD_Base_Pos;
oldpos2=VD_Elbow_Pos;

VD_Time=VD_Time+tsamp;
}

/*--------- OUTPUT ROUTINE ---------*/
output()
{
int Actual_Out;

Actual_Out=(int)(Safety*(-1*(u1*2047/10)/19.3));
if (Actual_Out>2047)
   Actual_Out=2047;
if (Actual_Out<-2047)
   Actual_Out=-2047;
Dac_Out(0, 0, Actual_Out);

Actual_Out=(int)(Safety*(-1*(u2*2047/10)/3.12));
if (Actual_Out>2047)
   Actual_Out=2047;
if (Actual_Out<-2047)
   Actual_Out=-2047;
Dac_Out(0, 1, Actual_Out);
}

/*--------- END_CONTROL ROUTINE ---------*/
end_control()
{
Dac_Out(0, 0, 0);
Dac_Out(0, 1, 0);
Dac_Out(1, 0, 0);
}
```

D.2 Flexible Rotor

```
/*=======================================================
Note: All measurements are in the rotated frame
The force is calculated w.r.t the rotating frame,
but applied by the magnets in the fixed frame,
hence use sin-cos transformation.
The LED position is measured by the cameras in
```

```
the fixed frame; therefore use the inverse transformation.
==========================================================*/

#include ''qc.def''
#include <math.h>

extern float tsamp;

typedef unsigned char BYTE;
typedef unsigned int  WORD;

#define  ECL_CARD_ADDR1  0x340
#define  ECL_CARD_ADDR2  0x360
#define  OUT_OF_RANGE         ((WORD)0x800)
#define  NOT_DETECTED         ((WORD)0x1000)
#define  MULTIPLE             ((WORD)0x2000)
#define PI 3.14159265358979323846264338

/*--- Variables and Functions for Camera Operation ---*/

void CcdLedEnable ( BYTE, BYTE );
void CcdThresholdSet1 ( WORD );
WORD CcdPixelCnt1 ( void );
void CcdThresholdSet2 ( WORD );
WORD CcdPixelCnt2 ( void );

BYTE led_id_on=1;  /* LED number 1 ON permanently */

WORD w_pixel_cnt1,
     led_bias1;
WORD w_pixel_cnt2,
     led_bias2;
int relative_pixel1;
int relative_pixel2;

WORD CCD_THRESHOLD,
     LED_ON_TIME_MASK;

float m_per_pixel,focal_length,p1,p2,focal_gain,max_def;

/*--- VD_ and VF_ variables for the camera ---*/
float VF_led_dist,VF_Delay;
float VD_pixel1,VD_pixel2,VD_Time;
```

```
/*--- VD_ and VF_ variables for control ---*/
float
 VD_Position_M,
 VD_Position_X, VD_Position_Y,
 VD_Shear_X, VD_Shear_Y,
 VD_ShearDer_X, VD_ShearDer_Y,
 VD_Id_X1, VD_Id_X2, VD_Id_Y1, VD_Id_Y2,
 VD_Current_X1, VD_Current_X2,
 VD_Current_Y1, VD_Current_Y2,
 VD_Current_M;

float
 VD_Error,
 VD_Fd_X, VD_Fd_Y,
 VD_Fddot_X, VD_Fddot_Y,
 VD_Fd_X1, VD_Fd_X2, VD_Fd_Y1, VD_Fd_Y2,
 VD_Torque, VD_Voltage_M,
 VD_EtaX, VD_EtaY,
 VD_Voltage_X1, VD_Voltage_X2,
 VD_Voltage_Y1, VD_Voltage_Y2,
 VD_Voltage_M;

float
 VD_Temp1, VD_Temp2, VD_Temp3,
 VD_Velocity_M,
 VD_Mhat, VD_EIhat,
 VD_Velocity_X, VD_Velocity_Y;

float
 VF_GammaEI, VF_GammaM,
 VF_Gamma0, VF_Alpha,VF_Phi,
 VF_Ks, VF_Kr, VF_Ke, VF_Init, VF_Km,
 VF_DesVel_M,
 VF_VelFilter_X, VF_VelFilter_Y, VF_VelFilter_M,
 VF_FddotFil_X, VF_FddotFil_Y,
 VF_ShearOffset_X, VF_ShearOffset_Y,
 VF_ShearGain_X, VF_ShearGain_Y,
 VF_ShearFil_X, VF_ShearFil_Y,
 VF_ShearDerFil_X, VF_ShearDerFil_Y;

int Safety = 1;

struct filter *filter_vel_x;
struct filter *filter_vel_y;
```

```
struct filter *filter_vel_m;
struct filter *filter_fddot_x;
struct filter *filter_fddot_y;
struct filter *filter_shear_x;
struct filter *filter_shear_y;
struct filter *filter_shearder_x;
struct filter *filter_shearder_y;

float
 unfilt_vel_x, unfilt_vel_y, unfilt_vel_m,
 unfilt_fddot_x, unfilt_fddot_y,
 unfilt_shear_x, unfilt_shear_y,
 unfilt_shearder_x, unfilt_shearder_y;

float
 M, EI, Ktau,
 Determ,
 aux_X, aux_Y,
 X1, X2, Y1, Y2,
 x1dot, x2dot, y1dot, y2dot,
 expX1, expX2, expY1, expY2,
 R, L1, L0, Beta,
 Lx1, Lx2, Ly1, Ly2,
 RootFdX1, RootFdX2, RootFdY1, RootFdY2,
 RootLexpX1, RootLexpX2, RootLexpY1, RootLexpY2,
 FdX1dot, FdX2dot, FdY1dot, FdY2dot,
 P_FdX1, P_FdX2, P_FdY1, P_FdY2,
 P_IdX1FdX1, P_IdX2FdX2, P_IdY1FdY1, P_IdY2FdY2,
 P_IdX1X1, P_IdX2X2, P_IdY1Y1, P_IdY2Y2,
 IddotX1, IddotX2, IddotY1, IddotY2,
 etaiX1, etaiX2, etaiY1, etaiY2,
 oldforce_X, oldforce_Y,
 oldshear_X, oldshear_Y,
 Fd_X, Fd_Y,
 Shear_X, Shear_Y,
 Position_X, Position_Y,
 Omega_X, Omega_Y, dMhat[2], dEIhat[2],
 oldpos_X, oldpos_Y, oldpos_M;


/*--------- CONTROL ROUTINE ---------*/
control()
{
if(VD_Time < VF_Delay) return;
```

```
X1 =  VD_Position_X;
X2 = -VD_Position_X;
Y1 =  VD_Position_Y;
Y2 = -VD_Position_Y;

x1dot =  VD_Velocity_X;
x2dot = -VD_Velocity_X;
y1dot =  VD_Velocity_Y;
y2dot = -VD_Velocity_Y;

expX1 = exp(Beta*X1);
expX2 = exp(Beta*X2);
expY1 = exp(Beta*Y1);
expY2 = exp(Beta*Y2);

Lx1 = L1*expX1 + L0;
Lx2 = L1*expX2 + L0;
Ly1 = L1*expY1 + L0;
Ly2 = L1*expY2 + L0;

/*--- Error Terms ---*/
VD_Error = VD_Velocity_M - VF_DesVel_M;
VD_EtaX = VD_Velocity_X
   - VD_Velocity_M*VD_Position_Y - VD_Shear_X;
VD_EtaY = VD_Velocity_Y
   + VD_Velocity_M*VD_Position_X - VD_Shear_Y;

/*--- Auxiliary Terms for Adaptive Laws ---*/
Omega_X = VD_Velocity_M*(VD_Velocity_M
   *VD_Position_X + VD_Velocity_Y)-VD_ShearDer_X;
Omega_Y = VD_Velocity_M*(VD_Velocity_M
   *VD_Position_Y - VD_Velocity_X)-VD_ShearDer_Y;

/*--- Adaptive Update Laws ---*/
dEIhat[1] = VF_GammaEI*(VD_Shear_X*VD_EtaX
   + VD_Shear_Y*VD_EtaY);
VD_EIhat = VD_EIhat + 0.5*tsamp*(dEIhat[1] + dEIhat[0]);
dEIhat[0] = dEIhat[1];

dMhat[1] = VF_GammaM*(Omega_X*VD_EtaX
   + Omega_Y*VD_EtaY);
VD_Mhat = VD_Mhat + 0.5*tsamp*(dMhat[1] + dMhat[0]);
dMhat[0] = dMhat[1];
```

```
if (VD_Mhat < 0.2) VD_Mhat=0.2;
if (VD_Mhat > 0.3) VD_Mhat=0.3;

if (VD_EIhat < 3.8E-07) VD_EIhat=3.8E-07;
if (VD_EIhat > 4.7E-07) VD_EIhat=4.7E-07;


/*--- Auxiliary Terms for Desired Force ---*/
Determ = 1+VD_Position_M*VD_Position_M;

Fd_X =  VF_Ks*VD_EtaX + VD_EIhat*VD_Shear_X
        + VD_Mhat*Omega_X;

Fd_Y =  VF_Ks*VD_EtaY + VD_EIhat*VD_Shear_Y
        + VD_Mhat*Omega_Y;

M = VD_Mhat;
EI= VD_EIhat;

/*--- Desired Force on each axis ---*/
/*---------------------------------------------------
The Fd_X and Fd_Y forces computed are w.r.t the rotating
frame.
The magnets which apply the force are a part of the fixed
frame; hence we decompose the forces as shown below
---------------------------------------------------*/

VD_Fd_X = cos(VD_Position_M)*Fd_X - sin(VD_Position_M)*Fd_Y;
VD_Fd_Y = sin(VD_Position_M)*Fd_X + cos(VD_Position_M)*Fd_Y;

/*--- Desired Torque applied to the driving motor ---*/
VD_Torque = - VF_Kr*VD_Error
   *(1-exp(-VF_Alpha*(VD_Time-VF_Delay)));

/*--- Desired Force applied by magnet ---*/
VD_Fd_X1 = 0.5*( VD_Fd_X + sqrt(VD_Fd_X*VD_Fd_X + VF_Gamma0));
VD_Fd_X2 = 0.5*(-VD_Fd_X + sqrt(VD_Fd_X*VD_Fd_X + VF_Gamma0));

VD_Fd_Y1 = 0.5*( VD_Fd_Y + sqrt(VD_Fd_Y*VD_Fd_Y + VF_Gamma0));
VD_Fd_Y2 = 0.5*(-VD_Fd_Y + sqrt(VD_Fd_Y*VD_Fd_Y + VF_Gamma0));

RootFdX1 = sqrt(2*VD_Fd_X1);
RootFdX2 = sqrt(2*VD_Fd_X2);
```

```
RootFdY1 = sqrt(2*VD_Fd_Y1);
RootFdY2 = sqrt(2*VD_Fd_Y2);

RootLexpX1 = sqrt(Beta*L1*expX1);
RootLexpX2 = sqrt(Beta*L1*expX2);
RootLexpY1 = sqrt(Beta*L1*expY1);
RootLexpY2 = sqrt(Beta*L1*expY2);

/*--- Filtering the forces ---*/
unfilt_fddot_x = (VD_Fd_X - oldforce_X)/tsamp;
VD_Fddot_X = filter(unfilt_fddot_x, filter_fddot_x);
oldforce_X = VD_Fd_X;

unfilt_fddot_y = (VD_Fd_Y - oldforce_Y)/tsamp;
VD_Fddot_Y = filter(unfilt_fddot_y, filter_fddot_y);
oldforce_Y = VD_Fd_Y;


/*--- Derivative of the force on each magnet ---*/
FdX1dot = 0.5*( VD_Fddot_X+((VD_Fd_X*VD_Fddot_X)
   /sqrt(VD_Fd_X*VD_Fd_X + VF_Gamma0)));
FdX2dot = 0.5*(-VD_Fddot_X+((VD_Fd_X*VD_Fddot_X)
   /sqrt(VD_Fd_X*VD_Fd_X + VF_Gamma0)));
FdY1dot = 0.5*( VD_Fddot_Y+((VD_Fd_Y*VD_Fddot_Y)
   /sqrt(VD_Fd_Y*VD_Fd_Y + VF_Gamma0)));
FdY2dot = 0.5*(-VD_Fddot_Y+((VD_Fd_Y*VD_Fddot_Y)
   /sqrt(VD_Fd_Y*VD_Fd_Y + VF_Gamma0)));


/*--- Desired Currents ---*/
VD_Id_X1 = RootFdX1/RootLexpX1;
VD_Id_X2 = RootFdX2/RootLexpX2;
VD_Id_Y1 = RootFdY1/RootLexpY1;
VD_Id_Y2 = RootFdY2/RootLexpY2;

/*--- Partial Derivative of Fd w.r.t. X ---*/
P_FdX1 = Beta*Beta*expX1*VD_Id_X1*VD_Id_X1;
P_FdX2 = Beta*Beta*expX2*VD_Id_X2*VD_Id_X2;
P_FdY1 = Beta*Beta*expY1*VD_Id_Y1*VD_Id_Y1;
P_FdY2 = Beta*Beta*expY2*VD_Id_Y2*VD_Id_Y2;

/*--- Partial Derivative of Id w.r.t. Fd ---*/
P_IdX1FdX1 = 1/(RootLexpX1*RootFdX1);
P_IdX2FdX2 = 1/(RootLexpX2*RootFdX2);
```

```
P_IdY1FdY1 = 1/(RootLexpY1*RootFdY1);
P_IdY2FdY2 = 1/(RootLexpY2*RootFdY2);

/*--- Partial Derivative of Id w.r.t. X  ---*/
P_IdX1X1 =  (P_FdX1 - Beta*VD_Fd_X1)
   /(RootLexpX1*RootFdX1);
P_IdX2X2 =  (P_FdX2 - Beta*VD_Fd_X2)
   /(RootLexpX2*RootFdX2);
P_IdY1Y1 =  (P_FdY1 - Beta*VD_Fd_Y1)
   /(RootLexpY1*RootFdY1);
P_IdY2Y2 =  (P_FdY2 - Beta*VD_Fd_Y2)
   /(RootLexpY2*RootFdY2);

/*--- Iddot ---*/
IddotX1 = P_IdX1FdX1*FdX1dot + P_IdX1X1*x1dot;
IddotX2 = P_IdX2FdX2*FdX2dot + P_IdX2X2*x2dot;
IddotY1 = P_IdY1FdY1*FdY1dot + P_IdY1Y1*y1dot;
IddotY2 = P_IdY2FdY2*FdY2dot + P_IdY2Y2*y2dot;

/*--- Current Tracking Errors ---*/
etaiX1 = VD_Id_X1 - VD_Current_X1;
etaiX2 = VD_Id_X2 - VD_Current_X2;
etaiY1 = VD_Id_Y1 - VD_Current_Y1;
etaiY2 = VD_Id_Y2 - VD_Current_Y2;


/*--- Desired  Voltages to the magnets ---*/
VD_Voltage_M = VF_Km*(VD_Torque-Ktau*VD_Current_M);

VD_Position_M = VD_Position_M/(2*PI)*360;
VD_Error = VD_Error/(2*PI)*360;

if (VD_Time < (VF_Delay + 3.0)) return;

VD_Voltage_X1 = VF_Init*(VF_Ke*etaiX1
  + R*VD_Current_X1 + Lx1*IddotX1
  + (L1*x1dot*expX1)*(VD_Current_X1-etaiX1/2));
VD_Voltage_X2 = VF_Init*(VF_Ke*etaiX2
  + R*VD_Current_X2 + Lx2*IddotX2
  + (L1*x2dot*expX2)*(VD_Current_X2-etaiX2/2));
VD_Voltage_Y1 = VF_Init*(VF_Ke*etaiY1
  + R*VD_Current_Y1 + Ly1*IddotY1
  + (L1*y1dot*expY1)*(VD_Current_Y1-etaiY1/2));
VD_Voltage_Y2 = VF_Init*(VF_Ke*etaiY2
```

```
  + R*VD_Current_Y2 + Ly2*IddotY2
  + (L1*y2dot*expY2)*(VD_Current_Y2-etaiY2/2));
}


/*--------- INIT_CONTROL ROUTINE ---------*/
init_control()
{
SetMultiQConfig(6, 6, 2, 0, 0);
VD_Time = -tsamp;

/*--- Initialize  Filters ---*/
filter_vel_x = (struct filter *)
   malloc(sizeof(struct filter));
initfilter(VF_VelFilter_X, filter_vel_x);

filter_vel_y = (struct filter *)
   malloc(sizeof(struct filter));
initfilter(VF_VelFilter_Y, filter_vel_y);

filter_vel_m = (struct filter *)
   malloc(sizeof(struct filter));
initfilter(VF_VelFilter_M, filter_vel_m);

filter_fddot_x = (struct filter *)
   malloc(sizeof(struct filter));
initfilter(VF_FddotFil_X, filter_fddot_x);

filter_fddot_y = (struct filter *)
   malloc(sizeof(struct filter));
initfilter(VF_FddotFil_Y, filter_fddot_y);

filter_shear_x = (struct filter *)
   malloc(sizeof(struct filter));
initfilter(VF_ShearFil_X, filter_shear_x);

filter_shear_y = (struct filter *)
   malloc(sizeof(struct filter));
initfilter(VF_ShearFil_Y, filter_shear_y);

filter_shearder_x = (struct filter *)
   malloc(sizeof(struct filter));
initfilter(VF_ShearFil_X, filter_shearder_x);
```

```
filter_shearder_y = (struct filter *)
   malloc(sizeof(struct filter));
initfilter(VF_ShearFil_Y, filter_shearder_y);

ResetEncoder(0);

focal_length = 0.085;     /* 85mm Nikon Lens */
CCD_THRESHOLD    = 0x180;
LED_ON_TIME_MASK = 0xA0;

CcdThresholdSet1 ( CCD_THRESHOLD );
CcdThresholdSet2 ( CCD_THRESHOLD );
led_id_on = (BYTE)1;
p1 = VF_led_dist;
CcdLedEnable ( led_id_on, LED_ON_TIME_MASK );

/* lens to image distance */
p2 = 1.0 / (1.0/focal_length - 1.0/p1);

focal_gain = p1 / p2;

/* Active line scan length is 26.624mm */
max_def = 0.026624 * focal_gain / 2.0;

/* 13um per pixel */
m_per_pixel = 0.000013 * focal_gain;
L1 = 0.0002;
L0 = 0.0552;
Beta = 60;
R = 10.0;
VD_Mhat = 0.25;

/* Bending Stiffness of a hollow
cylindrical plasticised vinyl pipe */
VD_EIhat = 3.786E-07;

/* Torque constant of the motor */
Ktau = 0.056492666;

oldpos_X = 0.0;
oldpos_Y = 0.0;
oldpos_M = 0.0;
oldforce_X = 0.0;
oldforce_Y = 0.0;
```

```
oldshear_X = 0.0;
oldshear_Y = 0.0;

etaiX1 = 0.0;
etaiX2 = 0.0;
etaiY1 = 0.0;
etaiY2 = 0.0;

VD_Position_X = 0.0;
VD_Position_Y = 0.0;
VD_Position_M = 0.0;
VD_Velocity_X = 0.0;
VD_Velocity_Y = 0.0;
VD_Velocity_M = 0.0;
VD_Shear_X    = 0.0;
VD_Shear_Y    = 0.0;
VD_ShearDer_X = 0.0;
VD_ShearDer_Y = 0.0;
VD_Fd_X = 0.0;
VD_Fd_Y = 0.0;

VD_Voltage_M  = 0.0;
VD_Voltage_X1 = 0.0;
VD_Voltage_X2 = 0.0;
VD_Voltage_Y1 = 0.0;
VD_Voltage_Y2 = 0.0;

VD_Current_M = 0.0;
VD_Current_X1 = 0.0;
VD_Current_X2 = 0.0;
VD_Current_Y1 = 0.0;
VD_Current_Y2 = 0.0;
}

/*--------- INPUT ROUTINE ---------*/
input()
{
long pos;

VD_Time=VD_Time+tsamp;

/*--- Read in the Positions of both axes ---*/
while(VD_Time < VF_Delay) {
    led_bias1 = CcdPixelCnt1();
```

```
    led_bias2 = CcdPixelCnt2();
    return;
}

w_pixel_cnt1 = CcdPixelCnt1();
w_pixel_cnt2 = CcdPixelCnt2();


if ( (w_pixel_cnt1 & 0x3800) == 0 )
  relative_pixel1 = w_pixel_cnt1 - led_bias1;
if ( (w_pixel_cnt2 & 0x3800) == 0 )
  relative_pixel2 = w_pixel_cnt2 - led_bias2;

/*-------------------------------------------------------
Note : Position Conventions
Variables Position_X and Position_Y reflect this sign
convention.

+ve Rotation is from +X to +Y, i.e. clockwise in
   the top view.
+ve Forces Fd_X and Fd_Y act along the positive
   X and Y axes respt.
-------------------------------------------------------*/

Position_X = -(relative_pixel1 * m_per_pixel);
Position_Y =  (relative_pixel2 * m_per_pixel);

VD_pixel1 = w_pixel_cnt1;
VD_pixel2 = w_pixel_cnt2;

/*--- Read in the Angular Position
of the motor in radians ---*/
pos = G_ENC[0];
VD_Position_M = (float)(pos)/4000*2*PI;

/*-------------------------------------------------------
The X and Y positions read are w.r.t the fixed frame.
The control uses the x and y positions of the rotating frame
Hence we reverse decompose it, as shown below
-------------------------------------------------------*/
VD_Position_X =  cos(VD_Position_M)*Position_X
   + sin(VD_Position_M)*Position_Y;
VD_Position_Y = -sin(VD_Position_M)*Position_X
   + cos(VD_Position_M)*Position_Y;
```

```
/*--- Backwards difference to
obtain velocites of X and Y axes ---*/

unfilt_vel_x = (VD_Position_X - oldpos_X)/tsamp;
VD_Velocity_X = filter(unfilt_vel_x, filter_vel_x);
oldpos_X = VD_Position_X;

unfilt_vel_y = (VD_Position_Y - oldpos_Y)/tsamp;
VD_Velocity_Y = filter(unfilt_vel_y, filter_vel_y);
oldpos_Y = VD_Position_Y;


/*--- Read in the currents of the magnets and the motor ---*/
VD_Current_X1= IToV (G_ADC[0]) * 2.0 + 0.1363;
VD_Current_X2= IToV (G_ADC[1]) * 2.0 + 0.1587;
VD_Current_Y1= IToV (G_ADC[2]) * 2.0 + 0.1954;
VD_Current_Y2= IToV (G_ADC[4]) * 2.0 + 0.2540;
VD_Current_M = IToV (G_ADC[5]) * 2.0 + 0.2466;

if(  VD_Current_X1 < -5.0 || VD_Current_X1 > 5.0 ||
     VD_Current_X2 < -5.0 || VD_Current_X2 > 5.0 ||
     VD_Current_Y1 < -5.0 || VD_Current_Y1 > 5.0 ||
     VD_Current_Y2 < -5.0 || VD_Current_Y2 > 5.0 ||
     VD_Current_M  < -4.5 || VD_Current_M  > 4.5 ) Safety=0;

/*--- Backwards difference to obtain velocity of motor ---*/
unfilt_vel_m = (VD_Position_M - oldpos_M)/tsamp;
VD_Velocity_M = filter(unfilt_vel_m, filter_vel_m);
oldpos_M = VD_Position_M;

/*--- Read in and filter the shear values ---*/
unfilt_shear_x = IToV(G_ADC[3]);
Shear_X = filter(unfilt_shear_x, filter_shear_x);

unfilt_shear_y = IToV(G_ADC[7]);
Shear_Y = filter(unfilt_shear_y, filter_shear_y);

VD_Shear_X = VF_ShearGain_X * (Shear_X - VF_ShearOffset_X);
VD_Shear_Y = VF_ShearGain_Y * (Shear_Y - VF_ShearOffset_Y);

/*--- Filtered derivative of the Shear ---*/
unfilt_shearder_x = (VD_Shear_X - oldshear_X)/tsamp;
VD_ShearDer_X = filter(unfilt_shearder_x, filter_shearder_x);
```

```
oldshear_X = VD_Shear_X;

unfilt_shearder_y = (VD_Shear_Y - oldshear_Y)/tsamp;
VD_ShearDer_Y = filter(unfilt_shearder_y, filter_shearder_y);
oldshear_Y = VD_Shear_Y;
}

/*--------- OUTPUT ROUTINE ---------*/
output ()
{
if(  VD_Voltage_X1 < -70.0 || VD_Voltage_X1 > 70.0 ||
     VD_Voltage_Y1 < -70.0 || VD_Voltage_Y1 > 70.0 ||
     VD_Voltage_X2 < -70.0 || VD_Voltage_X2 > 70.0 ||
     VD_Voltage_Y2 < -70.0 || VD_Voltage_Y2 > 70.0 ||
     VD_Voltage_M  < -20.0 || VD_Voltage_M  > 20.0 ) Safety=0;

G_DAC[0] = VToI(VD_Voltage_X1/55.0*(float)Safety);
G_DAC[1] = VToI(VD_Voltage_Y1/55.0*(float)Safety);
G_DAC[2] = VToI(VD_Voltage_X2/55.0*(float)Safety);
G_DAC[3] = VToI(VD_Voltage_Y2/55.0*(float)Safety);
G_DAC[6] = VToI(VD_Voltage_M/55.0*(float)Safety);
}

/*--------- END_CONTROL ROUTINE ---------*/
end_control()
{
  CcdLedEnable ( led_id_on, 0 );
  free(filter_vel_x);
  free(filter_vel_y);
  free(filter_vel_m);
  free(filter_fddot_x);
  free(filter_fddot_y);
  free(filter_shear_x);
  free(filter_shear_y);
  free(filter_shearder_x);
  free(filter_shearder_y);
}

/*-------------------------------------------------------
Functions   : CcdLedEnable(), CcdThresholdSet1(), CcdPixelCnt1(),
                     CcdThresholdSet2(), CcdPixelCnt2().
Description : Camera Routines Only

-------------------------------------------------------*/
```

```
void CcdLedEnable ( BYTE b_led_id, BYTE b_led_on_time )
{
  outp ( (ECL_CARD_ADDR1 + 0x00), ((BYTE)1 << b_led_id) );
  outp ( (ECL_CARD_ADDR1 + 0x02), b_led_on_time );
}


void CcdThresholdSet1 ( WORD w_threshold )
{
  BYTE b_valu;

/* isolate bit-11...4 */
  b_valu = (BYTE)(w_threshold >> 4);
  outp ( (ECL_CARD_ADDR1 + 0x05), b_valu );

/* isolate bit-3...0 */
  b_valu = (BYTE)w_threshold & (BYTE)0x0F;
  outp ( (ECL_CARD_ADDR1 + 0x04), b_valu );
}


WORD CcdPixelCnt1 (void)
{
  union
    {
      WORD w;
      BYTE b[2];
    } u_old;

  union
    {
      WORD w;
      BYTE b[2];
    } u_new;

/* begin with unlikely value */
  u_new.w = (WORD)0xFFFF;

  do
  {
    u_old.w = u_new.w;

/* high byte */
    u_new.b[1] = inp ( ECL_CARD_ADDR1 + 0x02 );
```

```
/* low byte  */
    u_new.b[0] = inp ( ECL_CARD_ADDR1 + 0x00 );

    u_new.w = u_new.w & 0x3FFF;
  } while ( u_new.w != u_old.w );

  return ( u_new.w );
}


void CcdThresholdSet2 ( WORD w_threshold )
{
  BYTE b_valu;

/* isolate bit-11...4 */
  b_valu = (BYTE)(w_threshold >> 4);
  outp ( (ECL_CARD_ADDR2 + 0x05), b_valu );

/* isolate bit-3...0 */
  b_valu = (BYTE)w_threshold & (BYTE)0x0F;
  outp ( (ECL_CARD_ADDR2 + 0x04), b_valu );
}


WORD CcdPixelCnt2 (void)
{
  union
    {
      WORD w;
      BYTE b[2];
    } u_old;

  union
    {
      WORD w;
      BYTE b[2];
    } u_new;

/* begin with unlikely value */
  u_new.w = (WORD)0xFFFF;

  do
  {
    u_old.w = u_new.w;
```

```
/* high byte */
    u_new.b[1] = inp ( ECL_CARD_ADDR2 + 0x02 );

/* low byte  */
    u_new.b[0] = inp ( ECL_CARD_ADDR2 + 0x00 );

    u_new.w = u_new.w & 0x3FFF;
  } while ( u_new.w != u_old.w );

  return ( u_new.w );
}
```

Index

MIX
Papier aus verantwortungsvollen Quellen
Paper from responsible sources
FSC® C105338

If you have any concerns about our products,
you can contact us on
ProductSafety@springernature.com

In case Publisher is established outside the EU,
the EU authorized representative is:
Springer Nature Customer Service Center GmbH
Europaplatz 3, 69115 Heidelberg, Germany

Printed by Libri Plureos GmbH
in Hamburg, Germany